BARRON'S

HOW TO PREPARE FOR THE

SAT II

CHEMISTRY

6TH EDITION

Joseph A. Mascetta

Former Principal

Chemistry Teacher and Coordinator, Science Department,
Mount Lebanon High School, Pittsburgh, Pennsylvania

Science Educational Consultant

BARRON'S

All inquiries should be addressed to:
Barron's Educational Series, Inc.
250 Wireless Boulevard
Hauppauge, New York 11788
http://www.barronseduc.com

Permission to reprint the material on pages xi and xii does not constitute review or endorsement by Educational Testing Service or the College Board of this publication as a whole or of any other questions or testing information it may contain.

International Standard Book No. 0-7641-0331-8

Library of Congress Catalog Card No. 98-6584

Library of Congress Cataloging-in-Publication Data
Mascetta, Joseph A.
    How to prepare for SAT II. Chemistry / Joseph A. Mascetta.—6th ed.
        p.   cm.
    Includes index.
    ISBN 0-7641-0331-8
    1. Chemistry—Examinations, questions, etc.   2. College entrance
achievement tests—United States—Study guides.   I. Title.
QD42.M3335   1998
540'.76—dc21
                                                                98-6584
                                                                CIP

PRINTED IN THE UNITED STATES OF AMERICA
9 8 7 6

Contents

Introduction   **viii**

  General Information about the Test   viii

    Is SAT II: Chemistry Required?   viii

    When Should You Take the Test?   viii

    When Is the Test Offered?   ix

    How Do You Register?   ix

    How Should You Prepare for the Test?   ix

    SAT II: Chemistry Dates   ix

    What Topics Appear on the Test, and to What Extent?   x

    How Can You Use This Book to Prepare for the Test?   xi

    How Can You Improve Your Problem-Solving Skills?   xii

A Diagnostic Test   **3**

  Answers and Explanations   16

  Diagnosing Your Needs   22

  Planning Your Study   23

  Final Preparation—The Day before the Test   25

    (Including a checklist of things to remember)

Chapter 1   **Introduction to Chemistry**   **27**

  Matter   27

    Definition of Matter 27   States of Matter 27   Composition of Matter 27
    Chemical and Physical Properties 28   Chemical and Physical Changes 28
    Conservation of Mass 29

  Energy   29

    Definition of Energy 29   Forms of Energy 30   Types of Reactions
    (Exothermic versus Endothermic) 30   Conservation of Energy 31

  Conservation of Mass and Energy   31

  Measurements and Calculations   31

    Metric System 31   Temperature Measurements 33
    Heat Measurements 34   Scientific Notation 34
    Factor-Label Method of Conversion (Dimensional Analysis) 35
    Precision, Accuracy, and Uncertainty 35   Significant Figures 36
    Calculations with Significant Figures 37

Chapter 2   **Atomic Structure**   **42**

  History   42

  Electric Nature of Atoms   42

    Basic Electric Charges 42   Bohr Model 44   Components of Atomic
    Structure 45   Oxidation Number and Valence 46   Metallic,
    Nonmetallic, and Noble Gas Structures 47   Reactivity 47

  Atomic Spectra   47

    Spectroscopy 48   Mass Spectroscopy 50

  The Wave-Mechanical Model   50

    Quantum Numbers 51   Hund's Rule of Maximum Multiplicity 53

  Sublevels and Electron Configuration   54

    Order of Filling and Notation 54   Electron Dot Notation (Lewis Dot
    Structures) 55

  Transition Elements and Variable Oxidation Numbers   55

  Periodic Table   56

    History 56   Periodic Law 57   The Table 57

  Properties Related to the Periodic Table   58

    Radii of Atoms 58   Atomic Radii in Periods 59
    Atomic Radii in Groups 59   Ionic Radius Compared to Atomic Radius 59
    Electronegativity 59   Ionization Energy 61

Chapter 3 **Bonding**   **64**

Types of Bonds   65
Ionic Bonds 65   Covalent Bonds 66   Metallic Bonds 68

Intermolecular Forces of Attraction   68
Dipole-Dipole Attraction 68   London Forces 68   Hydrogen Bonds 68

Double and Triple Bonds   69

Resonance Structures   70

Electrostatic Repulsion (VSEPR) and Hybridization   70
Electrostatic Repulsion 70   Hybridization 71

Sigma and Pi Bonds   74

Properties of Ionic Substances   75

Properties of Molecular Crystals and Liquids   75

Chapter 4 **Chemical Formulas**   **77**

Writing Formulas   77

More About Oxidation Numbers   79

Naming Compounds   81

Chemical Formulas   82

Laws of Definite Composition and Multiple Proportions   85

Writing and Balancing Simple Equations   85

Showing Phases in Chemical Equations   86

Writing Ionic Equations   87

Chapter 5 **Gases**   **89**

Some Representative Gases   89
Oxygen 89   Hydrogen 91

General Characteristics of Gases   94
Measuring the Pressure of a Gas 94   Kinetic-Molecular Theory 96
Some Particular Properties of Gases 96

Gas Laws and Related Problems   97
Graham's Law 97   Charles's Law 97   Boyle's Law 98   Combined
Gas Law 99   Pressure versus Temperature 99   Dalton's Law of Partial
Pressures 100   Corrections of Pressure 100   Ideal Gas Law 101
Ideal Gas Deviations 103

Chapter 6 **Chemical Calculations (Stoichiometry) and
the Mole Concept**   **106**

The Mole Concept   106

Molar Mass and Moles   106

Mole Relationships   108

Gas Volumes and Molar Mass   108

Density and Molar Mass   109

Mass-Volume Relationships   109

Mass-Mass Problems   111

Volume-Volume Problems   113

Problems with an Excess of One Reactant   115

Chapter 7 **Liquids, Solids, and Phase Changes**   **119**

Liquids   119
Importance of Intermolecular Interaction 119   Kinetics of Liquids 119

Phase Equilibrium   120

Boiling Point   121

Critical Temperature and Pressure   121

Solids 121

Phase Diagrams 121

Water 122
Purification of Water 122   Composition of Water 123   Properties and
Uses of Water 125   Water Calorimetry Problems 125   Water's Reactions
with Anhydrides 126

Polarity and Hydrogen Bonding 127

Solubility 128
General Rules of Solubility 129   Factors That Affect Rate of Solubility 129
Summary of Types of Solutes and Relationships of Type to Solubility 129

Water Solutions 129

Continuum of Water Mixtures 130

Expressions of Concentration 131

Dilution 134

Colligative Properties of Solutions 135

Crystallization 137

Chapter 8   Chemical Reactions   141

Types of Reactions 141

Predicting Reactions 141
Combination 142   Decomposition 143   Single Replacement 143
Double Replacement 143   Hydrolysis Reactions 145   Entropy 145

Origin of Chemical Energy 146

Changes in Enthalpy 146

Additivity of Reaction Heats 147

Bond Dissociation Energy 149

Enthalpy from Bond Energies 150

Chapter 9   Rates of Chemical Reactions   152

Measurements of Reaction Rates 152

Factors Affecting Reaction Rates 152

Collision Theory of Reaction Rates 153

Activation Energy 154

Reaction Rate Law 154
Reaction Mechanism and Rates of Reaction 155

Chapter 10   Chemical Equilibrium   157

Reversible Reactions and Equilibrium 157

Le Châtelier's Principle 160

Effects of Changing Conditions 161
Effect of Changing the Concentrations 161   Effect of Temperature on
Equilibrium 161   Effect of Pressure on Equilibrium 161

Equilibria in Heterogeneous Systems 162
Equilibrium Constant for Systems Involving Solids 162
Ionization Constants 162   Ionization Constant of Water 163
Solubility Products 164

Common Ion Effect 165

Factors Related to the Magnitude of K_e 166
Relation of Minimum Energy (Enthalpy) to Maximum Disorder (Entropy) 166
Change in Free Energy of a System 167

Chapter 11   Acids, Bases, and Salts   169

Definitions and Properties 169
Acids 169   Bases 170   Broader Acid-Base Theories 171
Conjugate Acids and Bases 171   Acid Concentration Expressed as pH 172

Indicators 173
Volumetric Analysis 173
Buffer Solutions 176
Salts 176
Amphoteric Substances 177

Chapter 12 **Oxidation-Reduction and Electrochemistry** **179**
Ionization 179
Oxidation-Reduction 180
Non-Standard-State Cell Potentials 183
Electrolytic Cells 184
Applications of Electrochemical Cells (Commercial Voltaic Cells) 186
Quantitative Aspects of Electrolysis 186
Relationship Between Quantity of Electricity and Amount of Product 186
Balancing Redox Equations Using Oxidation Numbers 187

Chapter 13 **Some Representative Groups and Families** **193**
Sulfur Family 193
Sulfuric Acid 194 Other Important Compounds of Sulfur 195
Halogen Family 196
Some Important Halides and Their Uses 196
Nitrogen Family 196
Nitric Acid 197 Other Important Compounds of Nitrogen 197
Other Members of the Nitrogen Family 199
Metals 199
Properties of Metals 199 Some Important Reduction Methods 201
Alloys 203 Metalloids 203

Chapter 14 **Carbon and Organic Chemistry** **205**
Carbon 205
Forms of Carbon 205 Carbon Dioxide 206
Organic Chemistry 208
Hydrocarbons 208 Isomers 215 Hydrocarbon Derivatives 215

Chapter 15 **Nucleonics** **221**
Radioactivity 221
The Nature of Radioactive Emissions 221 Methods of Detection of
Alpha, Beta, and Gamma Rays 223 Source of Radioactivity 224
Radioactive Dating 226
Nuclear Energy 226
Conditions for Fission 226 Methods of Obtaining Fissionable
Material 227 Fusion 228
New Subatomic Particles 228

Chapter 16 **Representative Laboratory Setups** **231**
Some Basic Setups 231
Summary of Qualitative Tests 238

Practice SAT II: Chemistry Tests 241
General Information 241
Basic Abilities Tested For 243
Practice Test 1 247
Answers and Explanations 260
Diagnosing Your Needs 264
Planning Your Study 265

Practice Test 2   269
    Answers and Explanations 282
    Diagnosing Your Needs 287
    Planning Your Study 288

Practice Test 3   293
    Answers and Explanations 305
    Diagnosing Your Needs 309
    Planning Your Study 310

Practice Test 4   315
    Answers and Explanations 327
    Diagnosing Your Needs 331
    Planning Your Study 332

Tables for Reference   334

Glossary of Common Terms   342

Index   351

Introduction

GENERAL INFORMATION ABOUT THE TEST

The College Board, which is a nonprofit membership organization, sponsors the Admissions Testing Program (ATP). Educational Testing Service develops and administers the tests for the College Board and also prepares a booklet about the testing program, of which the SAT II: Subject Tests are a part. Copies of this booklet, which describes each of the SAT II: Subject Tests and gives sample questions, are automatically sent at the beginning of each academic year to secondary schools for free distribution to students who plan to register for the SAT II: Subject Tests. The name of this publication is *Taking the SAT II: Subject Tests*, and requests from schools or individuals about how to obtain free copies should be addressed to College Board SAT Program P.O. Box 6200, Princeton, NJ 08541-6200. The phone number (Monday–Friday) is (609) 771-7600, 8:30 A.M. to 9:30 P.M. Eastern Time and the e-mail address is sat@ets.org.

All of the SAT II: Subject Tests are contained in the same test booklet. Each takes 1 hour of testing time, and you may choose any one, two, or three tests to take at one sitting. They all consist of multiple-choice questions except the SAT II: Writing Test.

Some colleges require SAT II: Subject Tests for admission. The scores are used in conjunction with your high school record, results on the SAT I: Reasoning Test, teacher recommendations, and other background information to provide a reliable measure of your academic achievements.

In addition to obtaining a standardized assessment of your achievement from your scores, some colleges use the test results for placement into their particular programs in the freshman year. At others, advisers use the results to guide freshmen in the selection of courses.

Is SAT II: Chemistry Required?

The best information on whether SAT II Tests are required and, if so, which ones is found in the individual college catalogs or a directory of colleges. Some colleges specify which tests you must take, while others allow you to choose. Obviously, if you have a choice and you have done well in chemistry, you should pick SAT II: Chemistry as one of your tests. The College Board publishes *The College Handbook*, which is a good source of information about colleges' requirements with regard to taking SAT II Tests. A copy of this book should be in your school library. You can also order it by using your registration form or by visiting the College Board's online store at http://www.collegeboard.org.

When Should You Take the Test?

You will undoubtedly do best if you take the test after completing the high school chemistry course or courses that you plan to take. At this time, the material will be fresh in your mind. Forgetting begins very quickly after you are past a topic or have finished the course. You should plan a review program for at least the last 6 weeks before the test date. (A plan is provided later in this book for such a review.) Careful review definitely helps—cramming just will not do if you want to get the best score of which you are capable!

Colleges that use SAT II: Subject Test results as part of the admissions process usually require that you take the test no later than December or January of your senior year. For early-decision programs, the test time is June of your junior year. Since chemistry is often a junior year course, June of that year is the optimum time to take the test.

When Is the Test Offered?

The chemistry test is available every time the SAT II: Subject Tests are given, that is, in October, November, December, January, March, May, and June. Be sure that the testing site for which you plan to register offers the SAT II Tests on each of these seven times. Remember that you may choose to take one or two additional tests besides chemistry on any one test day. You don't have to specify in advance which tests you plan to take or how many.

How Do You Register?

The *Registration Bulletin for the SAT I and SAT II Tests* contains all the information you need to register and to have your scores sent to the college(s) of your choice. Copies of this publication should be available in every high school guidance office. If you have a problem getting one, write to College Board Admissions Testing Program (see the address in "General Information about the Test," on page viii).

How Should You Prepare for the Test?

Barron's How to Prepare for SAT II: Chemistry will be especially helpful. The more you know about the test, the more likely you are to get the best score possible for you. This book provides you with a diagnostic test and four practice tests that allow you to become familiar with the question types and the wording of directions, and to gain a feel for the degree of emphasis on particular topics and the ways in which information may be tested. Each of these aspects should be consciously pursued as you use this book.

SAT II: Chemistry Test
Projected Calendar for Test Dates

When the test is given on:	Registration deadline is:	Late registration is:
October 13, 2001	September 11, 2001	September 15, 2001
November 3, 2001	September 28, 2001	October 10, 2001
December 1, 2001	October 26, 2001	November 7, 2001
January 26, 2002	December 21, 2001	January 2, 2002
May 4, 2002	March 29, 2002	April 10, 2002
June 1, 2002	April 26, 2002	May 8, 2002

What Topics Appear on the Test, and to What Extent?

The following charts* show the content of the test:

Topics Covered	Approximate Percentage of Test	Approximate Number of Questions
I. ATOMIC AND MOLECULAR STRUCTURE: • Atomic Theory and Structure, including periodic relationships • Nuclear Reactions • Chemical Bonding and Molecular Structure	23	20
II. STATES OF MATTER: • Kinetic Molecular Theory of Gases • Solutions, including concentration units, solubility, conductivity, and colligative properties	14	12
III. REACTION TYPES: • Acids and Bases • Oxidation-reduction, including electrochemical cells • Precipitation	14	12
IV. STOICHIOMETRY: Including the mole concept, Avogadro's number, empirical and molecular formulas, percentage composition, stoichiometric calculations, and limiting reagents	14	12
V. EQUILIBRIUM AND REACTION RATES: Equilibrium, including mass action expressions, ionic equilibria, and Le Châtelier's principle; factors affecting rates of reaction	7	6
VI. THERMODYNAMICS: Including energy changes in chemical reactions, and physical processes, Hess's Law, and randomness	7	6
VII. DESCRIPTIVE CHEMISTRY: Physical and chemical properties of elements and their more familiar compounds, including simple examples from organic chemistry, and environmental chemistry	14	12
VIII. LABORATORY: Equipment, measurement, procedures, observations, safety, calculations, and interpretation of results	7	6

Note: Every edition contains approximately five questions on equation balancing and/or predicting products of chemical reactions. These are distributed among the various content categories.

*Reprinted by permission of Educational Testing Service.

Skills Specifications	Approximate Percentage of Test
Recall of Knowledge: remembering fundamental concepts and specific information; demonstrating familiarity with terminology	20
Application of Knowledge: applying a single principle to unfamiliar and/or practical situations to obtain a qualitative result or solve a quantitative problem	45
Synthesis of Knowledge: inferring and deducing from qualitative and/or quantitative data; integrating two or more relationships to draw conclusions or solve problems	35

*Reprinted by permission of Educational Testing Service.

The above charts give you a general overview of the content of the test. Your knowledge of the topics and your skills in recalling, applying, and synthesizing this knowledge are evaluated through 85 multiple-choice questions. This material is that generally covered in an introductory course in chemistry at a level suitable for college preparation. While every test covers the topics listed, different aspects of each topic are stressed from year to year. Add to this the differences that exist in high school courses with respect to the percentage of time devoted to each major topic and to the specific subtopics covered, and you may find that there are questions on topics with which you have little or no familiarity.

Each of the sample tests in this book is constructed to match closely the distribution of topics shown in the preceding chart so that you will gain a feel for the makeup of the actual test. After each test, a chart will show you which questions relate to each topic. This will be very helpful to you in planning your review because you can identify the areas on which you need to concentrate in your studies. Another chart enables you to see which chapters correspond to the various topic areas.

A periodic chart is provided in SAT II: Chemistry as a resource and as the source of atomic numbers and atomic masses of elements that are needed in solving problems. You will not be allowed to use an electronic calculator during the test. Numerical calculations are limited to simple arithmetic. You should be familiar with the concepts of ratios and direct and inverse proportions, scientific notation, and exponential functions. In this test, the metric system of units is used.

The test is composed of three types of questions. The directions and formats are identical to those used in all of the sample tests in this book. The questions are all of the multiple-choice type, in which you must choose from the five choices given.

How Can You Use This Book to Prepare for the Test?

The best way to use this book is a two-stage approach, and the next sections are arranged accordingly. First, you should take the diagnostic pretest. This will give you a preliminary exposure to the type of test you are planning to take, as well as a measure of how you achieve on each of the areas of the test. You will also become aware of the types of questions that the test includes.

Having taken the diagnostic test, you should then follow a study program. A study plan covering the 6 weeks before the test has been developed for you and is given in detail on page 24. It requires a minimum of 1 or 2 hours per night on weekdays but leaves your weekends free.

How Can You Improve Your Problem-Solving Skills?*

Chemistry is a subject that deals with many problem situations that you, the student, must be able to solve. Solving problems may seem to be a natural process when the degree of difficulty is not very great, and you may not need a structured method to attack these problems. However, for complex problems an orderly process is required.

The following is such a problem-solving process. Each step is vital to the next step and to the final solution of the problem.

Step 1. Clarify the problem: to separate the problem into the facts, the conditions, and the questions that need to be answered, and to establish the goal.

Step 2. Explore: to examine the sufficiency of the data, to organize the data, and to apply previously acquired knowledge, skills, and understanding.

Step 3. Select a strategy: to choose an appropriate method to solve the problem.

Step 4. Solve: to apply the skills needed to carry out the strategy chosen.

Step 5. Review: to examine the reasonableness of the solution and to evaluate the effectiveness of the process.

The steps of the problem-solving process listed above should be followed in sequence. The subskills listed below for each step, however, are not in sequence. The order in which subskill patterns are used will differ with the nature of the problem and/or with the ways in which the individual problem solver thinks. Also, not every subskill need be employed in solving every problem.

Clarify the Problem

a. Identify the facts. What is known about the problem?

b. Identify the conditions. What is the current situation?

c. Identify the questions. What needs to be answered before the problem can be solved?

d. Visualize the problem.

    1. Make mental images of the problem.

    2. If desirable or necessary, draw a sketch or diagram, make an outline, write down symbols or equations that correspond to the mental images.

e. Establish the goal. The goal defines the specific result to be accomplished through the problem-solving process. It defines the purpose or function the solution is expected to achieve and serves as the basis for evaluating the solution.

Explore

a. Review previously acquired knowledge, skills, and understanding. Determine whether the current problem is similar to a previously seen type.

b. Estimate the sufficiency of the data. Does there seem to be enough information to solve the problem?

c. Organize the data. There are many ways in which data can be organized. Some examples are outline, written symbols and equations, chart, table, graph, map, diagram, and drawing. Determine whether the data organized in the way(s) you have chosen will enable you to partially or completely solve the problem.

*Adapted with permission from *Thinking Skills Resource Guide*, a noncopyrighted publication of Mount Lebanon School District.

d. Determine what new data, if any, need to be collected. What new information may be needed to solve the problem? Can the existing data be reorganized to generate new information? Do other resources need to be consulted? This step may suggest possible strategies to be used to solve the problem.

Select a Strategy

A strategy is a goal-directed sequence of mental operations. Selecting a strategy is the most important and also the most difficult step in the problem-solving process. Although there may be several strategies that will lead to the solution of a problem, the skilled problem solver uses the most efficient strategy. The choice of the most efficient strategy is based on knowledge and experience as well as a careful application of the clarify and explore steps of the problem-solving method. Some problems may require the use of a combination of strategies.

The following search methods may help you to select a strategy. They do not represent all of the possible ways in which this can be done. Other methods of strategy selection are related to specific content areas.

a. Trial-and-error search: Such a search either doesn't have or doesn't use information that indicates that one path is more likely to lead to the goal than any other path.

Trial-and-error search comes in two forms, blind and systematic. In *blind search*, the searchers pick paths to explore blindly, without considering whether they have already explored these paths. A preferable method is *systematic search*, in which the searchers keep track of the paths they have already explored and do not duplicate them. Because this method avoids multiple searches, systematic search is usually twice as efficient as blind search.

b. Reduction method: This involves breaking the problem into a sequence of smaller parts by setting up subgoals. Subgoals make problem solving easier becaue they reduce the amount of search required to find the solution.

You can set up subgoals by working part way into a problem and then analyzing the partial goal to be achieved. In doing this, you can drop the problem restrictions that do not apply to the subgoal. By adding up all the subgoals, you can solve the "abstracted" problem.

c. Working backward: When you have trouble solving a problem head-on, it is often useful to try to work backward. Working backward involves a simple change in representation or point of view. Your new starting point is the original goal. Working backward can be helpful because problems are often easier to solve in one direction than another.

d. Knowledge-based method: This strategy uses information stored in the problem solver's memory, or newly acquired information, to guide the search for the solution. The problem solver may have solved a similar problem and can use this knowledge in a new situation. In other cases, problem solvers may have to acquire needed knowledge. For example, they may solve an auxiliary problem to learn how to solve the one they are having difficulty with.

Searching for analogous (similar) problems is a very powerful problem-solving technique. When you are having difficulty with a problem, try to pose a related, easier one and hope thereby to learn something that will help you solve the harder problem.

Solve

Use the strategy chosen to actually solve the problem. Executing the solution provides you with a very valuable check on the adequacy of your plan. Sometimes students will look at a problem and decide that, since they know how to solve it, they need not bother with the drudgery of actually executing the solution. Sometimes the students are right, but at other times they miss an excellent opportunity to discover that they were wrong.

Review

a. Evaluation. The critical question in evaluation is this: "Does the answer I propose meet all of the goals and conditions set by the problem?" Thus, after the effort of finding a solution, you must turn back to the problem statement and check carefully to be sure your solution satisfies it.

With easy problems there is a strong temptation to skip evaluation because the probability of error seems small. In some cases, however, this can be costly. Evaluation may prove that errors were present.

b. Consolidation. Here the basic question to be answered is: "What can I learn from the experience of solving this problem?" The following more specific questions may help you to answer this general one:

1. Why was this problem difficult?
2. Was it difficult to follow a plan?
3. Was it difficult to decide on a plan? If so, why?
4. Did I take the long way to the answer?
5. Can I use this plan again in similar problems?

The important thing is to reflect on the process that you used in order to make future problem solving easier.

CHEMISTRY TEST

Material in the following table may be useful in answering the questions in this examination.

Periodic Table of the Elements

A DIAGNOSTIC TEST

The following test is a sample of the actual test you will take to measure your chemistry achievement. It has basically the same distribution of topics, directions, and number of questions. Before taking the actual test, follow the advice given in the section entitled "Final Preparation—The Day before the Test" (p. 25). Use the answer sheet provided below, and limit the test time to 1 hour.

A periodic chart has been included for your use on problems requiring this source of information. Use this chart with the practice tests, too.

Turn now to the test.

ANSWER SHEET FOR DIAGNOSTIC TEST

Determine the correct answer for each question. Then, using a No. 2 pencil, blacken completely the oval containing the letter of your choice.

CHEMISTRY* Fill in oval CE only if II is a correct explanation of I.

		I	II	CE*
101.		T F	T F	
102.		T F	T F	
103.		T F	T F	
104.		T F	T F	
105.		T F	T F	
106.		T F	T F	
107.		T F	T F	
108.		T F	T F	
109.		T F	T F	
110.		T F	T F	
111.		T F	T F	
112.		T F	T F	
113.		T F	T F	
114.		T F	T F	

1. A B C D E
2. A B C D E
3. A B C D E
4. A B C D E
5. A B C D E
6. A B C D E
7. A B C D E
8. A B C D E
9. A B C D E
10. A B C D E
11. A B C D E
12. A B C D E
13. A B C D E
14. A B C D E
15. A B C D E

16. A B C D E
17. A B C D E
18. A B C D E
19. A B C D E
20. A B C D E
21. A B C D E
22. A B C D E
23. A B C D E
24. A B C D E
25. A B C D E

ON THE ACTUAL CHEMISTRY TEST, THE FOLLOWING TYPE OF QUESTION MUST BE ANSWERED ON A SPECIAL SECTION (LABELED "CHEMISTRY") AT THE LOWER LEFT-HAND CORNER OF PAGE 2 OF YOUR ANSWER SHEET. THESE QUESTIONS WILL BE NUMBERED BEGINNING WITH 101 AND MUST BE ANSWERED ACCORDING TO THE DIRECTIONS.

ON THE ACTUAL CHEMISTRY TEST, THE REMAINING QUESTIONS MUST BE ANSWERED BY RETURNING TO THE SECTION OF YOUR ANSWER SHEET YOU STARTED FOR CHEMISTRY.

40. A B C D E
41. A B C D E
42. A B C D E
43. A B C D E
44. A B C D E
45. A B C D E
46. A B C D E
47. A B C D E
48. A B C D E
49. A B C D E
50. A B C D E
51. A B C D E
52. A B C D E
53. A B C D E
54. A B C D E
55. A B C D E

56. A B C D E
57. A B C D E
58. A B C D E
59. A B C D E
60. A B C D E
61. A B C D E
62. A B C D E
63. A B C D E
64. A B C D E
65. A B C D E
66. A B C D E
67. A B C D E
68. A B C D E
69. A B C D E
70. A B C D E

71. A B C D E
72. A B C D E
73. A B C D E
74. A B C D E
75. A B C D E
76. A B C D E
77. A B C D E
78. A B C D E
79. A B C D E
80. A B C D E
81. A B C D E
82. A B C D E
83. A B C D E
84. A B C D E
85. A B C D E

A Diagnostic Test

Note: For all questions involving solutions and/or chemical equations, assume that the system is in water unless otherwise stated.

Part A

Directions: Each set of lettered choices refers to the numbered statements or formulas immediately following it. Select the one lettered choice that best fits each statement or formula and then fill in the corresponding oval on the answer sheet. A choice may be used once, more than once, or not at all in each set.

Questions 1–4 refer to the periodic group shown, with letter choices substituted for the respective elements.

Halogen Group

(A)
(B)
(C)
(D)
(E)

1. The element that is most active chemically
2. The element with the smallest ionic radius
3. The element with the lowest first ionization potential
4. The element that first shows some metallic properties at room temperature

Questions 5–7

(A) X^+
(B) X^{2+}
(C) X^{3+}
(D) XO_3^{2-}
(E) XO_4^{2-}

5. A type of ion found in sodium acetate
6. A type of ion found in aluminum oxide
7. A type of ion found in lead(IV) carbonate

Questions 8–12

(A) Avogadro's Constant
(B) Boyle's Law
(C) Charles' Law
(D) Dalton's Theory
(E) Gay-Lussac's Law

8. Proposed basic postulates concerning elements and atoms
9. Proposed a relationship between the combining volumes of gases with respect to the reactants and gaseous products
10. Proposed a temperature-volume relationship of gases
11. Proposed a concept regarding the number of particles in a mole
12. Proposed a volume-pressure relationship of gases

Questions 13–16

(A) R–OH

(B) R–O–R*

(C) R–C$\diagup$$\!\!^{O}_{\diagdown H}$

(D) R–C$\diagup$$\!\!^{H}_{\diagdown OH}$

(E) R–$\overset{\overset{\text{O}}{\|}}{\text{C}}$–O–R*

(* Alkyl group that is not necessarily the same as R.)

13. The organic structure designation that includes the functional group of an aldehyde
14. The organic structure designation that includes the functional group of an acid
15. The organic structure designation that includes the functional group of an ester
16. The organic structure designation that includes the functional group of an ether

Questions 17–21

(A) $H_2(g)$
(B) $CO_2(g)$
(C) $2N_2O(g)$
(D) $2NaCl\,(aq)$
(E) $H_2SO_4(aq)$

17. The expression that can be used to designate a linear nonpolar molecule
18. The expression that can be used to designate 2 moles of atoms
19. The expression that can be used to designate 3 moles of atoms
20. The expression that can be used to designate 3 moles of ions
21. The expression that can be used to designate 6 moles of atoms

Questions 22–25

(A) Ionic substance
(B) Metallic substance
(C) Polar covalent molecule
(D) Nonpolar covalent molecule
(E) Aromatic organic structure

22. Carbon tetrachloride

23. Cesium chloride

24. Hydrogen chloride

25. Benzene

Part B

ON THE ACTUAL CHEMISTRY TEST, THE FOLLOWING TYPE OF QUESTION MUST BE ANSWERED ON A SPECIAL SECTION (LABELED "CHEMISTRY") AT THE LOWER LEFT-HAND CORNER OF PAGE 2 OF YOUR ANSWER SHEET. THESE QUESTIONS WILL BE NUMBERED BEGINNING WITH 101 AND MUST BE ANSWERED ACCORDING TO THE FOLLOWING DIRECTIONS.

Directions: Each question below consists of two statements, I in the left-hand column and II in the right-hand column. For each question, determine whether statement I is true or false <u>and</u> if statement II is true or false and fill in the corresponding T or F ovals on your answer sheet. <u>Fill in oval CE only if statement II is a correct explanation of statement I.</u>

Sample Answer Grid:

CHEMISTRY *Fill in oval CE only if II is a correct explanation of I.

	I	II	CE*
101.	ⓣ Ⓕ	ⓣ Ⓕ	◯

101. A catalyst can accelerate a chemical reaction | BECAUSE | a catalyst can decrease the activation energy required for the reaction.

102. Molten sodium chloride is a good electrical conductor | BECAUSE | sodium chloride is an ionic substance.

103. Ice is less dense than liquid water | BECAUSE | water molecules are nonpolar.

104. ^{14}C is an isotope of ^{14}N | BECAUSE | ^{14}C and ^{14}N have the same mass number.

105. A saturated solution can be classified as dilute | BECAUSE | the solute can have a very low solubility in that solvent.

106. Two liters of CO_2 can be produced by 1 gram of carbon burning completely | BECAUSE | the amount of gas evolved by this chemical reaction can be solved by using the mole relationship of the coefficients in the balanced equation.

107. A reaction is at equilibrium when it reaches completion | BECAUSE | the concentrations of the reactants in a state of equilibrium equal the concentrations of the products.

108. The anions in an electrochemical reaction migrate to the cathode | BECAUSE | positively charged ions are attracted to the negatively charged cathode.

109. A solution with pH = 5 has a higher concentration of hydronium ions than a solution with a pH = 3   BECAUSE   pH is defined as $-\log[H^+]$.

110. When $\Delta H_{\text{reaction}}$ is a negative quantity, the reaction is exothermic   BECAUSE   heat energy is released.

111. Weak acids have a small value for the equilibrium constant, K_a BECAUSE   the concentration of the hydronium ion is in the numerator of the expression.

112. One mole of NaCl contains 2 moles of ions   BECAUSE   NaCl is a stable salt at room temperature.

113. A pi bond is formed between the lobes of adjacent p orbitals in the same plane of two atoms   BECAUSE   both lobes of a single p orbital can hold two electrons of opposite spin.

114. H_2S and H_2O have a significant difference in their boiling points   BECAUSE   hydrogen sulfide has a higher degree of hydrogen bonding.

Part C

<u>Directions</u>: Each of the questions or incomplete statements below is followed by five suggested answers or completions. Select the one that is best in each case and then fill in the corresponding oval on the answer sheet.

40. Two immissible liquids, when shaken together vigorously, may form
    (A) a solution
    (B) a tincture
    (C) a sediment
    (D) a hydrated solution
    (E) a colloidal dispersion

41. During which interval on the cooling graph above is the system undergoing a change of state from a liquid to a solid?

42. If a principal energy level of an atom contains 18 electrons, they will be arranged in orbitals according to the pattern
    (A) $s^6p^6d^6$
    (B) $s^2p^6d^{10}$
    (C) $s^2d^6f^{10}$
    (D) $s^2p^6f^{10}$
    (E) $s^2p^2f^{14}$

43. For a particular organic compound, which of the following pairs can represent the empirical and the molecular formula, respectively?
    (A) CH and CH_4
    (B) CH and C_6H_6
    (C) CH_2 and C_2H_2
    (D) CH_2 and C_2H_3
    (E) CH_3 and C_3H_6

44. A liter of hydrogen is at 5°C temperature and under 640 torr pressure. If the temperature were raised to 60°C and the pressure decreased to 320 torr, how would the liter volume be modified?

    (A) $1 \text{ liter} \times \dfrac{5}{60} \times \dfrac{640}{32}$

    (B) $1 \text{ liter} \times \dfrac{60}{5} \times \dfrac{320}{640}$

    (C) $1 \text{ liter} \times \dfrac{278}{333} \times \dfrac{640}{320}$

    (D) $1 \text{ liter} \times \dfrac{333}{278} \times \dfrac{640}{320}$

    (E) $1 \text{ liter} \times \dfrac{333}{278} \times \dfrac{320}{640}$

45. Of the following statements about the number of subatomic particles in an ion of $^{32}_{16}S^{2-}$, which is (are) true?
    I. 16 protons
    II. 14 neutrons
    III. 18 electrons
       (A) II only
       (B) III only
       (C) I and II only
       (D) I and III only
       (E) I, II, and III

46. The most active metallic elements are found in
    (A) the upper right corner of the periodic chart
    (B) the lower right corner of the periodic chart
    (C) the upper left corner of the periodic chart
    (D) the lower left corner of the periodic chart
    (E) the middle of the table, just beyond the transition elements

47. If 1 mole of each of these substances were dissolved in 1,000 grams of water, which solution would have the highest boiling point?
    (A) NaCl
    (B) KCl
    (C) $CaCl_2$
    (D) $C_6H_{10}O_5$
    (E) $C_{12}H_{22}O_{11}$

48. A tetrahedral molecule, XY_4, would be formed if X were using the orbital configuration
    (A) p^2
    (B) s^2
    (C) sp
    (D) sp^2
    (E) sp^3

49. In the following reaction, how many liters of SO_2 will result from the complete burning of pure sulfur in 8 liters of oxygen?
$$S(s) + O_2(g) \rightarrow SO_2(g)$$
    (A) 1
    (B) 4
    (C) 8
    (D) 16
    (E) 32

50. In the above laboratory setup for the preparation of chlorine, where has a hazardous mistake been made?
    (A) A
    (B) B
    (C) C
    (D) D
    (E) E

51. In the laboratory preparation of a halogen, a halide, manganese dioxide, and sulfuric acid were heated in a retort. The product was condensed as a red-orange liquid. Which halogen was produced?
    (A) fluorine
    (B) chlorine
    (C) bromine
    (D) iodine
    (E) astatine

52. Which term is applied to any reaction between an Arrhenius acid and an Arrhenius base?
    (A) hydration
    (B) neutralization
    (C) single replacement
    (D) double replacement
    (E) oxidation-reduction

53. Standard conditions (STP) are
    (A) 0°C and 2 atm
    (B) 32°F and 76 torr
    (C) 273 K and 760 mm Hg
    (D) 4°C and 7.6 cm Hg
    (E) 0 K and 760 mm Hg

54. Laboratory results showed the composition of a compound to be 58.81% barium, 13.73% sulfur, and 27.46% oxygen. What is the empirical formula of the compound?
    (A) $BaSO_4$
    (B) BaS_2O
    (C) Ba_2SO_3
    (D) BaS_2O_4
    (E) Ba_2SO_4

55. What is the percentage composition of calcium in calcium hydroxide, $Ca(OH)_2$?
    (A) 40%
    (B) 43%
    (C) 54%
    (D) 69%
    (E) 74%

56. How many grams of hydrogen gas can be produced from the following reaction if 65 grams of zinc and 65 grams of HCl are present in the reaction?
$$Zn + 2HCl \rightarrow ZnCl_2 + H_2$$
    (A) 1
    (B) 1.8
    (C) 3.6
    (D) 7
    (E) 58

57. The following statements were recorded while preparing carbon dioxide gas in the laboratory. Which one involves an interpretation of the data rather than an observation?
    (A) No liquid was transfered from the reaction bottle to the beaker.
    (B) The quantity of solid minerals decreased.
    (C) The cloudiness in the last bottle of limewater was caused by the product of the reaction of the colorless gas and the limewater.
    (D) The bubbles of gas rising from the mineral remained colorless throughout the experiment.
    (E) There was a 4°C rise in temperature in the reaction vessel during the experiment.

58. The above laboratory setup can be used to prepare
    I. CO_2
    II. H_2
    III. NO
    (A) I only
    (B) III only
    (C) I and III only
    (D) II and III only
    (E) I, II, and III

59. In the experiment diagrammed above, the radioactive source emitted probably is
    (A) $_{-1}^{0}e$
    (B) $_{2}^{4}He$
    (C) $_{2}^{1}H$
    (D) a gamma ray
    (E) $_{1}^{0}e$

Questions 60–61 pertain to the reaction represented by the following equation:

$$2NO(g) + O_2(g) \qquad 2NO_2(g) + 150 \text{ kilojoules}$$

60. The equilibrium constant for the reaction is given by the expression:

(A) $\dfrac{[2NO_2]}{[2NO][O_2]}$

(B) $\dfrac{[NO]^2[O_2]}{[NO_2]^2}$

(C) $\dfrac{[2NO_2]}{[2NO]+[O_2]}$

(D) $\dfrac{[NO_2]^2}{[NO]^2[O_2]}$

(E) $\dfrac{[2NO][O_2]}{[2NO_2]}$

61. Suppose that 0.8 mole of NO is converted to NO_2. What amount of heat will be evolved?
    (A) 30 kJ
    (B) 60 kJ
    (C) 80 kJ
    (D) 130 kJ
    (E) 150 kJ

62. How does a Brønsted-Lowry acid differ from its conjugate base?
    (A) The acid has one more proton.
    (B) The acid has one less proton.
    (C) The acid has one more electron.
    (D) The acid has one less electron.
    (E) The acid and base have the same number of protons and electrons.

63. Two containers having 1 mole of hydrogen gas and 1 mole of oxygen gas, respectively, are opened. What will be the ratio of the rate of effusion of the hydrogen to that of the oxygen?
    (A) $\sqrt{2}:1$
    (B) $4:1$
    (C) $8:1$
    (D) $16:1$
    (E) $\sqrt{32}:1$

64. A molecule in which the electron configuration is a resonance hybrid is
    (A) SO_2
    (B) C_2H_6
    (C) Cl_2
    (D) HBr
    (E) NaCl

65. What is the pH of a solution in which the [OH⁻] is 1.0×10^{-4}?
    (A) −4
    (B) +4
    (C) +7
    (D) −10
    (E) +10

66. If 0.365 gram of hydrogen chloride is dissolved to make 1 liter of solution (Cl = 35.5 and H = 1), the pH concentration of the solution is
    (A) 0.001
    (B) 0.01
    (C) 1
    (D) 2
    (E) 12

67. A sample of hydrated salt was heated at 110°C for 30 minutes until all the water was driven off. The data were as follows:

    Mass of the hydrate before heating = 250 grams
    Mass of the hydrate after heating = 160 grams

    From these data, what was the percent of water by mass in the original sample?
    (A) 26.5
    (B) 36
    (C) 47
    (D) 56
    (E) 90

68. Which of the following oxides dissolves in water to form an acidic solution?
    (A) Na_2O
    (B) CaO
    (C) Al_2O_3
    (D) ZnO
    (E) SO_3

69. In the laboratory, 20.0 milliliters of an aqueous solution of calcium hydroxide, $Ca(OH)_2$, was used in a titration. A drop of phenolphthalein was added to it to indicate the end point. The solution turned colorless after 20.0 milliliters of a standard solution of 0.050 M HCl solution was added. What was the molarity of the $Ca(OH)_2$?
    (A) 0.01M
    (B) 0.025M
    (C) 0.50M
    (D) 0.75M
    (E) 1.00M

70. Which of the following reactions will NOT go to completion?
    (A) $Zn + S \rightarrow ZnS(s) + 48.5$ kcal
    (B) $CaCO_3 + 2HCl \rightarrow CaCl_2 + H_2O + CO_2(g)$
    (C) $Ag^+ + HCl \rightarrow AgCl(s) + H^+$
    (D) $Cu + 2H^+ \rightarrow Cu^{2+} + H_2;\ E^0 = -0.34$ V
    (E) $H_2SO_4 + 2NaOH \rightarrow 2Na_2SO_4 + 2H_2O$

71. For a laboratory experiment, a student placed sodium hydroxide crystals on a watch glass, assembled the titration equipment, and prepared a solution of 1.0 M sulfuric acid. Then he weighed 40 grams of sodium hydroxide and added it to enough water to make 1 liter of solution. What might be a source of error in the results of the titration?
    (A) Some sulfuric acid evaporated.
    (B) The sulfuric acid became more concentrated.
    (C) The NaOH solution gained weight, thus increasing its molarity.
    (D) The NaOH crystals gained H_2O weight, thus making the solution less than 1 molar.
    (E) The evaporation of sulfuric acid solution countered the absorption of H_2O by the NaOH solution.

72. If 60 grams of NO is reacted with sufficient O_2 to form NO_2 that is removed during the reaction, how many grams of NO_2 can be produced? (Molar mass: NO = 30 grams, NO_2 = 46 grams)
    (A) 46
    (B) 60
    (C) 92
    (D) 120
    (E) 180

73. Based on the information shown, each of the following equations represents a reaction in which ΔS is positive except:
    (A) $CaCO_3(s) \rightarrow CaO(s) + CO_2(g)$
    (B) $Zn(s) + 2H^+(aq) \rightarrow H_2(g) + Zn^{2+}(aq)$
    (C) $2C_2H_6(g) + 7O_2(g) \rightarrow 4CO_2(g) + 6H_2O(g)$
    (D) $NaCl(s) \rightarrow Na^+(aq) + Cl^-(aq)$
    (E) $N_2(g) + 3H_2(g) \rightarrow 2NH_3(g)$

74. $Cl_2(g) + 2Br^-$ (excess) $\rightarrow$?
    When 1 mole of chlorine gas reacts completely with excess KBr solution, as shown above, the products obtained are
    (A) 1 mole of Cl^- ions and 1 mole of Br^-
    (B) 1 mole of Cl^- ions and 2 moles of Br^-
    (C) 1 mole of Cl^- ions and 1 mole of Br_2
    (D) 2 moles of Cl^- ions and 1 mole of Br_2
    (E) 2 moles of Cl^- ions and 2 moles of Br_2

75. Two water solutions are made in the laboratory, one of glucose (molar mass = 180 grams), the other of sucrose (molar mass = 342 grams). If the glucose solution had 180 grams in 1,000 grams of water and the sucrose had 342 grams in 1,000 grams of water, which statement about the freezing points of the solutions is the most accurate?
    (A) The glucose solution would have the lower freezing point.
    (B) The sucrose solution would have the lower freezing point.
    (C) The freezing point of the sucrose solution would be lowered twice as much as that of the glucose solution.
    (D) Both solutions would have the same freezing point.
    (E) The freezing point of the solutions would not be affected, because the solutes are both nonpolar.

76. Which K_a value indicates the strongest acid?
    (A) 1.3×10^{-2}
    (B) 6.7×10^{-5}
    (C) 5.7×10^{-10}
    (D) 4.4×10^{-7}
    (E) 1.8×10^{-16}

77. What mass of $CaCO_3$ is needed to produce 11.2 liters of CO_2 when the calcium carbonate is reacted with an excess amount of hydrochloric acid? (Molar mass of $CaCO_3$ = 100 grams, HCl = 36.5 grams, CO_2 = 44 grams.)
    (A) 25 g
    (B) 44 g
    (C) 50 g
    (D) 100 g
    (E) none of the above

78. By experimentation it is found that a saturated solution of $BaSO_4$ at 25°C contains 3.9×10^{-5} mole/liter of Ba^{2+} ions. What is the K_{sp} of the $BaSO_4$?
    (A) 1.5×10^{-4}
    (B) 1.5×10^{-9}
    (C) 1.5×10^{-10}
    (D) 3.9×10^{-10}
    (E) 39×10^{-9}

79. What is the ΔH^0 value for the decomposition of sodium chlorate, given the following information?
$$NaClO_3(s) \rightarrow NaCl(s) + \tfrac{3}{2}O_2(g)$$
    ΔH_f^0 values for the substances are as follows: $NaClO_3(s)$ = –85.7 kilocalories/mole, $NaCl(s)$ = –98.2 kilocalories/mole, $O_2(g)$ = 0 kilocalories/mole.
    (A) 12.5 kcal
    (B) –12.5 kcal
    (C) 173.9 kcal
    (D) –173. 9 kcal
    (E) $\tfrac{3}{2}$(173.9 kcal)

$$NaCl(s) \rightleftharpoons Na^+ + Cl^-$$

80. To the equilibrium reaction shown above, a beaker of concentrated HCl (12 M) is added. Which is the best description of what will occur?
    (A) More salt will go into solution, and K_e will remain the same.
    (B) More salt will go into solution, and K_e will increase.
    (C) Salt will come out of the solution, and K_e will remain the same.
    (D) Salt will come out of the solution, and K_e will decrease.
    (E) No change in concentration will occur, and K_e will increase.

81. Balancing the following redox reaction gives what coefficient for H_2O?
$$HCl + KMnO_4 \rightarrow H_2O + KCl + MnCl_2 + Cl_2$$
    (A) 1
    (B) 2
    (C) 5
    (D) 8
    (E) 16

82. Each of the following systems is at equilibrium in a closed container. A decrease in the total volume of each container increases the number of moles of product(s) for which system?

    (A) $2NH_3(g) \rightleftharpoons N_2(g) + 3H_2(g)$
    (B) $H_2(g) + Cl_2(g) \qquad 2HCl(g)$
    (C) $2NO(g) + O_2 \qquad 2NO_2(g)$
    (D) $CO(g) + H_2O(g) \qquad CO_2(g) + H_2(g)$
    (E) $Fe_3O_4(s) + 4H_2(g) \qquad 3Fe(s) + 4H_2O(g)$

83. Which element(s), when forming an ionic bond, has (have) the following orbital notation?

    $$1s^2, 2s^2, 2p^6, 3s^2, 3p^6$$

    I. Potassium
    II. Sulfur
    III. Chlorine

    (A) I only
    (B) III only
    (C) I and III only
    (D) II and III only
    (E) I, II, and III

84. Hydrogen gas is collected in a eudiometer tube over water as shown above. The water level inside the tube is 40.8 millimeters higher than that outside. The barometric pressure is 730 millimeters Hg. The water vapor pressure at the room temperature of 29°C is found in a handbook to be 30.0 millimeters Hg. What is the pressure of the dry hydrogen?

    (A) 659.2 mm Hg
    (B) 689.2 mm Hg
    (C) 697.0 mm Hg
    (D) 740.8 mm Hg
    (E) 800.8 mm Hg

85. How many moles of electrons are required to reduce 2.93 grams of nickel ions from melted $NiCl_2$?

    (A) 0.05
    (B) 0.10
    (C) 1.0
    (D) 1.5
    (E) 2.0

STOP

IF YOU FINISH BEFORE ONE HOUR IS UP, YOU MAY GO BACK TO CHECK YOUR WORK OR COMPLETE UNANSWERED QUESTIONS.

1. (A) In the halogen family, the most active nonmetal would be the top element, fluorine, because it has the highest electronegativity.

2. (A) As you proceed down a group, the ionic radius increases as additional energy levels are filled farther from the nucleus. Therefore fluorine, the top element, has the smallest ionic radius.

3. (E) Since astatine has the largest atomic radius and its outer electrons are shielded from the protons by a large number of interior electrons, it has the lowest first ionization potential.

4. (D) Some physical characteristics of metal are found in iodine, the fourth halogen down in the group.

5. (A) Sodium acetate has Na^+ and $C_2H_3O_2^-$ ions.

6. (C) Aluminum oxide has Al^{3+} and O^{2-} ions.

7. (D) Lead carbonate has Pb^{4+} and CO_3^{2-} ions.

8. (D) John Dalton is credited with the basic postulates of atomic theory.

9. (E) Gay-Lussac is credited with the statement that, when gases combine, they do so in ratios of small whole numbers that are in relationship to the volumes of the reactants and the volumes of the products under the same conditions.

10. (C) Charles is credited with this temperature-volume relationship of gases:
$$\frac{V_1}{V_2} = \frac{T_1}{T_2}.$$

11. (A) Avogadro is credited with the concept regarding the number of particles in a mole, 6.02×10^{23}, and this number bears his name.

12. (B) Boyle is credited with the $P_1V_1 = P_2V_2$ relationship of gases.

13. (C) This includes the functional group of an aldehyde.

14. (D) This includes the functional group of an organic acid.

15. (E) An ester is the equivalent of an organic salt since it is usually formed from an organic alcohol, R–OH, plus an organic acid, $R^*-\overset{\overset{O}{\|}}{C}-OH$. The bonding is $R-O\underset{\text{-----}}{|HHO|}-\overset{\overset{O}{\|}}{C}-R^*$, which gives $R-O-\overset{\overset{O}{\|}}{C}-R^* + H_2O$.

(R* indicates that this hydrogen branch need not be the same as R.)

16. (B) This includes the functional group of an ether. It can be formed by the dehydration of two alcohol molecules. The reaction is
$$R-O\underset{\text{------}}{|H + HO|}-R^* \rightarrow R-O-R^* + H_2O.$$

17. (B) The molecular structure of carbon dioxide is $O{=}C{=}O$, where the oxygens are 180° apart and, although the bonding is polar to the carbon, counteract each other to constitute a nonpolar molecule.

18. (A) A hydrogen gas molecule is diatomic; it has 2 moles of atoms in each mole of molecules, represented by H_2.

19. (B) Each CO_2 has three atoms per molecule; hence the expression can represent 3 moles of atoms in 1 mole of molecules.

20. (E) With complete ionization $H_2SO_4 \rightarrow 2H^+ + SO_4^{2-}$, or 3 moles of ions per mole of H_2SO_4.

21. (C) The expression $2N_2O$ represents two triatomic molecules or 6 moles of atoms.

22. (D) Carbon tetrachloride has four hydrogens, each bonded to an sp^3 hybridized orbital. The four sp^3 orbitals of carbon are in a tetrahedral orientation so that the resulting configuration balances the polarity of each bond, forming a nonpolar covalent molecule.

23. (A) Because cesium is a very active metal, the chloride formed has an ionic bond and is an ionic substance.

24. (C) Chlorine's stronger electronegativity draws the hydrogen electron closer to itself more often, so that the charge distribution in the molecule is uneven, causing a negative and

positive polarity. Therefore hydrogen chloride is a polar covalent molecule.

25. (E) Benzene is an aromatic (ring) structure classified as an organic substance.

101. (T, T, CE) A catalyst can accelerate a chemical reaction by lowering the activation energy required for the reaction to occur.

102. (T, T) Sodium chloride is an ionic substance and when molten is a good electrical conductor, but the reason is that in the molten state the ions are free to migrate to the anode and cathode.

103. (T, F) Ice is less dense than liquid water. Water molecules, however, are polar, not nonpolar, and water expands as these molecules arrange themselves into a crystal lattice.

104. (F, T) It is true that the carbon and nitrogen have the same mass number, but isotopes are atoms of the same element (and atomic number) with different mass numbers.

105. (T, T, CE) A solute that has a very low solubility in a particular solvent can reach the saturated point in a solution with that solvent, but the solution can still be classified as dilute. The term *dilute* merely refers to a small amount of solute in a solvent.

106. (T, T, CE) The reaction is:

$$C + O_2 \rightarrow CO_2$$

and shows 1 mole or 12g of carbon produces 1 mole or 22.4L of CO_2. Then

$$2\cancel{L\ CO_2} \times \frac{12g\ C}{22.4\cancel{L\ CO_2}} = 1.07g\ C.$$

So the statement is correct, and the assertion explains it correctly.

107. (F, F) A reaction at equilibrium has reached a point where the forward and reverse reactions are occurring at an equal rate. The concentrations of the reactants and products are not necessarily equal and are described by the K_e value at the temperature of the reaction.

108. (F, T) In an electrochemical reaction, anions migrate to the anode. It is true that positively charged ions (cations) are attracted to the negatively charged cathode.

109. (F, T) The value pH = 5 can be expressed as

$$[H^+] = 1 \times 10^{-5} \frac{mole}{liter}$$

and pH = 3 as

$$[H^+] = 1 \times 10^{-3} \frac{mole}{liter}.$$

Thus pH = 3 represents a larger concentration of hydronium ions, H^+.

110. (T, T, CE) It is true that a negative $\Delta H_{reaction}$ indicates an exothermic reaction because heat energy is released.

111. (T, T, CE) In the expression for the equilibrium constant of an acid, $[H^+]$ is in the numerator:

$$K_a = \frac{[H^+][Y^-]}{[HY]}.$$

As $[H^+]$ decreases in the weak acids, the numerator becomes smaller as the denominator gets larger, therefore giving smaller K_a values.

112. (T, T) Both statements are true but are not related.

113. (T, F) A pi bond is formed between p lobes of adjacent atoms and in the same plane:

p Orbital   *p* Orbital

However, each p orbital consisting of two lobes can hold a total of two electrons, so the reason is false.

114. (T, F) The assertion that the boiling points of H_2S and H_2O are significantly different is true, but the reason is false. Water has the higher degree of hydrogen bonding.

40. (E) Two immiscible liquids, when shaken together vigorously, may each have small, colloidal-size particles, dispersed in the other liquid.

41. (D) In the chart, the first plateau must represent the condensation from gas to liquid because there is a second, lower plateau, which would represent the second change of state, from liquid to solid.

42. (B) If the electrons have the same principal energy level, they will fill the s^2, p^6, then the d^{10} level. This progression is from the lowest energy sublevel to the highest, to accommodate 18 electrons.

43. (B) The empirical formula identifies the atoms present and the lowest-whole-number ratio of their occurrence, as CH. The molecular formula, which gives the actual composition of the compound, must then be a multiple of the empirical formula: C_6H_6.

44. (D) Considering the concepts behind Charles's Law and Boyle's Law, you can arrive at the fraction to be used in kelvins and torrs. The volume must increase with an increase in temperature and also increase with a decrease in pressure. Therefore the fractions would be $\dfrac{333\ K}{278\ K}$ and $\dfrac{640\ torr}{320\ torr}$. Using the combined gas law equation,

$$\frac{V_1 P_1}{T_1} = \frac{V_2 P_2}{T_2}$$

and solving for V_2, you get

$$V_2 = \frac{T_2}{T_1} \times \frac{P_1}{P_2}$$

which is the same answer.

45. (D) The atomic mass, 32, is the total number of neutrons and protons. Since the atomic number, 16, gives the number of protons, $32 - 16 = 16$, or the number of neutrons, statement II is false. Since this ion has a charge of -2, it has two more electrons than protons, or 18 electrons. Statements I and III are true.

46. (D) The most active metallic elements are found in the lower left corner of the periodic table.

47. (C) The rise in the boiling point depends on the number of particles in solution. One mole of $CaCl_2$ gives 3 moles of ions, more than any other substance listed:

$$CaCl_2 \rightarrow Ca^{2+} + 2Cl^-.$$

The number of moles of ions given by the other substances are as follows: (A) = 2, (B) = 2, (D) = 1, (E) = 1.

48. (E) The sp^3 hybrid has the tetrahedron configuration. The sp^2 (D) is trigonal planar. The sp (C) is linear. The

s (B) and p (A) are the usual orbital structure.

49. (C) The reaction is

$$\overset{8\ L}{S(s)} + O_2(g) \rightarrow \overset{x\ L}{SO_2(g)}$$

The given (8 L) and the unknown (x L) are shown above. Since the equation, according to Gay-Lussac's Law, shows that 1 volume of oxygen yields 1 volume of sulfur dioxide, then

$$8\ \cancel{L\,O_2} \times \frac{1\ L\ SO_2}{1\ \cancel{L\,O_2}} = 8\ L\ SO_2$$

50. (A) In the laboratory setup shown in the question for the preparation of chlorine gas, the thistle tube at (A) is not below the liquid level. Therefore the chlorine gas could escape into the room.

51. (C) A physical property of bromine gas is that it condenses into an orange-red liquid.

52. (B) The reaction $H_3O^+ + OH^- \rightarrow H_2O$, from the Arrhenius acid and base theory, is neutralization.

53. (C) Standard conditions are 273 K and 760 mm Hg.

54. (A) Dividing the percentage of each element by the atomic mass of that element gives the basic ratio of atoms, but not necessarily in whole numbers. Thus, (Ba) $58.8 \div 137 = 0.43$, (S) $13.7 \div 32 = 0.43$, (O) $27.5 \div 16 = 1.72$.

Because atoms occur in whole numbers, you now must manipulate these numbers mathematically to get whole numbers. Usually dividing each number by the smallest one helps to accomplish this: (Ba) $0.43 \div 0.43 = 1$, (S) $0.43 \div 0.43 = 1$, (O) $1.72 \div .43 = 4$. The empirical formula is $BaSO_4$.

55. (C) The percentage composition can be found by dividing each total atomic mass in the formula by the molar mass of the compound.

$$Ca = 40 \qquad 40 \div 74 \times 100\% = 54\%\ Ca$$
$$O = 32 \qquad 32 \div 74 \times 100\% = 43\%\ O$$
$$2H = \frac{2}{74} \qquad 2 \div 74 \times 100\% = 2.7\%\ H$$

56. (B) The solution setup is:

$$\overset{65\ g}{\underset{65\ g}{Zn}} + \overset{65\ g}{\underset{73\ g}{2HCl}} \rightarrow ZnCl_2 + \overset{x\ g}{\underset{2\ g}{H_2(g)}}$$

Note that the equation mass is calculated under the substances that have mass units above them. According to the calculated equation masses, 73 g of HCl would be needed to react with 65 g of zinc. Since there is only 65 g of HCl, we use this and disregard the 65 g of Zn.

$$65 \text{ g HCl} \times \frac{2 \text{ g H}_2}{73 \text{ g HCl}}$$

$$= 1.78 \text{ or } 1.8 \text{ g H}_2$$

57. **(C)** All the other statements represent observations because they merely record what was seen.

58. **(E)** The setup is appropriate for the collection of a basically nonsoluble gas by the displacement of water. All three gases fit this description.

59. **(A)** Because the paths of the particles from the radioactive source bend toward the positive plate, the source must be negatively charged. The only radioactive particle that fits this criterion is the beta particle, $_{-1}^{0}e$.

60. **(D)** The correct K_e is

$$K_e = \frac{[NO_2]^2}{[NO]^2 [O_2]}$$

Note that the product concentration is in the numerator, with the coefficient as the exponent, while the reactants are in the denominator, again with the coefficient as the exponent.

61. **(B)** Since the equation shows that 2 moles of NO react to release 150 kJ, the solution is

$$0.8 \text{ mole NO} \times \frac{150 \text{ kJ}}{2 \text{ moles NO}} = 60 \text{ kJ}$$

This problem would be solved in the same manner if the heat had been expresssed in kilocalories. To convert one unit to the other, use 4.18×10^3 J = kcal.

62. **(A)** In the Brønsted-Lowry acid-base theory:

HCl + NH$_3$ NH$_4$ + Cl$^-$
acid   base   conjugate   conjugate
                 acid   base

63. **(B)** According to Graham's Law, the rates of effusion of two gases are inversely proportional to the square roots of their molar masses. Since 1 mol $O_2 = 32$ and 1 mol $H_2 = 2$,

$$\frac{H_2 \text{ rate}}{O_2 \text{ rate}} = \frac{\sqrt{32}}{\sqrt{2}} = \sqrt{\frac{16}{1}} = \frac{4}{1}$$

Therefore, the effusion rate of hydrogen is four times faster than that of oxygen.

64. **(A)** The structure of SO_2 is a resonance hybrid, shown as follows:

$$S{\overset{\textstyle O}{\underset{\textstyle O}{}}} \longleftrightarrow S{\overset{\textstyle O}{\underset{\textstyle O}{}}}$$

65. **(E)** The K_w of water is

$$K_w = [H^+][OH^-] = 10^{-14}$$

If $[OH^-] = 1.0 \times 10^{-4}$, then

$$[H^+] = \frac{10^{-14}}{10^{-4}} = 10^{-10}$$

and

$$pH = -\log [H^+] = 10$$

66. **(D)** $0.365 \text{ g HCl} \times \dfrac{1 \text{ mole HCl}}{36.5 \text{ g HCl}}$

= 0.01 mole HCl

0.01 mole HCl →0.01 mole H$^+$ + 0.01 mole Cl$^-$

If $[H^+] = 0.01 = 1 \times 10^{-2}$ mole/liter, then pH = 2.

67. **(B)** Hydrate mass before heating = 250 g
− Hydrate mass after heating = 160 g
_____
Water loss = 90 g

$$\frac{90 \text{ g mass loss}}{250 \text{ g original mass}} \times 100\%$$

= 36% by mass

68. **(E)** Since metallic oxides are basic anhydrides, the only nonmetallic oxide is SO_3. The reaction is as follows

$$SO_3 + H_2O \rightarrow H_2SO_4 \text{ (sulfuric acid)}$$

69. **(B)** In the titration, the reaction is:

$$2HCl + Ca(OH)_2 \rightarrow CaCl_2 + 2H_2O$$

The acid to base ratio is 2:1, or (moles acid used) = 2(moles base used), so

$M_a V_a = 2 M_b V_b$, where M is the molarity and V is the volume expressed in liters, then

$$M_b = \frac{M_a V_a}{2 V_b}$$

$$M_b = \frac{0.05 \text{ M} \times 0.02 \text{ L}}{2 (0.02 \text{ L})} = 0.025 \text{ M}$$

70. **(D)** All the reactions will go to completion except (D), which will not occur at all. If E^0 had been positive, however, this redox reaction would occur.

71. (D) Sodium hydroxide is hygroscopic and will attract water to its surface area. This water will influence its mass; consequently there will be less sodium hydroxide in the mass used.

72. (C) $\overset{60\ g}{\underset{60\ g}{2NO}} + O_2 \rightarrow \overset{x\ g}{\underset{92\ g}{2NO_2}}$

This is a mass stoichiometry problem. The equation masses are placed beneath the substances that have quantities above them.

Using the equation relationship gives

$$60\ \cancel{g\ NO} \times \frac{92\ g\ NO_2}{60\ \cancel{g\ NO}} = 92\ g\ NO_2$$

Using the mole method gives

$$60\ \cancel{g\ NO} \times \frac{1\ mole\ NO}{30\ \cancel{g\ NO}} = 2\ mole\ NO$$

Using the equation coefficients gives

$$2\ \cancel{moles\ NO} \times \frac{2\ moles\ NO_2}{2\ \cancel{moles\ NO}}$$
$$= 2\ moles\ NO_2$$

then

$$2\ \cancel{moles\ NO_2} \times \frac{46\ g\ NO_2}{1\ \cancel{mole\ NO_2}}$$
$$= 92\ g\ NO_2$$

73. (E) Since in this reaction 4 volumes of gases are forming 2 volumes of product, the randomness of the system is decreasing. Therefore, entropy is decreasing and ΔS is negative. In all the other reactions randomness is increasing.

74. (D) The reaction is

$$Cl_2(g) + 2\ Br^- (excess) \rightarrow 2Cl^- + Br_2(g)$$

Here 2 moles of chloride ions and 1 mole of Br_2 molecules are produced.

75. (D) Since both solutions were 1 molal and neither compound ionized into more particles, the freezing points would be the same.

76. (A) $$HY \leftrightharpoons H^+ + Y^-$$

Since $[H^+]$ is in the numerator of K_a:

$$K_a = \frac{[H^+][Y^-]}{[HY]}$$

The stronger the acid, the greater is $[H^+]$ and K_a value. Of the choices given, 1.3×10^{-2} is the largest.

77. (C) $$\overset{x\ g}{\underset{100\ g}{CaCO_3}} + 2HCl \rightarrow CaCl_2 + H_2O + \overset{11.2\ L}{\underset{22.4\ L}{CO_2}}$$

This is a mass-volume problem with the equation mass and volume indicated below the equation.

Using the factor-label method gives

$$11.2\ \cancel{L\ CO_2} \times \frac{100\ g\ CaCO_3}{22.4\ \cancel{L\ CO_2}}$$
$$= 50\ g\ CaCO_3$$

Using the mole method gives

$$11.2\ \cancel{L\ CO_2} \times \frac{1\ mole\ CO_2}{22.4\ \cancel{L\ CO_2}}$$
$$= 0.5\ mole\ CO_2$$

$$0.5\ \cancel{mole\ CO_2} \times \frac{1\ mole\ CaCO_3}{1\ \cancel{mole\ CO_2}}$$
$$= 0.5\ mole\ CaO_3$$

$$0.5\ \cancel{mole\ CaCO_3} \times \frac{100\ g\ CaCO_3}{1\ \cancel{mole\ CaCO_3}}$$
$$= 50\ mole\ CaO_3$$

78. (B) $$K_{sp} = [Ba^{2+}][SO_4^{2-}]$$

If $[Ba^{2+}] = 3.9 \times 10^{-5}$, then $[SO_4^{2-}]$ must also equal the same amount, so

$$K_{sp} = [3.9 \times 10^{-5}][3.9 \times 10^{-5}]$$
$$= 15.2 \times 10^{-10}\ or\ 1.5 \times 10^{-9}.$$

79. (B) $\Delta H^0_{reaction} = \Delta H^0_f (products) - \Delta H^0_f$ (reactants)

$$\Delta H^0_{reaction} = -98.2\ kcal - (-85.7\ kcal)$$
$$\Delta H^0_{reaction} = -12.5\ kcal$$

80. (C) Since the HCl solution will add to the chloride ion concentration, according to Le Châtelier's principle the equilibrium will shift in the direction to reduce this disturbance, so K_e will remain the same but salt will come out of solution. This process:

$$NaCl \leftrightharpoons Na^+ + Cl^-$$

will continue until K_e is reestablished. This phenomenon is called the "common ion effect."

81. (D) $HCl + KMnO_4 \rightarrow H_2O + KCl + MnCl_2 + Cl_2$

oxidation: $2Cl^- - 2e^- \rightarrow Cl_2$

reduction: $8H^+ + MnO_4^- + 5e^- \rightarrow Mn^{2+} + 4H_2O$

5 (oxidation reaction) and 2 (reduction reaction) will balance the e^- gain and loss, giving

$$10Cl^- - \cancel{10e^-} \rightarrow 5Cl_2$$
$$16H^+ + 2MnO_4^- + \cancel{10e^-} \rightarrow 2Mn^{2+} + 8H_2O$$

Therefore

$$16HCl + 2KMnO_4 \rightarrow 8H_2O + 2KCl + 2MnCl_2 + 5Cl_2$$

82. (C) A decrease in volume will cause the equilibrium to shift in the direction that has less volume(s) of gas(es). In every case except (C) this is the reverse reaction, which decreases the product. The coefficients give the volume relationships.

83. (E) Each of the elements when becoming ions has this configuration of 18 electrons.

K with 19 electrons $- e^- \rightarrow K^+$ with 18 electrons

S with 16 electrons $+ 2e^- \rightarrow S^{2-}$ with 18 electrons

Cl with 17 electrons $+ e^- \rightarrow Cl^-$ with 18 electrons

84. (C) Change the water height to the equivalent Hg height:

$$40.8 \text{ mm } H_2O \times \frac{1 \text{ mm Hg}}{13.6 \text{ mm } H_2O}$$

$$= 3 \text{ mm Hg}$$

Adjust for the difference in height to get the gas pressure. Pressure on the gas is

730 mm Hg – 3 mm Hg

= 727 mm Hg.

Vapor pressure of H_2O at 29°C accounts for 30 mm Hg pressure. Therefore

727 mm Hg – 30 mm Hg

= 697 mm Hg

85. (B) The reaction is

$$Ni^{2+} + 2e^- \rightarrow Ni \text{ (s)}$$

Since 2 moles of electrons are required to form 1 mole of nickel, and

$$2.9 \text{ g Ni} \times \frac{1 \text{ mole Ni}}{58.7 \text{ g Ni}} = 0.05 \text{ mole Ni}$$

then

$$0.05 \text{ mole Ni} \times \frac{2 \text{ moles electrons}}{1 \text{ mole Ni}}$$

$$= 0.1 \text{ mole of electrons}$$

DIAGNOSING YOUR NEEDS

This section will help you to diagnose your need to review the various categories tested by SAT II: Chemistry.

After taking the diagnostic test, check your answers against the correct ones. Then fill in the chart below. In the space under each question number, place a check (✔) if you answered that question correctly.

Next, total the checkmarks for each section and insert the number in the designated block. Now do the arithmetic indicated, and insert your percentage for each area.

SUBJECT AREA*   (✔) QUESTIONS ANSWERED CORRECTLY

I. Atomic Theory and Structure, including periodic relationships	1	2	3	5	6	7	8	42

No. of checks ÷ 8 × 100 = _____ %

II. Nuclear Reactions	104	45	59

No. of checks ÷ 3 × 100 = _____ %

III. Chemical Bonding and Molecular Structure	17	20	22	23	24	113	114	48	83

No. of checks ÷ 9 × 100 = _____ %

IV. States of Matter and Kinetic Molecular Theory of Gases	9	10	12	41	44	49	53	63

No. of checks ÷ 8 × 100 = _____ %

V. Solutions, including concentration units, solubility, and colligative properties	105	40	47	69	75

No. of checks ÷ 5 × 100 = _____ %

VI. Acids and Bases	109	111	52	62	65	66	68	76

No. of checks ÷ 8 × 100 = _____ %

VII. Oxidation-reduction and Electrochemistry	102	108	70	74	81	85

No. of checks ÷ 6 × 100 = _____ %

VIII.	**Stoichiometry**	11	18	19	106	112	43	54	55	56	72	77

☐ No. of checks ÷ 11 × 100 = _____%

IX.	**Reaction Rates**	101	107

☐ No. of checks ÷ 2 × 100 = _____%

X.	**Equilibrium**	60	78	80	82

☐ No. of checks ÷ 4 × 100 = _____%

XI.	**Thermodynamics:** energy changes in chemical reactions, randomness, and criteria for spontaneity	110	61	73	79

☐ No. of checks ÷ 4 × 100 = _____%

XII.	**Descriptive Chemistry:** physical and chemical properties of elements and their familiar compounds; organic chemistry; periodic properties	4	13	14	15	16	21	25	103
							46	51	64

☐ No. of checks ÷ 11 × 100 = _____%

XIII.	**Laboratory:** equipment, procedures, observations, safety, calculations, and interpretation of results	50	57	58	67	71	84

☐ No. of checks ÷ 6 × 100 = _____%

* The subject areas have been expanded to identify specific areas in the text.

PLANNING YOUR STUDY

The percentages give you an idea of how you have done on the various major areas of the test. Because of the limited number of questions on some parts, these percentages may not be as reliable as the percentages for parts with larger numbers of questions. However, you should now have at least a rough idea of the areas in which you have done well and those in which you need more study. (There are four more practice tests in the back of this book which may be used in a diagnostic manner as well.)

Start your study with the areas in which you are the weakest. The corresponding chapters are indicated on the next page.

Subject Area	Chapters to Review
I. Atomic Theory and Structure, including periodic relationships	2, 15
II. Nuclear Reactions	15
III. Chemical Bonding and Molecular Structure	3, 4
IV. States of Matter and Kinetic Molecular Theory of Gases	1
V. Solutions, including concentration units, solubility, and colligative properties	7
VI. Acids and Bases	11
VII. Oxidation-reduction and Electrochemistry	12
VIII. Stoichiometry	5, 6
IX. Reaction Rates	9
X. Equilibrium	10
XI. Thermodynamics, including energy changes in chemical reactions, randomness, and criteria for spontaneity	8
XII. Descriptive Chemistry: physical and chemical properties of elements and their familiar compounds; organic chemistry; periodic properties	1, 2, 13, 14
XIII. Laboratory: equipment, procedures, observations, safety, calculations, and interpretation of results.	All lab diagrams, 16

After you have spent some time reviewing your weaker areas, plan a schedule of work that spans the 6 weeks before the test. Unless you set up a regular study pattern and goals, you probably will not prepare sufficiently.

The following schedule provides such a plan. Note that weekends are left free, and the time spans are held to 1- or 2-hour blocks. This will be time well spent!

	Monday	Tuesday	Wednesday	Thursday	Friday
First Week:	Ch. 1: 1 hr	Ch. 2: 2 hr	Ch 2: 1 hr.	Ch. 3: 2 hr	Ch. 3: 1 hr
Second Week:	Ch. 4: 1 hr	Ch. 4: 1 hr	Ch. 5: 2 hr	Ch. 5: 1 hr	Ch. 6: 1 hr
Third Week:	Ch. 6: 2 hr	Ch. 7: 2 hr	Ch. 8: 1 hr	Ch. 9: 2 hr	Ch. 10: 1 hr
Fourth Week:	Ch. 10: 1 hr	Ch. 11: 2 hr	Ch. 12: 2 hr	Ch. 13: 1 hr	Ch. 14: 2 hr
Fifth Week:	Chs. 15, 16: 1 hr	Take Practice Test 1: 1 hr Study Ans.: 1 hr	Review weakest areas: 1 hr	Take Practice Test 2: 1 hr Study Ans.: 1 hr	Review weakest areas: 1 hr
Sixth Week:	Take Practice Test 3: 1 hr Study Ans.: 1 hr	Review weakest areas: 1 hr	Take Practice Test 4: 1 hr Study Ans.: 1 hr	Review weakest areas: 1 hr	Review a practice test already taken: 1 hr *Go to bed early.*

FINAL PREPARATION—THE DAY BEFORE THE TEST

The day before the test, review one of the practice tests you have already taken. Study again the directions for each type of question. Long hours of study at this point will probably only heighten your anxiety, so just look over the answer section of the practice test and refer to any chapter in the book if you need more information. This type of limited, relaxed review will probably make you feel more comfortable and better prepared.

Get together the materials that you will need. They are:

- Your admission ticket.
- Your identification. (You will not be admitted without some type of positive identification such as a student I.D. card with picture or a driver's license.)
- Two No. 2 pencils with erasers.

You should also go over this checklist:

A. Plan your activities so that you will have time for a good night's sleep.
B. Lay out comfortable clothes for the next day.
C. Review the following helpful tips about taking the test:

    1. Read the directions carefully.
    2. In each group of questions, answer first those that you know. Temporarily skip difficult questions, but mark them in the margin so you can go back if you have time. Keep in mind that an easy question answered correctly counts as much as a difficult one.
    3. Avoid haphazard guessing since this will probably lower your score. However, if you can eliminate one or more of the choices to a question, it will generally be to your advantage to guess which of the remaining answers is correct. Your score will be based on the number right minus a fraction of the number answered incorrectly.
    4. Remember that you neither gain nor lose credit for questions you do not answer.
    5. Pace your time throughout the hour.
    6. Mark the answer grid clearly and correctly. Be sure each answer is placed in the proper space and within the oval. *Erase all stray marks completely.*
    7. You may write as much as you like in the test booklet. Use it as a scratch pad. Only the answers on the answer sheet are scored for credit.

D. Set your alarm clock so as to allow plenty of time to dress, eat your usual (or even a better) breakfast, and reach the test center without haste or anxiety.

Introduction to Chemistry

MATTER

Definition of Matter

Matter is defined as anything that occupies space and has mass. *Mass* is the quantity of matter which a substance possesses and, depending on the gravitational force acting on it, has a unit of weight assigned to it. Although the *weight* then can vary, the mass of the body is a constant and can be measured by its resistance to a change of position or motion. This property of mass to resist a change of position or motion is called *inertia*. Since matter does occupy space, we can compare the masses of various substances that occupy a particular unit volume. This relationship of mass to a unit volume is called the *density* of the substance. It can be shown in a mathematical formula as $D = \dfrac{m}{V}$. The basic unit of mass (m) in chemistry is the gram (g), and of volume (V) is the cubic centimeter (cm^3) or milliliter (mL).

An example of how density varies can be shown by the difference in the volume occupied by 1 gram of a metal, such as gold, to 1 gram of Styrofoam. Both have the same mass, 1 gram, but the volume occupied by the Styrofoam is much larger. Therefore, the density of the metal will be much larger than that of the Styrofoam. When dealing with gases in chemistry, the standard units for density of gases is given in grams/liter at a standard temperature and pressure. This aspect of density of gases is dealt with in Chapter 6. Basically then, density can be defined as the mass per unit volume.

States of Matter

Matter occurs in three states: solid, liquid, and gas. A *solid* has both a definite size and shape. A *liquid* has a definite volume but takes the shape of the container and a *gas* has neither definite shape nor definite volume. These states of matter can often be changed by the addition of heat energy. An example of this is ice changing to liquid water and finally steam.

Composition of Matter

Matter can be subdivided into two general categories: distinct substances and mixtures. The distinct substances are substances that can be subdivided into the smallest particle that still has the properties of that substance. At that point, if the substance is made up of only one kind of atom, it is called an *element*. Atoms are considered to be the basic building blocks of matter that cannot be easily created nor destroyed. The word *atom* comes from the Greeks and means the smallest possible piece of something. Today, scientists recognize approximately 109 different kinds of atoms each with their own unique composition. These atoms then are the building blocks of elements when only one kind of atom makes up the substance. If, however, two or more kinds of atoms join together in a definite grouping, this distinct substance is called a *compound*. Compounds are made by combining elements in a definite proportion (or ratio) by mass and are made up of two or more kinds of atoms. This is called the *Law of Definite Composition (or Proportions)*. The smallest naturally occurring unit of a compound is called a *molecule* of that compound. A molecule of a compound has a definite shape that is determined by how the atoms are bonded to or combined with each other. This bonding is described in Chapter 3, Bonding. An example is the compound water: it always occurs in a two hydrogen atoms to one oxygen atom relationship. *Mixtures*, however, can vary in their composition.

In general, then:

Mixtures	*Distinct Substances*
	ELEMENTS
1. Composition is indefinite (generally heterogeneous).* (Example: marble)	1. Composition is made up of one kind of atom. (Examples: nitrogen, gold, neon)
2. Properties of the constituents are retained.	2. All parts are the same throughout (homogeneous).
3. Parts of the mixture react differently to changed conditions.	COMPOUNDS
	1. Composition is definite (homogeneous). (Examples: water, carbon dioxide)
*Solutions are mixtures, such as sugar in water, but since the substance, like sugar, is distributed evenly throughout the water, it can be said to be a homogeneous mixture.	2. All parts react the same.
	3. Properties of the compound are distinct and different from the properties of the individual elements that are combined in its make-up.

Chemical and Physical Properties

Physical properties of matter are those properties that can usually be observed with our senses. They include everything about a substance that can be noted when no change is occurring in the type of structure that makes up its smallest component. Some common examples are physical state, color, odor, solubility in water, density, melting point, taste, boiling point, and hardness.

Chemical properties are those properties that can be observed in regard to whether or not a substance reacts with other substances. Some common examples are: iron rusts in moist air, nitrogen does not burn, gold does not rust, sodium reacts with water, silver does not react with water, and water can be decomposed by an electric current.

Chemical and Physical Changes

The changes matter undergoes are classified as either physical or chemical. In general, a *physical change* alters the physical properties of matter, but the composition remains constant.

The most often altered properties are form and state. Some examples are: breaking glass, cutting wood, melting ice, and magnetizing a piece of metal. In some cases, the process that caused the change can be easily reversed and the substance regains its original form.

Chemical changes are changes in the composition and structure of a substance. They are always accompanied by energy changes. If the energy released in the formation of a new structure exceeds the chemical energy in the original substances, energy will be given off, usually in the form of heat or light or both. This is called an *exothermic reaction*. If, however, the new structure needs to absorb more energy than is available from the reactants, the result is an *endothermic reaction*. This can be shown graphically.

Notice that in Figures 1 and 2 the term *activation energy* is used. The activation energy is the energy necessary to get the reaction going by increasing the energy of the reactants so they can combine. You know you have to heat paper before it burns. This heat raises the energy of the reactants so that the burning can begin; then enough energy is given off from the burning so that an external source of energy is no longer necessary.

FIGURE 1. *An Exothermic Reaction*   FIGURE 2. *An Endothermic Reaction*

Conservation of Mass

When ordinary chemical changes occur, the mass of the reactants equals the mass of the products. This can be stated another way: In a chemical change, matter can neither be created nor destroyed, but only changed from one form to another. This is referred to as the *Law of Conservation of Matter* (Lavoisier—1785). This law is extended by the Einstein mass-energy relationship, which states that matter and energy are interchangeable (see page 31).

ENERGY

Definition of Energy

The concept of energy plays an important role in all of the sciences. In chemistry, all physical and chemical changes have energy considerations associated with them. To understand how and why these changes happen, an understanding of energy is required.

Energy is defined as the capacity to do work. Work is done whenever a force is applied over a distance. Therefore, anything that can force matter to move, to change speed, or to change direction has energy. The following example will help you understand this definition of energy. When you charge a battery with electricity, you are storing energy in the form of chemical energy. The charged battery has a capacity to do work. If you use the battery to operate a toy car, the stored energy is transformed into mechanical energy that exerts a force on the mechanism that turns the wheels and makes the car move. This continues until the *charge* or stored energy is completely used. In its uncharged condition, the battery no longer has the capacity to do work.

Just as work itself is measured in *joules*, so is energy. In some problems, energy may be expressed in *kilocalories*. The relationship between these two units is that 4.18×10^8 joules (J) equals 1 kilocalorie (kcal).

Forms of Energy

Energy may appear in a variety of forms. Most commonly, energy in reactions is evolved as *heat*. Some other forms of energy are *light, sound, mechanical energy, electrical energy*, and *chemical energy*. Energy can be converted from one form to another, as when the heat from burning fuel is used to vaporize water to steam. The energy of the steam is used to turn the wheels of a turbine to produce mechanical energy. The turbine turns the generator armature to produce electricity, which is then available in homes for use as light or heat, or in the operation of many modern appliances.

Two general classifications of energy are *potential energy* and *kinetic energy*. Potential energy is due to position; kinetic energy is energy of motion. The difference can be illustrated by a boulder sitting on the side of a mountain. It has a high potential energy due to its position above the valley floor. If it falls, however, its potential energy is converted to kinetic energy. This illustration is very similar to the situation of electrons cascading to lower energy levels in the atomic model.

This concept can also be applied at the molecular level. For example, the molecule of N_2 in the diagram below shows three variations of kinetic energy. The N_2 molecule also possesses potential energy because of the chemical bond within the molecule. To break this bond would require energy. The potential energy in each isolated atom of nitrogen would then be greater. This would be similar to raising the boulder to a higher position on the mountainside. Both systems have a higher potential energy and have a tendency to fall back to the lower state, which is more stable.

| Translational | Vibrational | Rotational |

Types of Reactions (Exothermic versus Endothermic)

When physical or chemical changes occur, energy is involved. If the heat content of the product(s) is higher than that of the reactants, the reaction is *endothermic*. If, on the other hand, the heat content of the product(s) is less than that of the reactants, the reaction is called *exothermic*. This change of heat content can be designated as ΔH. The heat content (H) is sometimes referred to as the *enthalpy*. Every system has a certain amount of heat. This changes during the course of a physical or chemical change. The change in heat content, ΔH, is the difference between the heat content of the products and that of the reactants. The equation is:

$$\Delta H = H_{\text{products}} - H_{\text{reactants}}$$

If the heat content of the products is greater than the heat content of the reactants, ΔH is a positive quantity ($\Delta H > 0$) and the reaction is endothermic. If, however, the heat content of the products is less than the heat content of the reactants, ΔH is a negative quantity ($\Delta H < 0$) and the reaction is exothermic. This relationship is shown graphically in Figures 1 and 2 on page 29. This topic is developed in detail in Chapter 8.

Conservation of Energy

Experiments have shown that energy is neither gained nor lost in physical or chemical changes. This principle is known as the *Law of Conservation of Energy* and is often stated as follows: Energy is neither created nor destroyed in ordinary physical and chemical changes.

CONSERVATION OF MASS AND ENERGY

With the introduction of atomic theory and a more complete understanding of the nature of both mass and energy, it was found that a relationship exists between these two concepts. Einstein formulated the *Law of Conservation of Mass and Energy*. This states that mass and energy are interchangeable under special conditions. The conditions have been created in nuclear reactors and accelerators, and the law has been verified. This relationship can be expressed by Einstein's famous equation:

$$E = mc^2$$
$$\text{Energy} = \text{Mass} \times (\text{Velocity of light})^2$$

MEASUREMENTS AND CALCULATIONS

The student of chemistry must be able to use the correct measurement terms accurately and to solve problems that require mathematical skill, as well as proper terminology, for their correct solution. The following sections review these topics.

Metric System

It is important that scientists around the world use the same units when communicating information. For this reason, scientists use the modernized metric system, designated in 1960 by the General Conference on Weights and Measures as the International System of Units. This is commonly known as SI, an abbreviation for the French name Le Systeme International d'Unites. It is now the most common system of measurement in the world.

The reason it is so widely accepted is twofold. First, SI uses the decimal system as its base. Secondly, in many cases units for various quantities are defined in terms of units for simpler quantities.

There are seven basic units that can be used to express the fundamental properties of measurement. These are called the SI base units and are shown in the table that follows.

SI BASE UNITS		
Property	Unit	Abbreviation
mass	kilogram	kg
length	meter	m
time	second	s
electric current	ampere	A
temperature	kelvin	K
amount of substance	mole	mol
luminous intensity	candela*	cd

*The candela is rarely used in chemistry.

Other SI units are derived by combining prefixes with a root unit. The prefixes represent multiples or fractions of 10. The following table gives some basic prefixes used in the metric system.

Prefix	Multiple	Scientific Notation	Abbreviation
mega-	1,000,000	10^6	M
kilo-	1,000	10^3	k
hecto-	100	10^2	h
deka-	10	10^1	da
deci-	0.1	10^{-1}	d
centi-	0.01	10^{-2}	c
milli-	0.001	10^{-3}	m
nano-	0.000,000,001	10^{-9}	n

For an example of how the prefix works in conjunction with the root word, consider the term *kilometer*. The prefix *kilo-* means "multiply the root word by 1000," so a kilometer is 1,000 meters. By the same reasoning, a millimeter is 1/1,000 of a meter.

Because of the prefix system, all units and quantities can be easily related by some factor of 10. Here is a brief table of some metric unit equivalents.

Length
10 millimeters (mm) = 1 centimeter (cm)
100 cm = 1 meter (m)
1,000 m = 1 kilometer (km)

Volume
1,000 milliliters (mL) = 1 liter (L)
1,000 cubic centimeters (cm^3) = 1 liter
1 mL = 1 cm^3

Mass
1,000 milligrams (mg) = 1 gram (g)
1,000 g = 1 kilogram (kg)

A unit of length, used especially in expressing the length of light waves, is the angstrom, abbreviated as Å and equal to 10^{-8} cm.

Because measurements are occasionally reported in units of the English system, it is important to be able to convert them to metric units. Some common conversion factors are shown in the following table.

> 2.54 cm = 1 inch
> 1 meter = 39.37 inches (10% longer than a yard)
> 1 ounce = 28.35 grams
> 1 pound = 454 grams
> 2.2 pounds = 1 kilogram
> 1 quart = 0.946 liter
> 1.06 quarts = 1 liter (5% larger than a quart)

The metric system standards were chosen as natural standards. The meter was first described as the distance marked off on a platinum-iridium bar but now is defined as the length of the path traveled by light in a vacuum during a time interval of $1/2.99792458 \times 10^8$ second.

There are some interesting relationships between volume and mass units in the metric system. Since water is most dense at 4°C, the gram was intended to be 1 cubic centimeter of water at this temperature. This means, then, that:

1,000 cm^3 = 1 liter of water @ 4°C

1,000 cm^3 of water weighs 1,000 g @ 4°C

Therefore

1 liter of water @ 4°C weighs 1 kg

and

1 mL of water @ 4°C weighs 1 g

1 mL = 1 cm^3

10 cm

10 cm

10 cm

1 liter or 1,000 cm^3

When 1 liter is filled with water @ 4°C, it has a mass of 1 kilogram.

Temperature Measurements

The most commonly used temperature scale in scientific work is the Celsius scale. It gets its name from the Swedish astronomer Anders Celsius and dates back to 1742. For a long time it was called the centigrade scale because it is based on the concept of dividing the distance on a thermometer between the freezing point of water and its boiling point into 100 equal markings or degrees.

There is another scale that is based on the lowest theoretical temperature (called absolute zero). This temperature has never actually been reached, but scientists in laboratories have reached temperatures within about a thousandth of a degree above absolute zero. Lord William Kelvin proposed this scale, on which a degree is the same size as a Celsius degree and which is referred to as the Kelvin scale. Through experiments and calculations, it has been determined that absolute zero is 273.15 degrees below zero on the Celsius scale. This figure is usually rounded off to –273°C.

The diagram and formulas that follow give the graphic and algebraic relationships between the temperature scales encountered in chemistry.

Conversion Formulas

$$°F = \frac{9}{5}°C + 32°$$
$$°C = \frac{5}{9}(°F - 32°)$$
$$K = °C + 273°$$
$$°C = K - 273°$$

• **EXAMPLES:**

30°C = ?°F	$°F = \frac{9}{5}(30°) + 32° = 86°$
68°F = ?°C	$°C = \frac{5}{9}(68° - 32°) = 20°$
10°C = ? K	K = 10 + 273 = 283
200 K = ?°C	°C = 200 − 273° = −73°

Note: In Kelvin notation, the degree sign is omitted: 283 K. The unit is the kelvin, abbreviated as K.

Heat Measurements

The scales above are used to measure the degree of heat. A pail and a thimble can both be filled with water at $100°$ Celsius. The water in both measures the same degree of heat. However, the pail of water has a greater quantity of heat. This could be easily demonstrated by the amount of ice that could be melted by the water in these two containers. Obviously, the pail of water at $100°$ Celsius will melt more ice than will a thimble full of water at the same temperature. Therefore, the pail of water contains a greater number of *calories* of heat. The calorie unit is used to measure the quantity of heat. It is defined as the amount of heat needed to raise the temperature of 1 gram of water by 1 degree on the Celsius scale. This is a rather small unit to measure the quantities of heat involved in most chemical reactions. Therefore, the *kilocalorie* is more often used. The kilocalorie equals 1,000 calories. It is the quantity of heat that will increase the temperature of 1 kilogram of water by 1 degree on the Celsius scale. As a unit of heat energy, 1 calorie equals approximately 4.18 joules (J).

Problems involving heat transfers in water are called water calorimetry problems. These are explained on page 125.

Scientific Notation

When students must do mathematical operations with numerical figures, the *scientific notation system* is very useful. Basically this system uses the exponential means of expressing figures. With large numbers, such as 3,630,000., move the decimal point to the left until only one digit remains to the left (3.630000) and then indicate the number of moves of the decimal point as the exponent of 10 (3.63×10^6). With a very small number such as 0.000000123, move the decimal point to the right until only one digit is to the left (0000001.23) and then express the number of moves as the negative exponent of 10 (1.23×10^{-7}).

With numbers expressed in this exponential form, you can now use your knowledge of exponents in mathematical operations. An important fact to remember is that in multiplication you add the exponents of 10, and in division you subtract the exponents.

- **EXAMPLES:**

Multiplications:

$(2.3 \times 10^5)(5.0 \times 10^{-12})$. Multiplying the first number in each, you get 11.5 and the addition of the exponents gives 10^{-7}. Now changing to a number with only one digit to the left of the decimal point gives you 1.15×10^{-6} for the answer.

Try these:

$(5.1 \times 10^{-6})(2 \times 10^{-3}) = 10.2 \times 10^{-9} = 1.02 \times 10^{-8}$

$(3 \times 10^5)(6 \times 10^3) = 18 \times 10^6 = 1.8 \times 10^9$

Divisions:

$(1.5 \times 10^3) \div (5.0 \times 10^{-2}) = .3 \times 10^5 = 3 \times 10^4$

$(2.1 \times 10^{-2}) \div (7.0 \times 10^{-3}) = .3 \times 10^1 = 3$

(Notice that in division the exponents of 10 are subtracted.)

Factor-Label Method of Conversion (Dimensional Analysis)

When you are working problems that involve numbers with units of measurement, it is convenient to use this method so that you do not become confused in the operations of multiplication or division. For example, if you are changing 0.001 kilogram to milligrams, you set up each conversion as a fraction so that all the units will factor out except the one you want in the answer.

$$1 \times 10^{-3} \text{ kg} \times \frac{1 \times 10^3 \text{ g}}{1 \text{ kg}} \times \frac{1 \times 10^3 \text{ mg}}{1 \text{ g}} = 1 \times 10^3 \text{ mg}$$

Notice that the kilogram is made the denominator in the first fraction to be factored with the original kilogram unit. The numerator is equal to the denominator except that the numerator is expressed in smaller units. The second fraction has the gram unit in the denominator to be factored with the gram unit in the preceding fraction. The answer is in milligrams since this is the only unit remaining and it assures you that the correct operations have been performed in the conversion.

- **ANOTHER EXAMPLE:**

1 ft. = ? cm

$$1 \text{ ft.} \times \frac{12 \text{ in.}}{1 \text{ ft.}} \times \frac{2.54 \text{ cm}}{1 \text{ in.}} = 30.48 \text{ cm}$$

This method is used in examples throughout this book.

Precision, Accuracy, and Uncertainty

Two other factors to consider in measurement are *precision* and *accuracy*. Precision indicates the reliability or reproducibility of a measurement. Accuracy indicates how close a measurement is to its known or accepted value.

For example, suppose you were taking a reading of the boiling point of pure water at sea level. Using the same thermometer in three trials, you record 96.8, 96.9, and 97.0 degrees Celsius. Since these figures show a high reproducibility, you can say that they are precise. However, the values are considerably off from the accepted value of 100 degrees Celsius, so we say they are not accurate. In this example we probably would suspect that the inaccuracy was the fault of the thermometer.

Regardless of precision and accuracy, all measurements have a degree of *uncertainty*. This is usually dependent on two factors—the limitation of the measuring instrument and the skill of the person making the measurement. This can best be shown by example.

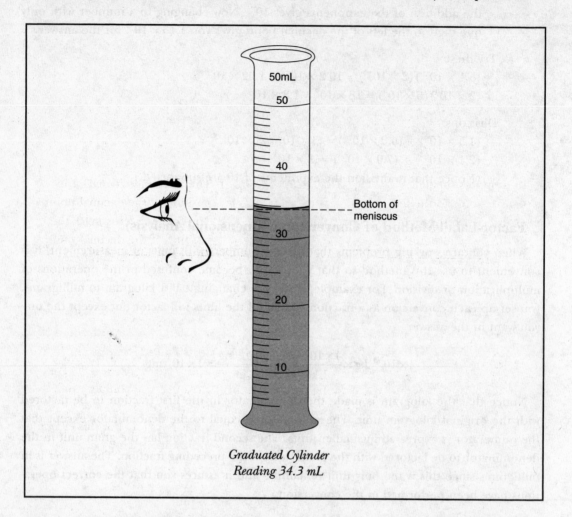

Graduated Cylinder
Reading 34.3 mL

The graduated cylinder in the illustration contains a quantity of water to be measured. It is obvious that the quantity is betwen 30 and 40 milliliters because the meniscus lies between these two marked quantities. Now, checking to see where the bottom of the *meniscus* lies with reference to the ten intervening subdivisions, we see that it is between the fourth and fifth. This means that the volume lies between 34 and 35 milliliters. The next step introduces the uncertainty. We have to guess how far the reading is between these two markings. We can make an approximate guess, or estimate, that the level is more than 0.2 but less than 0.4 of the distance. We therefore report the volume as 34.3 mL (milliliters). The last digit in any measurement is an estimate of this kind and is uncertain.

Significant Figures

Any time a measurement is recorded, it includes all the digits that are certain plus one uncertain digit. These certain digits plus the one uncertain digit are referred to as *significant figures*. The more digits you are able to record in a measurement, the less relative uncertainty there is in the measurement. The following table summarizes the rules of significant figures.

Rule	Example	Number of Significant Figures	
All digits other than zeros are significant.	25 g	2	
	5.471 g	4	
Zeros between nonzero digits are significant.	309 g	3	
	40.06 g	4	
Final zeros to the right of the decimal point are significant.	6.00 mL	3	
	2.350 mL	4	
In numbers smaller than 1, zeros to the left or directly to the right of the decimal point are not significant.	0.05 cm	1	The zeros merely mark the position of the decimal point.
	0.060 cm	2	The first two zeros mark the position of the decimal point. The final zero is significant.

One last rule deals with final zeros in a whole number. These zeros may or may not be significant, depending on the measuring instrument. For instance, if an instrument that measures to the nearest mile is used, the number 3,000 miles has four significant figures. If, however, the instrument in question records miles to the nearest thousands, there is only one significant figure. The number of significant figures in 3,000 could be one, two, three, or four, depending on the limitation of the measuring device.

This problem can be avoided by using the system of scientific notation. For this example, the following notations would indicate the numbers of significant figures:

$$3 \times 10^3 \quad \text{one significant figure}$$
$$3.0 \times 10^3 \quad \text{two significant figures}$$
$$3.00 \times 10^3 \quad \text{three significant figures}$$
$$3.000 \times 10^3 \quad \text{four significant figures}$$

Calculations With Significant Figures

When you do calculations involving numbers that do not have the same number of significant figures in each, keep the following two rules in mind.

First, in multiplication and division, the number of significant figures in a product or a quotient of measured quantities is the same as the number of significant figures in the quantity having the smaller number of significant figures.

• EXAMPLE 1

Problem	Unrounded answer	Answer rounded to the correct number of significant figures
4.29 cm × 3.24 cm =	13.8996 cm^2 =	13.9 cm^2

Explanation: Both measured quantities have three significant figures. Therefore, the answer should be rounded to three significant figures.

• EXAMPLE 2

Problem	Unrounded answer	Answer rounded to the correct number of significant figures
4.29 cm $\times$ 3.2 cm =	13.728 cm^2 =	14 cm^2

Explanation: One of the measured quantities has only two significant figures. Therefore, the answer should be rounded to two significant figures.

• EXAMPLE 3

Problem	Unrounded answer	Answer rounded to the correct number of significant figures
8.47 cm^2/4.26 cm =	1.9882629 cm =	1.99 cm

Explanation: Both measured quantities have three significant figures. Therefore, the answer should be rounded to three significant figures.

Second, when adding or subtracting measured quantities, the sum or difference should be rounded to the same number of decimal places as the quantity having the least number of decimal places.

• EXAMPLE 1

Problem	Unrounded answer	Answer rounded to the correct number of significant figures
3.56 cm		
2.6 cm		
+ 6.12 cm		
Total =	12.28 cm =	12.3 cm

Explanation: One of the quantities added has only one decimal place. Therefore, the answer should be rounded to only one decimal place.

• EXAMPLE 2

Problem	Unrounded answer	Answer rounded to the correct number of significant figures
3.514 cm		
− 2.13 cm		
Difference =	1.384 cm =	1.38 cm

Explanation: One of the quantities has only two decimal places. Therefore, the answer should be rounded to only two decimal places.

Chapter

1

Review

• QUESTIONS

1. 1.2 mg = _____ g

2. 6.3 cm = _____ mm

3. 5.12 km = _____ cm

4. 12 L = _____ qt.

5. 6.111 mL = _____ L

6. 1 km = _____ mm

7. 1.03 kg = _____ g

8. 0.003 g = _____ kg

9. 22.4 L = _____ mL

10. 10,013 cm = _____ km

11. The density of CCl_4 (carbon tetrachloride) is 1.58 g/mL. What will 100 mL of CCl_4 weigh?

12. A piece of sulfur weighs 8 oz. When it was submerged in a graduated cylinder containing 50 mL of H_2O, the level rose to 150 mL. What is the density (g/mL) of the sulfur? (16 oz. = 454 g)

13. (a) A box 20 cm × 20 cm × 5.08 inches has what volume in cubic centimeters? (b) What weight of H_2O @ 4°C will the box hold?

14. Set up the following using the *factor-label method:*

$$\frac{5 \text{ cm}}{\text{sec}} = \frac{\text{km}}{\text{hr}}$$

15. How many significant figures are in each of the following? (a) 1.01 (b) 200.0 (c) 0.0021   (d) 0.0230

16. If the graphic representation of the energy levels of the reactants and products in a chemical reaction looks like this:

(a) there is an exothermic reaction   (b) there is an endothermic reaction (c) the *a* portion is the energy given off   (d) the *b* portion is called the activation energy.

17. The amount of mass per unit volume refers to the (a) density   (b) specific weight   (c) volume   (d) weight.

18. A baking powder can carries the statement, "Ingredients: corn starch, sodium bicarbonate, calcium acid phosphate, and sodium aluminum sulfate." Therefore, this baking powder is (a) a compound   (b) a mixture   (c) a molecule (d) a mixture of elements.

19. Which of the following is a physical property of sugar? (a) It decomposes readily.   (b) Its composition is carbon, hydrogen, and oxygen.   (c) It turns black with concentrated H_2SO_4.   (d) It can be decomposed with heat.   (e) It is a white crystalline solid.

20. A substance that can be further simplified may be either (a) an element or a compound (b) an element or a mixture (c) a mixture or a compound (d) a mixture or an atom.

21. A substance composed of two or more elements chemically united is called (a) an isotope (b) a compound (c) an element (d) a mixture.

22. An example of a chemical change is the (a) breaking of a glass bottle (b) sawing of a piece of wood (c) rusting of iron (d) melting of an ice cube.

23. A substance that cannot be further decomposed by ordinary chemical means is (a) water (b) air (c) sugar (d) silver.

24. An example of a physical change is (a) the fermenting of sugar to alcohol (b) the rusting of iron (c) the burning of paper (d) a solution of sugar in water.

25. Chemical action may involve all of the following except (a) combining of atoms of elements to form a molecule (b) separation of the molecules in a mixture (c) breaking down compounds into elements (d) reacting a compound and an element to form a new compound and a new element.

26. The energy of a system can be (a) easily changed to mass (b) transformed into a different form (c) measured only as potential energy (d) measured only as kinetic energy.

27. If the ΔH of a reaction is a negative quantity, the reaction is definitely (a) endothermic (b) unstable (c) exothermic (d) reversible.

28. In the following list, write E for the elements, C for the compounds, and M for the mixtures.

Water	Aluminum oxide	Hydrochloric acid
Wine	Hydrogen	Nitrogen
Soil	Carbon dioxide	Tin
Silver	Air	Potassium chloride

• ANSWERS

1. 0.0012
2. 63
3. 512
4. 12.7
5. 0.006111
6. 1,000,000 or 1×10^6
7. 1,030 or 1.03×10^3
8. 0.000003 or 3×10^{-6}
9. 22,400 or 2.24×10^4
10. 0.10013

11. 158 grams

12. 2.27 g/mL

To find the volume of 8 oz. sulfur, subtract the volume of water before from the volume after immersion.

$$150 \text{ mL} - 50 \text{ mL} = 100 \text{ mL}$$

Then $8 \text{ oz.} \times \dfrac{454 \text{ g}}{8 \text{ oz.}} = 227 \text{ g}$

So, $\dfrac{227 \text{ g}}{110 \text{ mL}} = \underline{\underline{2.27 \text{ g/mL}}}$

13. (a) 5,160 cm³ (b) 5,160 g

(a) Converting 5.08 in. to cm,

$$5.08 \text{ in.} \times \frac{2.54 \text{ cm}}{1 \text{ in.}} = 12.9 \text{ cm}$$

Then 20 cm × 20 cm × 12.9 cm = $\underline{\underline{5,160 \text{ cm}^3}}$

(b) Since 1 cm³ of water at 4°C has a mass of 1 g,
then 5,160 cm³ = $\underline{\underline{5,160 \text{ g}}}$

14. $\dfrac{5 \, \cancel{\text{cm}}}{\cancel{\text{s}}} \times \dfrac{1 \, \cancel{\text{m}}}{100 \, \cancel{\text{cm}}} \times \dfrac{1 \text{ km}}{1,000 \, \cancel{\text{m}}} \times \dfrac{60 \, \cancel{\text{s}}}{1 \, \cancel{\text{min}}} \times \dfrac{60 \, \cancel{\text{min}}}{1 \text{ hr}} =$

15. (a) 3   (b) 4   (c) 2   (d) 3

16. (a)   20. (c)   24. (d)
17. (a)   21. (b)   25. (b)
18. (b)   22. (c)   26. (b)
19. (e)   23. (d)   27. (c)

28. C   C   C
    M   E   E
    M   C   E
    E   M   C

Terms You Should Know

accuracy	Fahrenheit	matter
activation energy	gas	mixture
angstrom	heterogeneous	physical change
calorie	homogeneous	physical property
Celsius	inertia	potential energy
chemical change	joule	precision
chemical property	Kelvin	scientific notation
compound	kilocalorie	significant figures
density	kinetic energy	SI units
element	liquid	solid
endothermic	mass	uncertainty
enthalpy	meniscus	weight
exothermic		

Law of Conservation of Energy
Law of Conservation of Matter
Law of Conservation of Mass and Energy
Law of Definite Composition or Proportion

2

Atomic Structure

HISTORY

The idea of small, invisible particles being the building blocks of matter can be traced back more than 2,000 years to the Greek philosophers Democritus and Leucippus. These particles were supposed to be so small and indestructible that they could not be divided into smaller particles. These particles were called *atoms,* the Greek word for indivisible. The English word *atom* comes from this Greek word. This early concept of atoms was not based upon experimental evidence but was simply a result of thinking and reasoning on the part of the philosophers. It was not until the eighteenth century that experimental evidence in favor of the atomic hypothesis began to accumulate. Finally, around 1805, John Dalton proposed some basic assumptions about atoms based on what was known through scientific experimentation and observation at that time. These assumptions were very closely related to what scientists presently know about atoms. Because of this, Dalton is often referred to as the father of modern atomic theory. Some of these basic ideas were:

1. All matter is made up of very small, discrete particles called atoms.
2. All atoms of an element are alike in weight and different from the weight of any other kind of atom.
3. Atoms unite in chemical changes to form compounds. The atom loses its characteristic properties when this combination occurs.
4. The atom is not entirely altered in the formation of a compound, and its weight remains constant.

ELECTRIC NATURE OF ATOMS

From around the beginning of the twentieth century, scientists have been gathering evidence about the structure of atoms and fitting the information into a model of the atomic structure.

Basic Electric Charges

The discovery of the electron as the first subatomic particle is credited to J. J. Thomson (England, 1897). He used an evacuated tube connected to a spark coil as shown in Figure 3. As the voltage across the tube was increased, a beam of light became visible. This was

42

referred to as a cathode ray. Thomson found that the beam was deflected by both electrical and magnetic fields. From this, he concluded that the cathode rays were made up of very small negatively charged particles; which he named *electrons*.

FIGURE 3. Electron Experiment

Further experimentation led Thomson to find the ratio of the electrical charge of the electron to its mass. This was a major step toward understanding the nature of the particle. He was awarded a Nobel Prize in 1906 for his accomplishment.

It was an American scientist, Robert Millikan, who in 1909 was able to measure the charge on an electron using the apparatus pictured below.

Millikan Oil Drop Experiment

Oil droplets were sprayed into the chamber and, in the process, became randomly charged by gaining or losing electrons. The electric field was adjusted so that a negatively charged drop would move slowly upward in front of the grid in the telescope. Knowing the rate at which the drop was rising, the strength of the field, and the weight of the drop, Millikan was able to calculate the charge on the drop. Combining the information with the results of Thomson, he was able to calculate a value for the mass of a single electron. Eventually, this number was found to be 9.11×10^{-28} g.

Ernest Rutherford (England, 1911) performed a gold foil experiment (Figure 4) that had tremendous implications for atomic structure.

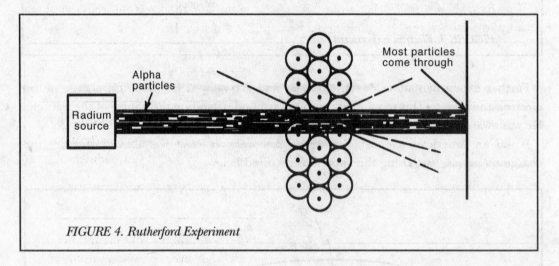

FIGURE 4. Rutherford Experiment

Alpha particles (helium nuclei) passed through the foil with few deflections. However, some deflections (1 per 8,000) were almost directly back toward the source. This was unexpected and suggested an atomic model with mostly empty space between a *nucleus*, in which most of the mass of the atom was located and which was positively charged, and the electrons that defined the volume of the atom.

Further experiments showed that the nucleus was made up of still smaller particles called *protons*. Rutherford realized, however, that protons, by themselves, could not account for the entire mass of the nucleus. He predicted the existence of a new nuclear particle that would be neutral and would account for the missing mass. In 1932, James Chadwick (England) discovered this particle, the *neutron*.

Today the number of subatomic particles identified as discrete units has risen to well over 90 named units. An update about the newest subatomic particles research is given in Chapter 15.

Bohr Model

In 1913, Neils Bohr (Denmark) proposed his model of the atom. This pictured the atom as having a dense, positively charged nucleus and negatively charged electrons in specific spherical orbits, also called energy levels or shells, around this nucleus. These shells are arranged concentrically around the nucleus, and each shell was designated by a letter: *K, L, M, N, O* . . . . The closer to the nucleus, the less energy an electron needs in one of these shells, but it has to gain energy to go from one shell to another that is farther away from the nucleus.

Because of its simplicity and general ability to explain chemical change, the Bohr model still has some usefulness today.

Shell	Maximum Number of Electrons ($2n^2$)
1	2
2	8
3	18
4	32
5	32

Components of Atomic Structure

The chart below lists the basic particles of the atom. For the latest information about subatomic particles, see "New Subatomic Particles" in Chapter 15.

Particle	Charge	Actual Mass	Relative Mass Compared to Proton	Discovery
Electron	– (e^-)	9.109×10^{-28} g	1/1,837	J. J. Thomson–1897
Proton	+ (p^+)	1.673×10^{-24} g	1	——— early 1900s
Neutron	0 (n^0)	1.675×10^{-24} g	1	J. C. Chadwick–1932

(There are now some 30 or more named particles or units of atomic structure, but the above are the most commonly used.)

When these components are used in the Bohr model, we show the protons and neutrons in the nucleus. These particles are known as *nucleons*. The electrons are shown in the shells.

The number of protons in the nucleus of an atom determines the *atomic number*. All atoms of the same element have the same number of protons and therefore the same atomic number; atoms of different elements have different atomic numbers. Thus, the atomic number identifies the element. An English scientist, Henry Moselely, first determined the atomic numbers of the elements through the use of X-rays.

Since the actual masses of subatomic particles and atoms themselves are very small numbers when expressed in grams, scientists use atomic mass units (amu) instead. An *atomic mass unit* is defined as 1/12 the mass of a carbon-12 atom. Thus, the mass of any atom is expressed relative to the mass of one atom of carbon-12, which is sometimes called the *unified mass unit* (u) or the *dalton*.

The sum of the number of protons and the number of neutrons in the nucleus is called the *mass number*.

Table 1 (see page 46) summarizes the relationships just discussed. Notice that the outermost energy level can contain no more than eight electrons. The explanation of this is given in the next section.

If atoms of the same element have different masses, they are called *isotopes*. The variance in mass and mass number is due to a different number of neutrons in the nucleus. Isotopes of an element occur in the same relative percentages wherever a natural sample of the element is found. The atomic mass (formerly atomic weight) of an element is defined as the weighted average of the masses of all its isotopes as they occur in a natural sample of the element. It is due to the occurrence of isotopes that the atomic mass of an element is not a whole number, but rather a number containing a decimal. When the atomic mass is rounded off to the nearest whole number, the resulting number represents the atomic mass of the most common isotope. This number of amu will be the same as the mass number of that isotope.

Element	Atomic No.	Mass No.	Number of Protons	Number of Neutrons	Number of Electrons	Electrons in Energy Levels			
						1	2	3	4
Hydrogen	1	1	1	0	1	1			
Helium	2	4	2	2	2	2			
Lithium	3	7	3	4	3	2	1		
Beryllium	4	9	4	5	4	2	2		
Boron	5	11	5	6	5	2	3		
Carbon	6	12	6	6	6	2	4		
Nitrogen	7	14	7	7	7	2	5		
Oxygen	8	16	8	8	8	2	6		
Fluorine	9	19	9	10	9	2	7		
Neon	10	20	10	10	10	2	8		
Sodium	11	23	11	12	11	2	8	1	
Magnesium	12	24	12	12	12	2	8	2	
Aluminum	13	27	13	14	13	2	8	3	
Silicon	14	28	14	14	14	2	8	4	
Phosphorus	15	31	15	16	15	2	8	5	
Sulfur	16	32	16	16	16	2	8	6	
Chlorine	17	35	17	18	17	2	8	7	
Argon	18	40	18	22	18	2	8	8	
Potassium	19	39	19	20	19	2	8	8	1
Calcium	20	40	20	20	20	2	8	8	2
Scandium	21	45	21	24	21	2	8	9	2

TABLE 1. *Table of the First 21 Elements**

*A complete list of the names and symbols of the known elements can be found in the Tables for Reference section.

Drawings like those in Figure 5 give valuable aid in determining oxidation numbers (see below); metallic, nonmetallic, and inert elements; chemical reactivity; and the periodic arrangement of the elements.

FIGURE 5. *Some Representative Atoms and Isotopes*

Oxidation Number and Valence

Each atom attempts to have its outer energy level complete and accomplishes this by borrowing, lending, or sharing its electrons. The electrons found in the outermost energy level are called *valence electrons*. The absolute number of electrons gained, lost, or borrowed is referred to as the valence of the atom. When valence electrons are lost or partially lost by sharing, the valence number is assigned a + sign for that element and is called its *oxidation number*. If valence electrons are gained or partially gained by an atom, its valence is assigned a – sign for its oxidation number.

- **EXAMPLE:**

$_{17}$Cl = nucleus ●)2)8)7← valence electrons

This picture can be simplified to ·C̈l: , showing only the valence electrons as dots. This is called the Lewis dot structure of the atom. To complete its outer orbit to eight electrons, Cl must borrow one from another atom. Its valence number then is 1. When electrons are gained, we assign a – sign to this number so the oxidation number of Cl is –1.

- **ANOTHER EXAMPLE:**

$_{11}$Na = nucleus ●)2)8)1← valence electrons

Na • (Lewis dot structure)

Since Na tends to lose this electron, its oxidation number is +1.

Some other general rules for oxidation numbers are:

1. Atoms of free elements have an oxidation number of zero.
2. Hydrogen has an oxidation number of +1 except in metallic hydrides, where it is –1.
3. Oxygen has an oxidation number of –2 except in peroxides, where it is –1. In combination with fluorine, it is +2.

A table of oxidation numbers is given on page 78. Information about oxidation numbers in chemical combinations is given on page 79.

Metallic, Nonmetallic, and Noble Gas Structures

On the basis of atomic structure, atoms are classified as *metals* if they tend to lend electrons; as *nonmetals* if they tend to borrow electrons; and as noble gases if they tend neither to borrow nor to lend electrons and have a complete outer energy level.

Reactivity

The fewer electrons an atom tends to borrow, or lend, or share to fill its outer energy level, the more reactive it tends to be in chemical reactions.

ATOMIC SPECTRA

The Bohr model was based on a simple postulate. Bohr applied to the hydrogen atom the concept that the electron can exist only in certain energy levels without an energy change but that, when the electron changes its state, it must absorb or emit the exact amount of energy which will bring it from the initial state to the final state. The *ground state* is the lowest energy state available to the electron. The *excited state* is any level higher than the ground state. The formula for changes in energy (ΔE) is:

$$\Delta E_{electron} = E_{final} - E_{initial}$$

When the electron moves from the ground state to an excited state, it must absorb energy. When it moves from an excited state to the ground state, it emits energy. This release of energy is the basis for *atomic spectra*. (See Figure 6.)

The energy values shown were calculated from Bohr's equation.

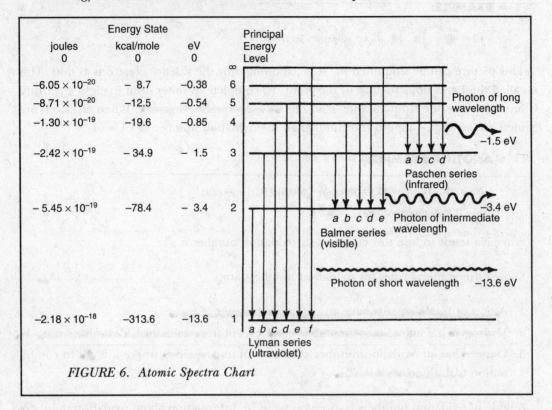

FIGURE 6. *Atomic Spectra Chart*

When energy is released in the "allowed" values, it is released in the form of discrete radiant energy called *photons*. Each of the first three levels has a particular name associated with the emissions that occur when an electron reaches its ground state on that level. The emissions, consisting of ultraviolet radiation, that occur when an electron cascades from a level higher than the first level down to $n = 1$ are known as the Lyman series. Note in Figure 6 that the next two higher levels have the names Balmer (for $n = 2$) and Paschen ($n = 3$) series, respectively.

Spectroscopy

When the light emitted by energized atoms is examined with an instrument called a *spectroscope*, the prism or diffraction grating in the spectroscope disperses the light to allow an examination of the *spectra* or distinct colored lines. Since only particular energy jumps are available in each type of atom, each element has its own unique emission spectra made up of only the lines of specific wavelength that correspond to its atomic structure.

- **EXAMPLE:**

    Hydrogen can have an electron drop from the n = 4 to $n = 2$ level. What visible spectral line in the Balmer series will result from this emission of energy?

$$\Delta E_{\text{evolved}} = E_{n=2} - E_{n=4}$$

From Figure 6, $E_2 = -5.45 \times 10^{-19}$ J and $E_4 = -1.36 \times 10^{-19}$ J. Then

$E_{\text{evolved}} = -5.45 \times 10^{-19}$ J $- (-1.36 \times 10^{-19}$ J$) = -4.09 \times 10^{-19}$ J

Since ΔE is negative, energy is released. The formula for the relationship of ΔE to the emission frequency is

$$\Delta E = \frac{h \text{ (Planck's constant) } c \text{ (velocity of light)}}{\lambda \text{ (wavelength)}}$$

Substituting, we get

$$-4.09 \times 10^{-19} \text{ J} = \frac{(6.626 \times 10^{-34} \text{ Js}) \ (2.9979 \times 10^{8} \text{ m / s})}{\lambda}$$

$$\lambda = \frac{(6.626 \times 10^{-34} \text{ Js}) \ (2.9979 \times 10^{8} \text{ m / s})}{4.09 \times 10^{-19} \text{ J}} =$$

$$\lambda = 4.87 \times 10^{-7} \text{m}$$

This line is in the blue-green sector of the spectrum.

A partial atomic spectrum for hydrogen would look like this:

The right-hand group is in the visible range and is part of the Balmer series. The left-hand group is in the ultraviolet region and belongs to the Lyman series.

Investigating spectral lines like these can be used in the identification of unknown specimens.

Mass Spectroscopy

Another tool used to identify specific atomic structures is mass spectroscopy, which is based on the concept that differences in mass cause differences in the degree of bending that occurs in a beam of ions passing through a magnetic field. This is shown in Figure 7.

FIGURE 7. *Mass Spectroscope*

The intensity on the photographic plate indicates the amount of each particular *isotope.* Other collectors may be used in place of the photographic plate to collect and interpret these data.

THE WAVE-MECHANICAL MODEL

In the early 1920s, some difficulties with the Bohr model of the atom were becoming apparent. Although Bohr used classical mechanics (which is the branch of physics that deals with the motion of bodies under the influence of forces) to calculate the orbits of the hydrogen atom, it could not be used to explain the ability of electrons to stay in only certain energy levels without the loss of energy. Nor could it explain that the only change of energy occurred when an electron "jumped" from one energy level to another and that the electron could not exist in the atom at any energy level between these levels. According to Newton's laws, the kinetic energy of a body always changes smoothly and continuously, not in sudden jumps. The idea of only certain *quantized* energy levels being available in the Bohr atom was a very important one. The energy levels explained the existence of atomic spectra, described in the previous sections.

Another difficulty with the Bohr model was that it worked well only for the hydrogen atom with its single electron. It did not work with atoms that had more electrons. A new approach to the laws governing the behavior of electrons inside the atom was needed, and such an approach was developed in the 1920s by the combined work of many scientists. Their work dealt with a more mathematical model usually referred to as *quantum mechanics* or *wave mechanics.* By this time, Albert Einstein had already proposed a relativity mechanics model to deal with the relative nature of mass as its speed approaches the speed of light. In the same manner, a quantum/wave mechanics model was now needed to fit the data of the atomic model. Max Planck suggested in his quantum theory of light that light has both particle-like properties as well as wavelike characteristics. In 1924, Louis de Broglie, a young French physicist, suggested that if light can have both wavelike and particle-like characteristics as Planck had suggested, then perhaps particles can also have wavelike characteristics. In 1927, de Broglie's ideas were proven to be true experimentally when investigators showed that electrons could produce diffraction patterns, a property associated with waves. Diffraction patterns are patterns produced by waves as they pass through small holes or narrow slits.

In 1927, Werner Heisenberg stated what is now called the *uncertainty principle*. This principle states that it is impossible to know both the precise location and precise velocity of a subatomic particle at the same time. Heisenberg, in conjunction with the Austrian physicist, Erwin Schrodinger, agreed with the de Broglie concept that the electron was bound to the nucleus in a manner similar to a standing wave. They developed the complex equations that describe the *wave mechanical model* of the atom. The solution of these equations gives specific wave functions called *orbitals*. These were not related at all to the Bohr orbits. The electron does not move in a circular orbit in this model. These orbitals give only a probability picture of where the electron can be found. It gives no information about the pathway. Notice the drawings in Figures 8a and 8b are only probability distribution representations of where electrons in these orbitals might be found.

Quantum Numbers

Each electron orbital of the atom may be described by a set of four quantum numbers in this model. They give the position with respect to the nucleus, the shape of the orbital, its spatial orientation, and the spin of the electron in the orbital.

Principal quantum number (n)
1, 2, 3, 4, 5, etc.

Refers to average distance of the orbital from the nucleus. 1 is closest to the nucleus and has the least energy. These numbers correspond to the orbits in the previous model. They are called energy levels.

Secondary or azimuthal (l)
quantum number
s, p, d, f
(in order of
increasing energy)

Refers to the shape of the orbital. The number of possible shapes is limited by the principal quantum number. The first energy level has only one possible shape, the s orbital. The second has two possible, the s and p shapes. See figures 8a and 8b for representations of these shapes.

Magnetic quantum number (m_l)
$s = 1$ space-oriented orbital
$p = 3$ space-oriented orbitals
$d = 5$ space-oriented orbitals
$f = 7$ space-oriented orbitals

The drawings in Figure 8a show the s-orbital shape, which is a sphere, and the p orbitals, which have dumbbell shapes with three possible orientations on the axis shown. The number of spatial orientations of orbitals is referred to as the magnetic quantum number. The possible orientations are listed. Figure 8b represents the d orbitals.

Spin quantum number (m_s)
+ spin − spin

Electrons are assigned one more quantum number called the spin quantum number. This describes the spin in either of two possible directions. Each orbital can be filled by only two electrons, each with an opposite spin. The main significance of electron spin is explained by the postulate of Wolfgang Pauli. It states that in a given atom no two electrons can have the same set of four quantum numbers (n, ℓ, m_ℓ, and m_s). This is referred to as the *Pauli Exclusion Principle*. Therefore each orbital in Figures 8a and 8b can hold only two electrons.

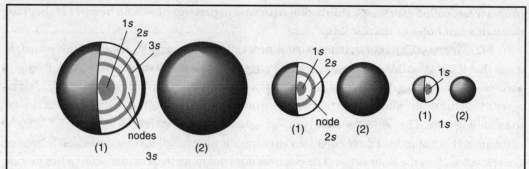

Two representations of the hydrogen $1s$, $2s$, and $3s$ orbitals. (a) The electron probability distribution; the nodes indicate regions of zero probability. (b) The surface that contains 90% of the total electron probability (the size of the orbital, by definition).

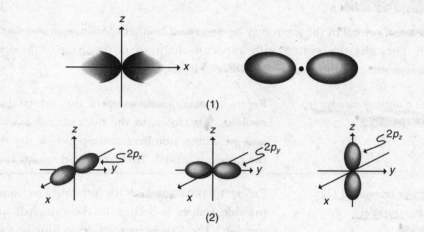

Representation of the $2p$ orbitals. (a) The electron probability distribution and (b) the boundary surface representations of all three orbitals.

FIGURE 8a. Representations of s *and* p *Orbitals*

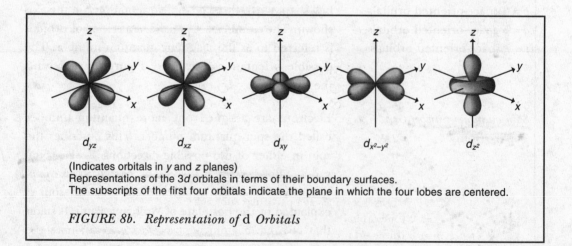

(Indicates orbitals in y and z planes)
Representations of the $3d$ orbitals in terms of their boundary surfaces.
The subscripts of the first four orbitals indicate the plane in which the four lobes are centered.

FIGURE 8b. Representation of d *Orbitals*

Quantum numbers are summarized in the table below.

SUMMARY OF QUANTUM NUMBERS FOR THE FIRST FOUR LEVELS OF ORBITALS IN THE HYDROGEN ATOM				
Principal Quantum No., n	Azimuthal Quantum No., ℓ	Orbital Designation	m_ℓ	Number of Orbitals
1	0	$1s$	0	1
2	0	$2s$	0	1
	1	$2p$	−1, 0, +1	3
3	0	$3s$	0	1
	1	$3p$	−1, 0, +1	3
	2	$3d$	−2, −1, 0, +1, +2	5
4	0	$4s$	0	1
	1	$4p$	−1, 0, +1	3
	2	$4d$	−2, −1, 0, +1, +2	5
	3	$4f$	−3, −2, −1, 0, +1, +2, +3	7
Limits of Quantum Numbers				
$n = 1, 2, 3, \ldots$	$\ell = 0, 1, \ldots (n-1)$		$m_\ell = -\ell, \ldots, 0, \ldots, +\ell$	

Hund's Rule of Maximum Multiplicity

It is important to remember that, when there is more than one orbital at a particular energy level, such as three p orbitals or five d orbitals, only one electron will fill each orbital until each has one electron. After this, pairing will occur with the addition of one more electron to each orbital. This is called *Hund's Rule of Maximum Multiplicity* and is shown in Table 2, where each slant line indicates an electron (∅).

Chemical Symbol	Atomic No.	Orbital Notation			Electron Configuration Notation
		$1s$	$2s$	$2p$	
H	1	⊘	◯	◯ ◯ ◯	$1s^1$
He	2	⊗	◯	◯ ◯ ◯	$1s^2$
Li	3	⊗	⊘	◯ ◯ ◯	$1s^2, 2s^1$
Be	4	⊗	⊗	◯ ◯ ◯	$1s^2, 2s^2$
B	5	⊗	⊗	⊘ ◯ ◯	$1s^2, 2s^2, 2p^1$
C	6	⊗	⊗	⊘ ⊘ ◯	$1s^2, 2s^2, 2p^2$
N	7	⊗	⊗	⊘ ⊘ ⊘	$1s^2, 2s^2, 2p^3$
O	8	⊗	⊗	⊗ ⊘ ⊘	$1s^2, 2s^2, 2p^4$
F	9	⊗	⊗	⊗ ⊗ ⊘	$1s^2, 2s^2, 2p^5$
Ne	10	⊗	⊗	⊗ ⊗ ⊗	$1s^2, 2s^2, 2p^6$

Maximum electrons in orbitals at a particular sublevel:

$s = $ 2 (one orbital)
$p = $ 6 (three orbitals)
$d = $ 10 (five orbitals)
$f = $ 14 (seven orbitals)

TABLE 2. Orbital Notations

If each orbital is indicated in an energy diagram as a circle (O), we can show the relative energies in a chart such as Figure 9. If this drawing represented a ravine with the energy levels as ledges onto which stones could come to rest only in numbers equal to the circles for orbitals, then pushing stones into the ravine would cause the stones to lose their kinetic energy as they dropped to the lowest level available to them. Much the same is true for electrons.

SUBLEVELS AND ELECTRON CONFIGURATION

Order of Filling and Notation

The sublevels do not fill up in numerical order, and the pattern of filling is shown on the right side of the approximate relative energy levels chart (Figure 9). In the first instance of failure to follow numerical order, the 4s fills before the 3d. (Study Figure 9 carefully before going on.)

FIGURE 9. *Approximate Relative Energy Levels of Subshells*

$_{19}$K $\quad 1s^2, 2s^2, 2p^6, 3s^2, 3p^6, 4s^1$

$_{20}$Ca $\quad 1s^2, 2s^2, 2p^6, 3s^2, 3p^6, 4s^2$

$_{21}$Sc $\quad 1s^2, 2s^2, 2p^6, 3s^2, 3p^6, 4s^2, 3d^1$ (note 4s filled before 3d)

There is a more stable configuration to a half-filled or filled sublevel, so at atomic number 24 the 3d sublevel becomes half-filled by taking a 4s electron;

$_{24}$Cr $\quad 1s^2, 2s^2, 2p^6, 3s^2, 3p^6, 3d^5, 4s^1$

and at atomic number 29 the 3d becomes filled by taking a 4s electron:

$_{29}$Cu $\quad 1s^2, 2s^2, 2p^6, 3s^2, 3p^6, 3d^{10}, 4s^1$

Table 3 shows the electron configurations of the elements. A triangular mark ▶ indicates the phenomenon of an outer-level electron dropping back to a lower unfilled orbital. By following the atomic numbers throughout this chart, you will get the same order of filling as shown in Figure 9.

TABLE 3. Electron Configurations of the Elements

Electron Dot Notation (Lewis Dot Structures)

In 1916 G. N. Lewis devised the electron dot notation, which may be used in place of the electron configuration notation. The electron dot notation shows only the chemical symbol surrounded by dots to represent the electrons in the incomplete outer level. Examples are:

$$\dot{K}, \cdot \ddot{As} \cdot, \ddot{Sr}, \ddot{:I:}, \text{and } \ddot{:Rn:}$$

The symbol denotes the nucleus and all electrons except the valence electrons. The dots are arranged at the four sides of the symbol and are paired when appropriate. In the examples above, the depicted electrons are the valence electrons found in the outer energy level orbitals.

$4s^1$ is shown for potassium (K)

$4s^2$, $4p^3$ are shown for arsenic (As)

$5s^2$ is shown for strontium (Sr)

$5s^2$, $5p^5$ are shown for iodine (I)

$6s^2$, $6p^6$ are shown for radon (Rn)

TRANSITION ELEMENTS AND VARIABLE OXIDATION NUMBERS

The elements involved with the filling of a *d* sublevel with electrons after two electrons are in the *s* sublevel of the next principal energy level are often referred to as the *transition elements*. The first examples of these are the elements between calcium, atomic number 20, and gallium, atomic number 31. Their electron configurations are the same in the 1*s*, 2*s*, 2*p*, 3*s*, and 3*p* sublevels. It is the filling of the 3*d* and changes in the 4*s* sublevels that are of interest. This is shown in the following table.

Element	Atomic No.	Electron Configuration $1s^2, 2s^2, 2p^6, 3s^2, 3p^6, 3d$	$4s$
Scandium	21	1	2
Titanium	22	All 2	2
Vanadium	23	3	2
Chromium	24	the 5	1*
Manganese	25	same 5	2
Iron	26	6	2
Cobalt	27	7	2
Nickel	28	8	2
Copper	29	10	1*
Zinc	30	10	2

The asterisk (*) shows where a $4s$ electron is promoted into the $3d$ sublevel. This is due to the fact that the $3d$ and $4s$ sublevels are very close in energy and that there is a state of greater stability in half-filled and filled sublevels. Therefore, chromium gains stability by the movement of an electron from the $4s$ sublevel into the $3d$ sublevel to give a half-filled $3d$ sublevel. It then has one electron in each of the five orbitals of the $3d$ sublevel. In copper, the movement of one $4s$ electron into the $3d$ sublevel gives the $3d$ sublevel a completely filled configuration.

The fact that the electrons in the $3d$ and $4s$ sublevels are so close in energy levels leads to the possibility of some or all the $3d$ electrons being involved in chemical bonding. With the variable number of electrons available for bonding, it is not surprising that transition elements can exhibit variable oxidation numbers. An example is manganese with possible oxidation numbers of +2, +3, +4, +6, and +7, which correspond, respectively, to the use of no, one, two, four, and five electrons from the $3d$ sublevel.

The transition elements in the other periods of the table show this same type of anomaly, as they have d sublevels filling in the same manner.

PERIODIC TABLE

History

The history of the development of a systematic pattern for the elements includes the work of a number of scientists such as John Newlands, who, in 1863, proposed the idea of repeating octaves of properties.

Dimitry I. Mendeleyev in 1869 proposed a table containing 17 columns and is usually given credit for the first periodic table since he arranged elements in groups according to their atomic weights and properties. It is interesting to note that Lothar Meyer proposed a similar arrangement about the same time. In 1871 Mendeleyev rearranged some elements and proposed a table of eight columns, obtained by splitting each of the long periods into a period of seven elements, an eighth group containing the three central elements (such as Fe, Co, Ni), and a second period of seven elements. The first and second periods of seven were later distinguished by use of the letters a and b attached to the group symbols, which are Roman numerals. This nomenclature of periods (Ia, IIa, etc.) appears slightly revised in the present periodic table, even in the extended form of assigning Arabic numbers from 1–18 as shown in Table 4.

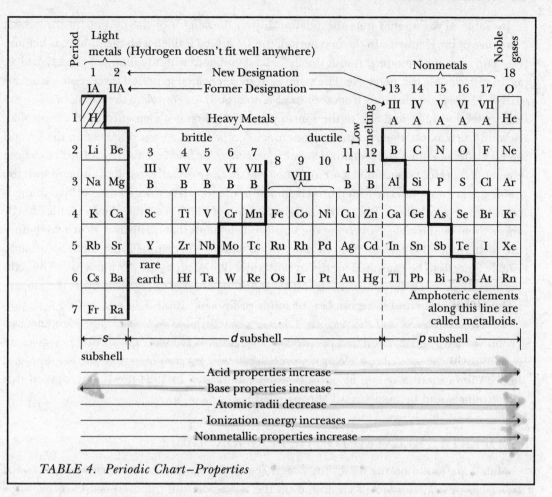

TABLE 4. *Periodic Chart—Properties*

Mendeleyev's table had the elements arranged by atomic weights with recurring properties in a periodic manner. Where atomic weight placement disagreed with the properties that should occur in a particular spot in the table, he gave preference to the element with the correct properties. He even predicted elements for places that were not yet occupied in the table. These predictions proved to be amazingly accurate.

Periodic Law

Henry Moseley stated, after his work with X-ray spectra in the early 1900s, that the properties of elements are a periodic function of their atomic numbers, thus changing the basis of the periodic law from atomic weight to atomic number. This is the present statement of the periodic law.

The Table

The horizontal rows of the periodic table are called *periods* or *rows*. There are seven periods, each of which begins with an atom having only one valence electron and ends with a complete outer shell structure of an inert gas. The first three periods are short, consisting of 2, 8, and 8 elements, respectively. Periods 4 and 5 are long, with 18 each, while period 6 has 32 elements, and period 7 is incomplete with 22 elements, most of which are radioactive and do not occur in nature.

In Table 4, you should note the relationship of the length of the periods to the orbital structure of the elements. In the first period, the $1s^2$ orbital is filled with the noble gas helium, He. The second period begins with the $2s^1$ orbital and ends with the filling of the $2p^6$ orbital, again with a noble gas, neon, Ne. The same pattern is repeated in period three, going from $3s^1$ to $3p^6$. These eight elements from sodium, Na, to argon, Ar, complete the filling of the $n = 3$ energy level with $3s^2$ and $3p^6$. In the fourth period, the first two elements fill the $4s^2$ orbital. Beyond calcium, Ca, the pattern becomes more complicated. As was discussed in the section about Order of Filling and Notation, the next orbitals to be filled are the five $3d$ orbitals whose elements represent transition elements. Then the three $4p$ orbitals are filled and end with the noble gas krypton, Kr. The fifth period is similar to the fourth period. The $5s^2$ orbital filling is represented by rubidium, Rb, and strontium, Sr, both of which resemble the elements directly above them on the table. Next come the transition elements that fill the five $4d$ orbitals before the next group of elements, from indium, In, to xenon, Xe, complete the three $5p$ orbitals. (Table 3 needs to be consulted for the irregularities that occur as the d orbitals fill.) The sixth period follows much the same pattern and has the filling order of $6s^2$, $4f^{14}$, $5d^{10}$, $6p^6$. Here again, irregularities occur and these can best be followed by using Table 3.

The vertical columns of the periodic table are called *groups* or *families*. The elements in a group exhibit similar or related properties. The Roman numeral group number gives an indication of the number of electrons probably found in the outer shell of the atom and thus an indication of one of its possible valence numbers. In 1984 the IUPAC agreed that the groups would be numbered 1 through 18.

PROPERTIES RELATED TO THE PERIODIC TABLE

Metals are found on the left of the chart (see Table 4) with the most active metal in the lower left corner. Nonmetals are found on the right side with the most active nonmetal in the upper right-hand corner. The noble or inert gases are on the far right. Since the most active metals react with water to form bases, the Group I metals are called alkali metals. As you proceed to the right, the base-forming property decreases and the acid-forming properties increase. The metals in the first two groups are the light metals, and those toward the center are heavy metals.

The elements found along the dark line in the periodic table (Table 4) are called metalloids. These elements have certain characteristics of metals and other characteristics of nonmetals. Some examples of metalloids are boron, silicon, arsenic, and tellurium.

Study Table 4 carefully because it summarizes many of these properties. For a more detailed description of metals, alloys, and metalloids see pages 199 through 203 in Chapter 13.

Radii of Atoms

The size of an atom is difficult to describe because atoms have no definite outer boundary. Unlike a volleyball, the atom does not have a definite circumference.

To overcome this problem, the size of an atom is estimated by describing its radius. In metals, this is done by measuring the distance between two nuclei in the solid state and dividing this distance by 2. Such measurements can be made with X-ray diffraction. For nonmetallic elements that exist in pure form as molecules, such as chlorine, measurements can be made of the distance between nuclei for two atoms covalently bonded together. Half of this distance is referred to as the *covalent radius*. The method for finding the covalent radius of the chlorine atom is illustrated on the following page.

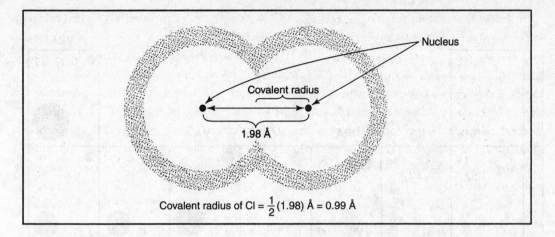

Covalent radius of $Cl = \frac{1}{2}(1.98)$ Å $= 0.99$ Å

Figure 10 shows the relative atomic and ionic radii for some elements. As you review this chart, you should note two trends:

1. Atomic radii decrease from left to right across a period in the periodic table (until the noble gases).
2. Atomic radii increase from top to bottom in a group or family.

The reason for these trends will become clear in the following discussions.

Atomic Radii in Periods

Since the number of electrons in the outer principal energy level increases as you go from left to right in each period, the corresponding increase in the nuclear charge due to the additional protons pulls the electrons more tightly around the nucleus. This attraction more than balances the repulsion between the added electrons and the other electrons, and the radius is generally reduced. The inert gas has a slight increase because of the electron repulsion in the filled outer principal energy level.

Atomic Radii in Groups

For a group of elements, the atoms of each successive member have another outer principal energy level in the electron configuration and the electrons there are held less tightly by the nucleus. This is so because of their increased distance from the nuclear positive charge and the shielding of this positive charge by all the intermediate electrons. Therefore the atomic radius increases down a group. See Figure 10.

Ionic Radius Compared to Atomic Radius

Metals tend to lose electrons in forming positive ions. With this loss of negative charge, the positive nuclear charge pulls in the remaining electrons closer and thus reduces the ionic radius below that of the atomic radius.

Nonmetals tend to gain electrons in forming negative ions. With this added negative charge, which increases the inner electron repulsion, the ionic radius is increased beyond the atomic radius. See Figure 10 for relative atomic and ionic radii values.

Electronegativity

The electronegativity of an element is a number that measures the relative strength with which the atoms of the element attract valence electrons in a chemical bond. This elec-

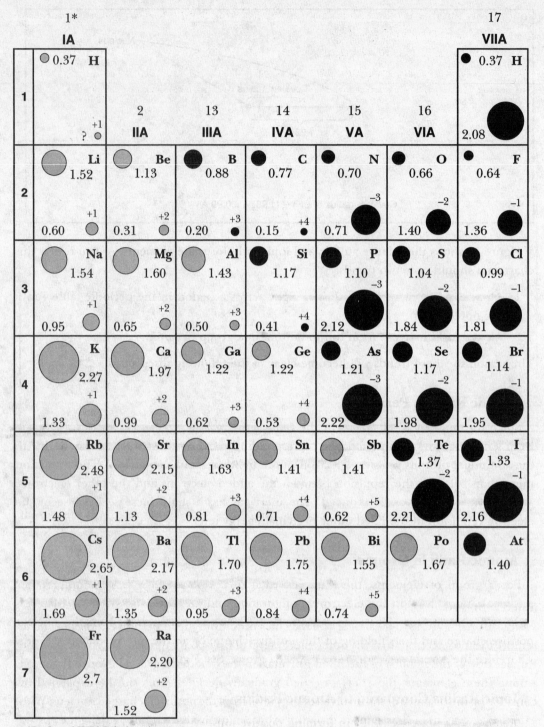

FIGURE 10. Radii of Some Atoms and Ions (in Angstroms)

Note: The atomic radius is usually given for metal atoms, which are shown in gray, and the covalent radius is usually given for atoms of nonmetals, which are shown in black.

tronegativity number is based on an arbitrary scale going from 0 to 4. In general, a value of less than 2 indicates a metal.

Notice in Table 5 that the electronegativity decreases down a group and increases across a period. The lower the electronegativity number the more electropositive an element is said to be. The most electronegative element is in the upper right corner—F. The most electropositive is in the lower left corner of the chart—Fr.

H						
2.1						
Li	Be	B	C	N	O	F
1.0	1.5	2.0	2.5	3.0	3.5	4.0
Na	Mg	Al	Si	P	S	Cl
0.9	1.2	1.5	1.8	2.1	2.5	3.0
K	Ca	Ga	Ge	As	Se	Br
0.8	1.0	1.7	1.8	2.0	2.4	2.8
Rb	Sr	In	Sn	Sb	Te	I
0.8	1.0	1.7	1.8	1.9	2.1	2.5
Cs	Ba	Tl	Pb	Bi	Po	At
0.7	0.9	1.8	1.8	1.9	2.0	2.2
Fr	Ra					
0.7	0.9					

TABLE 5. Electronegativities of the Elements

Ionization Energy

The atoms hold their valence electrons, then, with different amounts of energy. If enough energy is supplied to one outer electron to remove it from its atom, this amount of energy is called the *first ionization energy*. With the first electron gone, the removal of succeeding electrons becomes more difficult because of the imbalance between the positive nuclear charge and the remaining electrons. The lowest ionization energy would be found with the least electronegative atom.

Not surprisingly, the highest peaks on the chart occur with the ionization energy needed to remove the first electron from the outer energy level of the noble gases, He, Ne, Ar, Kr, Xe, and Rn, because of the stability of the filled p orbitals in the outer energy level. Notice that even among these elements, the energy needed gradually declines. This can be explained by considering the distance of the involved energy level from the positively charged nucleus. With each succeeding noble gas, a more distant p orbital is involved, therefore making it easier to remove an electron from the positive attraction of the nucleus. Besides this consideration, as more energy levels are added to the atomic structure as the atomic number increases, the additional negative fields associated with the additional electrons screen out some of the positive attraction of the nucleus. Within a period such as from Li to Ne, the ionization energy generally increases. The lowest occurs when a lone electron occupies the outer s orbital as in Li. As the s orbital fills with two electrons at atomic number 4, Be, the added stability of a filled $2s$ orbital explains the small peak at 4. At atomic number 5, B, one lone electron occupies the $2p$ orbital. This electron can be removed with less energy and, therefore, a dip occurs in the chart. With the $2p$ orbitals filling according to Hund's Rule (refer to Table 2, page 53), with only one electron in each orbital before pairing occurs, again a slightly more stable situation and, therefore, another small peak occur at atomic number 7. After this peak, a dip and continual increases occur until the 2*p* orbitals are completely filled with paired electrons at the noble gas Ne. As you continue to associate the atomic number with the line in the chart, you find peaks occurring in the same general pattern. These peaks are always related to the state of filling of the orbitals involved and the distance of these orbitals from the nucleus.

Element	Atomic Number	First Ionization Energy (eV)	Second Ionization Energy (eV)	
Group 1 or IA { Li	3	5.39	75.7	
K	19	4.34	31.8	Third Ionization Energy (eV)
Group 2 or IIA { Be	4	9.32	18.2	154
Mg	12	7.64	15.1	80.3

Sample Ionization Energies for
Second and Third Electron Removal

Chapter

2

Review

• **QUESTIONS**

1. The two main parts of an atom are (a) the principal energy levels and energy sublevels (b) nucleus and kernel (c) nucleus and energy levels (d) planetary electrons and energy levels.

2. The lowest principal quantum number that an electron can have is (a) 0 (b) 1 (c) 2 (d) 3.

3. The sublevel that has only one orbital is identified by the letter (a) *s* (b) *p* (c) *d* (d) *f*.

4. The sublevel that can be occupied by a maximum of 10 electrons is identified by the letter (a) *d* (b) *f* (c) *p* (d) *s*.

5. An orbital may never be occupied by (a) 1 electron (b) 2 electrons (c) 3 electrons (d) 0 electrons.

6. An atom of beryllium consists of 4 protons, 5 neutrons, and 4 electrons. The mass number of this atom is (a) 13 (b) 9 (c) 8 (d) 5.

7. The number of orbitals in the second principal energy level, *n* = 2, of an atom is (a) 1 (b) 9 (c) 16 (d) 4.

8. An electron-dot symbol consists of the symbol representing the element and an arran⟨...⟩ dots that usually shows (a) the atomic number (b) the atomic mass ⟨...⟩ neutrons (d) the electrons in the outermost energy level⟨...⟩

9. Chl⟨...⟩ by the electron-dot symbol $:\overset{\cdot}{\underset{\cdot}{Cl}}:$. The atom that would be ⟨...⟩ntical electron-dot arrangement has the atomic number ⟨...⟩ (d) 19.

Using t⟨...⟩er the following questions:

10. Na⟨...⟩e first 20 whose atom gives up an electron the most re⟨...⟩ c) Cl (d) K

11. N⟨...⟩hose atom shows the greatest affinity for an additional e⟨...⟩ F (c) Cl (d) O

12. N⟨...⟩e nonmetal in Period 3. (a) Na (b) Cl (c) Ar (d) S

13. ⟨...⟩ing elements is most likely to form covalent compounds? ⟨...⟩ Cs (d) C

14. ⟨...⟩probable oxidation number for silicon in a compound? ⟨...⟩ +3 (d) +4

15. ⟨...⟩c series do you find strong base-formers? (a) left (b) right ⟨...⟩rt gases

• AN⟨...⟩

⟨...⟩a)	7. (d)	10. (d)	13. (d)
	8. (d)	11. (b)	14. (d)
	9. (b)	12. (b)	15. (a)

Term⟨...⟩

atomic⟨...⟩	ionic radii	oxidation number
atomi⟨...⟩	ionization energy	Pauli Exclusion Principle
atomi⟨...⟩	isotopes	periodic law
Bohr moɑei	Lewis (electron-dot) structure	period or row
covalent radius	Mendeleyev	proton
Dalton's atomic theory	metallic atoms	quantum numbers
electron	Moseley	*s, p, d, f* orbitals
electronegativity	neutron	transition elements
group or family	nonmetallic atoms	valence electrons
Hund's Rule	nucleus	wave mechanical model
inert atoms		

(Handwritten notes on overlapping paper scrap:)
X BC
2) B
3) A
4) BA
X DC
6) B
7) D
8) D
9) B
10) A D
11) B
12) B
ok 7) D
A B D
15) A
Risp R

Bonding

Some elements show no tendency to combine with either like atoms or other kinds of elements. These elements are said to be monoatomic molecules; three examples are helium, neon, and argon. A *molecule* is defined as the smallest particle of an element or a compound that retains the characteristics of the original substance. Water is a triatomic molecule since two hydrogen atoms and one oxygen atom must combine to form the substance water with its characteristic properties. When atoms do combine to form molecules, there is a shifting of valence electrons, that is, the electrons in the outer energy level of each atom. Usually, this results in completion of the outer energy level of each atom. This more stable form may be achieved by the gain or loss of electrons or the sharing of pairs of electrons. The resulting attraction of the atoms involved is called a *chemical bond*. When a chemical bond forms, energy is released; when this bond is broken, energy is absorbed.

This relationship of bonding to the valence electrons of atoms can be further explained by studying the electron structure of the atoms involved. As already mentioned, the noble gases are monoatomic molecules. The reason for this can be seen in the electron distribution of these noble gases as shown below:

Noble Gas	Electron Distribution					Electrons in Valence Energy Level
helium	$1s^2$					2
neon	$1s^2$	$2s^2, 2p^6$				8
argon	$1s^2$	$2s^2, 2p^6$	$3s^2, 3p^6$			8
krypton	$1s^2,$	$2s^2, 2p^6$	$3s^2, 3p^6, 3d^{10}$	$4s^2, 4p^6$		8
xenon	$1s^2,$	$2s^2, 2p^6$	$3s^2, 3p^6, 3d^{10}$	$4s^2, 4p^6, 4d^{10}$	$5s^2, 5p^6$	8
radon	$1s^2,$	$2s^2, 2p^6$	$3s^2, 3p^6, 3d^{10},$	$4s^2, 4p^6, 4d^{10}, 4f^{14}$	$5s^2, 5p^6, 5d^{10}$ $6s^2, 6p^6$	8

The distinguishing factor in these very stable configurations is the arrangement of two *s* electrons and six *p* electrons in the valence energy level in five of the six atoms. (Note that helium, He, has only a single *s* valence energy level that is filled with two electrons, making He a very stable atom.) This arrangement is called a *stable octet*. All other elements, other than the noble gases, have one to seven electrons in their outer energy level. These elements are reactive to varying degrees. When they do react to form chemical bonds, usually the electrons shift in such a way that stable octets form. In other words, in bond formation, atoms usually attain the stable electron structure of one of the noble gases. The type of bond formed is directly related to whether this structure is achieved through gaining, losing, or sharing of electrons.

TYPES OF BONDS

Ionic Bonds

When the electronegativity values of two kinds of atoms differ by 1.7 or more (especially differences greater than 1.7), the more electronegative atom will borrow the electrons it needs to fill its energy level, and the other kind of atom will lend electrons until it, too, has a complete energy level. Because of this exchange, the borrower becomes negatively charged and the lender becomes positively charged. They are now referred to as *ions*, and the bond or attraction between them is called an *ionic* or *electrovalence bond*. These ions do not retain the properties of the original atoms. An example can be seen in Figure 11.

These ions do not form an individual molecule in the liquid or solid phase but are arranged into a crystal lattice or giant ion-molecule containing many of these ions. Ionic solids like this tend to have high melting points and will not conduct a current of electricity until they are in the molten state.

The reaction:

$$Li\ (atom) + F\ (atom) \longrightarrow Li^+ + F^-\ (Ionic\ compound\ formed)$$

$$e^-$$

Electron notation representations:

$$Li\ 1s^2, 2s^1 + F\ 1s^2, 2s^2, 2p^5 \longrightarrow Li^+\ 1s^2 + F^-\ 1s^2, 2p^2, 2p^6$$

Orbital representation:

FIGURE 11. Three Representations of the Ionic Bonding of LiF

Covalent Bonds

When the electronegativity difference between two or more atoms is zero or very small (not greater than about 0.5), they tend to share the valence electrons in the respective outer energy levels. This attraction is called a *nonpolar covalent bond.* Here is an example using electron-dot notation and orbital notation:

flourine atoms ──→ flourine molecule

These covalent bonded molecules do not have an electrostatic charge like the ionic bonded substances. In general, covalent compounds are either gases, or liquids having fairly low boiling points, or solids that melt at relatively low temperatures. Unlike ionic compounds, they do not conduct electric currents.

When the electronegativity difference is between 0.5 and 1.7, there will not be an equal sharing of electrons between the atoms involved. The shared electrons will be more strongly attracted to the atom of greater electronegativity. As the difference in the electronegativities of the two elements increases above 0.5, the polarity or degree of ionic character increases. At a difference of 1.7 or more, the bond has more than 50% ionic character. However, when the difference is between 0.5 and 1.7, the bond is called a *polar covalent bond.* An example:

<table>
<tr><td>H :Cl :   • hydrogen electron
         ○ chlorine electrons</td><td>H :O :   • hydrogen electrons
       ○ oxygen electrons</td></tr>
<tr><td>Hydrogen chloride</td><td>H
Water</td></tr>
</table>

Notice the electron pair in the bond is shown closer to the more electronegative atom. Because of this unequal sharing, the molecules shown are said to be polar molecules, or a *dipole.* There are cases of polar covalent bonds existing in nonpolar molecules. Some examples are CO_2, CH_4, and CCl_4. (See Figure 12.)

FIGURE 12. Covalent Bonds in Nonpolar Molecules

In all the examples in Figure 12 the bonds are polar covalent bonds, but the important thing is that they are symmetrically arranged in the molecule. This results in a nonpolar molecule.

In the *covalent* bonds described so far, the shared electrons in the pair were contributed one each from the atoms bonded. In some cases, however, both electrons for the shared pair are supplied by only one of the atoms. These bonds are called *coordinate covalent bonds.* Two examples are NH_4^+ and H_2SO_4. (See Figure 13.)

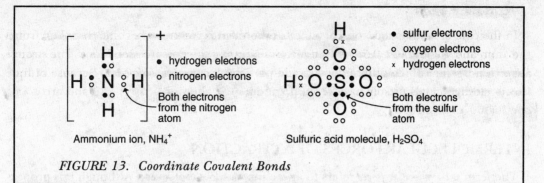

FIGURE 13. *Coordinate Covalent Bonds*

The formation of a covalent bond can be described in a graphic form and related to the potential energy of the atoms involved. Using the formation of the hydrogen molecule as an example, we can show how the potential energy changes as the two atoms approach and form a covalent bond. In the illustration that follows, frames (1), (2), and (3) show the effect on potential energy as the atoms move closer to each other. In frame (3), the atoms have reached the condition of lowest potential energy, but the inertia of the atoms pulls them even closer, as shown in frame (4). The repulsion between them then forces the two nucleii to a stable position, as shown in frame (5).

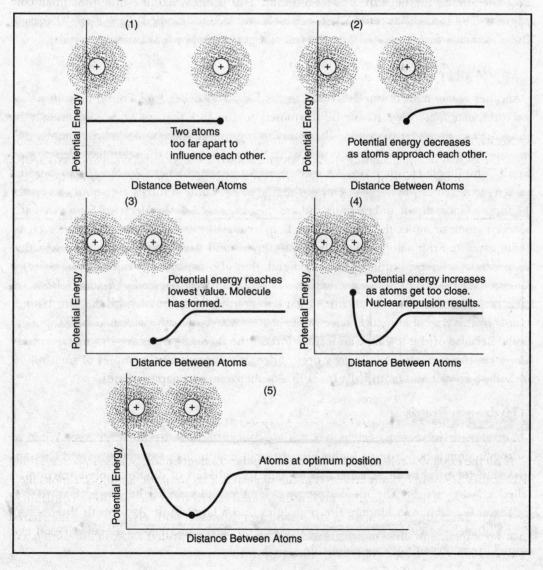

Metallic Bonds

In the case of most metals, one or more of the valence electrons become detached from the atom and migrate in a "sea" of free electrons among the positive metal ions. The strong attraction between these differently charged particles forms a *metallic bond*. Because of this firm bonding, metals usually have high melting points, show great strength, and are good conductors of electricity.

INTERMOLECULAR FORCES OF ATTRACTION

The term *intermolecular forces* refers to attractions *between* molecules. Although it is proper to refer to all intermolecular forces as *van der Waals forces*, this concept should be expanded for clarity.

Dipole-Dipole Attraction

One component of van der Waals forces is *dipole-dipole attraction*. It was shown in the discussion of polar covalent bonding that the unsymmetrical distribution of electronic charges leads to positive and negative charges in the molecules, which are referred to as *dipoles*. In polar molecular substances, the dipoles line up so that the positive pole of one molecule attracts the negative pole of another. This is much like the lineup of small bar magnets. The force of attraction between polar molecules is called *dipole-dipole attraction*. These attractive forces are less than the full charges carried by ions in ionic crystals.

London Forces

Another component of van der Waals forces is called *London force*. Found in both polar and nonpolar molecules, it can be attributed to the fact that an atom that usually is nonpolar sometimes becomes polar because the constant motion of its electrons may cause uneven charge distribution at any one instant. When this occurs, the atom has a temporary dipole. This dipole can then cause a second, adjacent atom to be distorted and to have its nucleus attracted to the negative end of the first atom. London forces are about one-tenth the force of most dipole interactions and are the weakest of all the electrical forces that act between atoms or molecules. These forces help to explain why nonpolar substances such as noble gases and the halogens condense into liquids and then freeze into solids when the temperature is lowered sufficiently. In general, they also explain why liquids composed of discrete molecules with no permanent dipole attraction have low boiling points relative to their molecular masses. It is also true that compounds in the solid state that are bound mainly by this type of attraction have rather soft crystals, are easily deformed, and vaporize easily. Because of the low intermolecular forces, the melting points are low and evaporation takes place so easily that it may occur at room temperature. Examples of such solids are iodine crystals and moth balls (paradichlorobenzene and naphthalene).

Hydrogen Bonds

A proton or hydrogen nucleus has a high concentration of positive charge. When a hydrogen atom is bonded to a highly electronegative atom, its positive charge will have an attraction for neighboring electron pairs. This special kind of dipole-dipole attraction is called a *hydrogen bond*. The more strongly polar the molecule is, the more effective the hydrogen bonding is in binding the molecules into a larger unit. As a result the boiling points of such molecules are higher than those of similar nonpolar molecules. Good examples are water and hydrogen fluoride.

Figure 14 shows that in the formation of H_2O, H_2S, H_2Se, and H_2Te, an unusual rise in the boiling point of H_2O occurs that is not in keeping with the typical slow increase of boiling point as the molecular mass increases. Instead of the expected slope line between H_2O and H_2S that is shown in Figure 14 as a dashed line, the actual boiling point of H_2O is quite a bit higher—$100°C$. The explanation is that hydrogen bonding occurs in H_2O but not to any significant degree in the other compounds.

FIGURE 14. Boiling Points of Hydrogen Compounds with Similar Electron Dot Structures

This same phenomena occurs with the hydrogen halides (HF, HCl, HBr, and HI). Note in Figure 14 that hydrogen fluoride, which has strong hydrogen bonding, shows an unexpectedly high boiling point.

Hydrogen bonding also explains why some substances have unexpectedly low vapor pressure, high heats of vaporization, and high melting points. In order for vaporization or melting to take place, molecules must be separated. Energy must be expended to break hydrogen bonds and thus break down the larger clusters of molecules into separate molecules. As with the boiling point, the melting point of H_2O is abnormally high when compared with the melting point of the hydrogen compounds of the other elements having six valence electrons, which are chemically similar but which have no apparent hydrogen bonding. The hydrogen bonding effect in water is covered on pages 127–128.

DOUBLE AND TRIPLE BONDS

To achieve the *octet* structure, which is an outer energy level resembling the noble gas configuration of eight electrons, it is necessary for some atoms to share two or even three pairs of electrons. Sharing two pairs of electrons produces a *double bond*. An example:

$\overset{x}{\underset{xx}{\overset{x}{O}}}\overset{x}{\underset{x}{\overset{\circ}{C}}}\overset{x}{\underset{xx}{\overset{x}{O}}}$, and by a line formula $O = C = O$

    carbon dioxide

In the line formula, only the shared pair of electrons is indicated by a bond (—). The sharing of three electron pairs results in a *triple* bond. An example:

$H\overset{x}{\underset{\circ}{C}}\overset{\circ\circ}{\underset{\circ\circ}{C}}\overset{x}{\underset{\circ}{C}}H$, and by a line formula $H — C \equiv C — H$

    acetylene

It can be assumed from these structures that there is a greater electron density between the nuclei involved and hence a greater attractive force between the nuclei and the shared electrons. Experimental data verify that greater energy is required to break double bonds than single bonds, and triple bonds than double bonds. Also, since these stronger bonds tend to pull atoms closer together, the atoms joined by double and triple bonds have smaller interatomic distances and greater bond strengths, respectively.

RESONANCE STRUCTURES

It is not always possible to represent the bonding structure by either the Lewis dot structure or the line drawing because data about the bonding distance and bond strength are between possible drawing configurations and really indicate a hybrid condition. To represent this, the possible alternatives are drawn with arrows in between. Classic examples are sulfur trioxide and benzene. These are shown in Chapters 13 and 14, respectively, but are repeated here as examples.

Sulfur trioxide resonance structures:

Benzene resonance structures:

ELECTROSTATIC REPULSION (VSEPR) AND HYBRIDIZATION

Electrostatic Repulsion

Data have shown that bond angles for atoms in molecules with p orbitals in the outer energy level do not conform to the expected 90° separation of an x, y, z axis orientation. This variation can be expressed by *electrostatic repulsion between valence electron charge clouds* or by the concept of *hybridization*. The valence shell electron pair repulsion is sometimes abbreviated as VSEPR model.

Electrostatic repulsion uses as its basis the fact that like charges will orient themselves in such a way as to diminish the repulsion between them.

1. Mutual repulsion of two electron clouds forces them to the opposite sides of a sphere. This is called a *linear* arrangement.

• **EXAMPLE:** BeF_2

2. Minimum repulsion between three electron pairs occurs when the pairs are at the vertices of an equilateral triangle inscribed in a sphere. This is called *trigonal planar*.

• **EXAMPLE:** BF_3

3. Four electron pairs are farthest apart at the vertices of a tetrahedron inscribed in a sphere. This is called a *tetrahedral* shape.

• **EXAMPLE:** CH_4

4. Mutual repulsion of six identical electron clouds directs them to the corners of an inscribed regular octahedron. This is said to have *octahedral* geometry.

• **EXAMPLE:** SF_6

Hybridization

These same configurations can also be arrived at through the concept of *hybridization.* Briefly stated, this means that two or more pure atomic orbitals (usually *s, p,* and *d*) can be mixed to form two or more hybrid atomic orbitals that are identical. This can be illustrated as follows:

1. *sp* Hybrid Orbitals
   Beryllium fluoride spectroscopic measurements reveal a bond angle of 180° and equal bond lengths.

   F—BE—F
     180°

   The ground state of beryllium is:

To accommodate the experimental data we theorize that a $2s$ electron is excited to a $2p$ orbital; then the two orbitals hybridize to yield two identical orbitals called *sp* orbitals. Each contains one electron but is capable of holding two electrons.

2. *sp²* Hybrid Orbitals

Boron trifluoride has bond angles of 120° of equal strength. To accommodate these data the boron atom hybridizes from its ground state of $1s^2 2s^2 2p^1$ to:

3 orbitals of *sp²*

Boron + 3 F

3. *sp³* Hybrid Orbitals

Methane, CH_4, can be used to illustrate this hybridization. Carbon has a ground state of $1s^2 2s^2 2p^2$. One 2s electron is excited to a 2p orbital, and the four involved orbitals then form four new identical *sp³* orbitals.

Tetrahedron
Four *sp3* hybrid orbitals

Carbon   +   4 H   =

In some compounds where only certain sp^3 orbitals are involved in bonding, distortion in the bond angle occurs because of unbonded electron repulsion. Examples:

a. Water, H_2O.

Water, H_2O

b. Ammonia, NH_3.

4. sp^3d^2 Hybrid Orbitals

These orbitals are formed from the hybridization of an s and a p electron promoted to d orbitals and transformed into six equal sp^3d^2 orbitals. The spatial form is:

• **EXAMPLE:** SF_6

SUMMARY OF HYBRIDIZATION					
Number of Bonds	*Number of Unused Electron Pairs*	*Type of Hybrid Orbital*	*Angle between Bonded Atoms*	*Geometry*	*Example*
2	0	sp	180°	Linear	BeF_2
3	0	sp^2	120°	Trigonal planar	BF_3
4	0	sp^3	109.5°	Tetrahedral	CH_4
3	1	sp^3	90° to 109.5°	Pyramidal	NH_3
2	2	sp^3	90° to 109.5°	Angular	H_2O
6	0	sp^3d^2	90°	Octahedral	SF_6

SIGMA AND PI BONDS

When bonding occurs between s and p orbitals, each bond is identified by a special term. A *sigma bond* is a bond between s orbitals, or an s orbital and another orbital such as a p orbital. It includes bonding between hybrids of s orbitals such as sp, sp^2, and sp^3.

In the methane molecule, the sp^3 orbitals are each bonded to hydrogen atoms. These are sigma bonds.

When two p orbitals share electrons in a covalent bond, this is called a *pi bond*. An example:

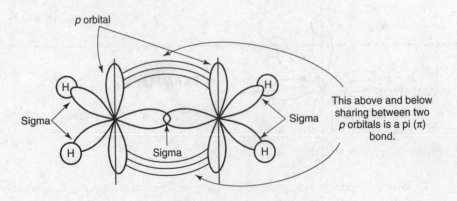

Chapter 14 gives more examples of sigma and pi bonding.

PROPERTIES OF IONIC SUBSTANCES

Laboratory experiments reveal that, in general, ionic substances are characterized by the following properties:

1. In the solid phase at room temperature they do not conduct appreciable electric current.

2. In the liquid phase they are relatively good conductors of electric current. The conductivity of ionic substances is much smaller than that of metallic substances.

3. They have relatively high melting and boiling points. There is a wide variation in the properties of different ionic compounds. For example, potassium iodide (KI) melts at 686°C and boils at 1,330°C, while magnesium oxide (MgO) melts at 2,800°C and boils at 3,600°C. Both KI and MgO are ionic compounds.

4. They have relatively low volatilities and low vapor pressures. In other words, they do not vaporize readily at room temperatures.

5. They are brittle and easily broken when a stress is exerted on them.

6. Those that are soluble in water form electrolytic solutions that are good conductors of electricity. There is, however a wide range in the solubilities of ionic compounds. For example, at 25°C, 92 g of sodium nitrate ($NaNO_3$) dissolves in 100 g of water, while only 0.0002 g of $BaSO_4$ dissolves in the same mass of water.

PROPERTIES OF MOLECULAR CRYSTALS AND LIQUIDS

Experiments have sh   llowing are general properties of these substances:

1. Neither the liquids   conduct electric current appreciably.

2. Many exist as gases   erature and atmospheric pressure, and many solids and liquids are relati

3. The melting points o   are relatively low.

4. The boiling points of   e relatively low.

5. The solids are gener   ve a waxy consistency.

6. A large amount of   equired to decompose the substance chemically into simpler substa

Chapter

3

Review

• QUESTIONS

Use the following   questions 1 through 7: (a) ionic   (b) covalent
(c) polar covalent   ordinate covalent   (e) metallic   (f) van der Waals forces   (g) hydroge   nding.

1. When the electronegativity difference between two atoms is 2, what type bond can be predicted?

2. If two atoms are bonded in such a way that one member of the pair is supplying both electrons that are shared, what is this type of bond called?

3. If the seven choices above were ordered from the strongest bond to the weakest, which would be the last one?

4. Which of the above bonds explains water's abnormally high boiling point?

5. When electron pairs are shared equally between two atoms, what is the bond called?

6. If the sharing of an electron pair is unequal, what is this sharing called?

7. The force of attraction of the electrons of one atom for the protons of another atom in close proximity is which choice above?

8. The VSEPR and model of the BF_3 molecule results in the same trigonal planar structure as that of the hybridized form represented by (a) sp (b) sp^2 (c) sp^3 (d) sp^3d^2.

• **ANSWERS:**

1. (a)   4. (g)   7. (f)
2. (d)   5. (b)   8. (b)
3. (f)   6. (c)

Terms You Should Know

covalent bond	ionic bond	stable octet
dipole-dipole attraction	London forces	sigma bond
electrostatic repulsion	metallic bond	van der Waals forces
hybridization	pi bond	VSEPR
hydrogen bond	resonance structure	

Chemical Formulas

WRITING FORMULAS

From the knowledge of oxidation numbers and valence and the understanding of atomic structure, it is now possible to write chemical formulas. Table 6 is a list of oxidation numbers often encountered in a first-year chemistry course.

• GENERAL OBSERVATIONS:

1. The symbols of the metals have + signs, while those of the nonmetals and all the polyatomic ions *except* ammonium have – signs.

2. When an element exhibits two possible oxidation states, the lower state can be indicated with the suffix *-ous* and the higher one with *-ic.* Another method of indicating this difference is to use the Roman numeral of the oxidation state in parentheses after the name of the element. For example: ferrous iron or iron(II) for the +2 oxidation state of iron.

3. A *polyatomic ion* is a group of elements that act like a single atom in the formation of a compound. The bonds within these polyatomic ions are predominantly covalent, but the groups of atoms as a whole have an excess of electrons when combined, and thus are usually negative ions.

When you attempt to write a formula, it is important to know whether the substance actually exists. For example, one can easily write the formula of carbon nitrate, but no chemist has ever prepared this compound. Here are the basic rules for writing formulas with three examples carried through each step:

1. Represent the symbols of the components, writing the positive part first, and then the negative part.

Sodium chloride	Calcium oxide	Ammonium sulfate
NaCl	CaO	NH_4SO_4

2. Indicate the oxidation number above and to the right of each symbol. (Enclose polyatomic ions in parentheses for the time being.)

$Na^{1+}Cl^{1-}$	$Ca^{2+}O^{2-}$	$(NH_4)^{1+}(SO_4)^{2-}$

Metals / Cations (+)

Monovalent I	Bivalent II	Trivalent III	Tetravalent IV	V
Hydrogen H	Barium Ba	Aluminum Al	Carbon C	Arsenic (ic) As
Potassium K	Calcium Ca	Gold (auric) Au	Silicon Si	Phosphorus (ic) P
Sodium Na	Cobalt Co	Arsenic (ous) As	Manganese (ic) Mn	Antimony (ic) Sb
Silver Ag	Magnesium Mg	Chromium Cr	Tin (stannic) Sn	Bismuth (ic) Bi
Mercury (ous) Hg	Lead Pb	Iron (ferric) Fe	Platinum Pt	
Copper (ous) Cu	Zinc Zn	Phosphorus (ous) P	Sulfur S	
Gold (aurous) Au	Mercury (ic) Hg	Antimony (ous) Sb		
*Ammonium (NH_4)	Copper (cupric) Cu	Bismuth (ous) Bi		
	Iron (ferrous) Fe			
	Manganese (ous) Mn			
	Tin (stannous) Sn			

Use of the Roman numeral, instead of the suffix, is now preferred, for example, iron(II) oxide.

*Polyatomic ion

Nonmetals† / Anions (−)

Monovalent I	Bivalent II	Trivalent III	Tetravalent IV
Fluorine F	Oxygen O	Nitrogen N	Carbon C
Chlorine Cl	Sulfur S	Phosphorus P	
Bromine Br			
Iodine I			

†Last syllable of nonmetal is changed to -ide in binary compound.

Polyatomic Ions (−)

Monovalent I	Bivalent II	Trivalent III	Tetravalent IV
Hydroxide (OH)	Carbonate (CO_3)	Borate (BO_3)	Ferrocyanide $[Fe(CN)_6]$
Bicarbonate (HCO_3)	Sulfite (SO_3)	Phosphate (PO_4)	
Nitrite (NO_2)	Sulfate (SO_4)	Phosphite (PO_3)	
Nitrate (NO_3)	Tetraborate (B_4O_7)	Ferricyanide $[Fe(CN)_6]$	
Hypochlorite (ClO)	Silicate (SiO_3)		
Chlorate (ClO_3)	Chromate (CrO_4)		
Chlorite (ClO_2)	Oxalate (C_2O_4)		
Perchlorate (ClO_4)			
Acetate $(C_2H_3O_2)$			
Permanganate (MnO_4)			
Bisulfate (HSO_4)			

TABLE 6. Table of Oxidation Numbers

3. Write a subscript number equal to the oxidation number of the other element or polyatomic ion. This is the same as the mechanical criss-cross method.

$$\text{Na}_1{}^{1+}\text{Cl}_1{}^{1-} \qquad\qquad \text{Ca}_2{}^{2+}\text{O}_2{}^{2-} \qquad\qquad (\text{NH}_4)_2{}^{1+}(\text{SO}_4)_1{}^{2-}$$

Since the positive oxidation number shows the number of electrons that may be lost or shared and the negative oxidation number shows the number of electrons that may be gained or shared, you must have just as many electrons lost (or partially lost in sharing) as are gained (or partially gained in sharing).

4. Now rewrite the formulas, omitting the subscript 1, the parentheses of the polyatomic ions that have the subscript 1, and the plus and minus numbers.

$$\text{NaCl} \qquad\qquad \text{Ca}_2\text{O}_2 \qquad\qquad (\text{NH}_4)_2\text{SO}_4$$

5. As a general rule, reduce the subscript numbers in the final formula to their lowest terms. There are, however, certain exceptions, such as hydrogen peroxide (H_2O_2), and acetylene (C_2H_2). For these exceptions, you must know more specific information about the compound.

The only way to become proficient at writing formulas is to memorize the oxidation numbers of common elements (or learn to use the period chart group numbers) and practice writing formulas. It is especially important to memorize the polyatomic ions and their oxidation numbers.

MORE ABOUT OXIDATION NUMBERS

Because it is important to know how to assign oxidation numbers in order to write formulas correctly, the basic rules concerning the assignment of oxidation numbers need to be listed. The oxidation number concept was introduced in Chapter 2, which dealt with atomic structure. The oxidation number can be defined as the charge that an atom is assumed to have in a particular molecule or ion.

The basic rules follow:

1. The oxidation number of a monoatomic ion is the same as the charge on the ion. Therefore, in NaCl, the oxidation number assigned to sodium is 1+, and the oxidation number of chlorine is 1–.

2. The oxidation number of any element is zero when the atoms of that element are in the elemental state. An atom is in its elemental state when its atoms are not chemically combined with atoms of other elements.

3. In a covalent compound of two elements, the more electronegative element has a negative oxidation number and the other element has a positive oxidation number. For example, in ammonia, NH_3, the nitrogen is more electronegative. Each of the valence electrons from the three hydrogen atoms is drawn nearer to the nitrogen atom. The oxidation number of the nitrogen in ammonia is therefore 3–. Since each hydrogen atom partially loses one electron, the oxidation number of hydrogen is 1+.

4. The oxidation number of oxygen in most of its compounds is 2–. In peroxides an exception occurs, and oxygen is assigned 1–.

5. The oxidation number of hydrogen in most compounds is 1+. In hydrides an exception occurs, and hydrogen is assigned 1–.

6. In neutral compounds, for all the atoms represented by the formula, the quantitative sum of the positive oxidation numbers must equal the quantitative sum of the negative oxidation numbers, and their sum must equal 0. This can best be illustrated in an example.

In Na_2SO_4, what is the oxidation number of sulfur?

$$Na_2SO_4$$

We know the oxidation number of one atom of Na is 1+. There are two atoms of Na.

So, $2 \times 1+ = 2+$

We know one atom of O is assigned 2–. There are four atoms of O.

So, $4 \times 2- = 8-$

Since the positive sum plus the negative sum must equal 0,

Then

$$(2+) + x + (8-) = 0$$
$$x = 6+$$

The sulfur must be using a 6+ oxidation state.

Another example: What is the oxidation number of chromium in $K_2Cr_2O_7$?

$$K = 2 \times (1+) = 2+$$
$$Cr = 2 \times (x) = 2x$$
$$O = 7 \times (2-) = 14-$$

$$(2+) + (2x) + (14-) = 0$$
$$2x = 12+$$
$$x = 6+$$

So, chromium is using a 6+ oxidation number.

7. In a polyatomic ion, the algebraic sum of the positive and negative oxidation numbers of all the atoms in the formula must equal the charge on the ion.

Example: What is the oxidation number of sulfur in SO_4^{2-}?

$$S = 1 \times (x) = 1x$$
$$O = 2 \times (2-) = 8-$$

$$(1x) + (8-) = -2 \text{ (the charge of the ion)}$$
$$x = 6+$$

The sulfur is using a 6+ oxidation state.

Another example: In the NO_3^- ion, find the oxidation number of N.

$$NO_3^-$$
$$(x) + 3(2-) = 1-$$
$$x = 5+$$

Nitrogen is using an oxidation number of 5+.

NAMING COMPOUNDS

A binary compound consists of two elements. The name of the compound also consists of the two elements, the second name having its ending changed to -ide, such as NaCl = sodium chloride; AgCl = silver chloride. If the metal has two different oxidation numbers, this can be indicated by the use of the suffix -ous for the lower one and -ic for the higher one. As previously stated, the preferred way is to use a Roman numeral after the name to indicate the oxidation state.

- **EXAMPLES:**

  $FeCl_2$ = ferrous chloride or iron(II) chloride

  $FeCl_3$ = ferric chloride or iron(III) chloride

If elements combine in varying proportions, thus forming two or more compounds of varying compositions, the name of the second element may be preceded by a prefix, such as mon(o)- (one), di- (two), tri- (three), or pent(a)- (five). (These prefixes do not use the letter in parentheses when preceding a vowel in the word to which it is attached.) Some examples are: carbon dioxide, CO_2; carbon monoxide, CO; phosphorous trioxide, P_2O_3; phosphorous pentoxide, P_2O_5. Notice that, when these prefixes are used, it is not necessary to indicate the oxidation state of the first element in the name since it is given indirectly by the prefixed second element.

A ternary compound consists of three components, usually an element and a polyatomic ion. To name the compound, you merely name each component in the order of positive first and negative second.

Binary acids use the prefix hydro- in front of the stem or full name of the nonmetallic element, and add the ending -ic. Examples are hydrochloric acid (HCl) and hydrosulfuric acid (H_2S).

Ternary acids usually contain hydrogen, a nonmetal, and oxygen. Since the amount of oxygen often varies, the name of the most common form of the acid in the series consists of merely the stem of the nonmetal with the ending -ic. The acid containing one less atom of oxygen than the most common acid is designated by the ending -ous. The name of the acid containing one more atom of oxygen than the most common acid has the prefix per- and the ending -ic ; that of the acid containing one less atom of oxygen than the -ous acid has the prefix hypo- and the ending -ous.

You can remember the names of the common acids and their salts by learning the following simple rules:

Rule	Example
-ic acids form -ate salts.	Sulfuric acid forms sulfate salts.
-ous acids form -ite salts.	Sulfurous acid forms sulfite salts.
hydro-(stem)-ic acids form -ide salts.	Hydrochloric acid forms chloride salts.

When the name of the ternary acid has the prefix hypo- or per-, that prefix is retained in the name of the salt (hypochlorous acid = sodium hypochlorite).

The names and formulas of some common acids and bases are listed in Table 7.

ACIDS, BINARY		ACIDS, TERNARY	
Name	*Formula*	*Name*	*Formula*
Hydrofluoric	HF	Nitric	HNO_3
Hydrochloric	HCl	Nitrous	HNO_2
Hydrobromic	HBr	Hypochlorous	HClO
Hydriodic	HI	Chlorous	$HClO_2$
Hydrosulfuric	H_2S	Chloric	$HClO_3$
		Perchloric	$HClO_4$
BASES		Sulfuric	H_2SO_4
		Sulfurous	H_2SO_3
Sodium hydroxide	NaOH	Phosphoric	H_3PO_4
Potassium "	KOH	Phosphorous	H_3PO_3
Ammonium "	NH_4OH	Carbonic	H_2CO_3
Calcium "	$Ca(OH)_2$	Acetic	$HC_2H_3O_2$
Magnesium "	$Mg(OH)_2$	Oxalic	$H_2C_2O_4$
Barium "	$Ba(OH)_2$	Boric	H_3BO_3
Aluminum "	$Al(OH)_3$	Silicic	H_2SiO_3
Ferrous "	$Fe(OH)_2$		
Ferric "	$Fe(OH)_3$		
Zinc "	$Zn(OH)_2$		
Lithium "	LiOH		

TABLE 7. Formulas of Common Acids and Bases

CHEMICAL FORMULAS

The chemical formula is an indication of the makeup of a compound in terms of the kinds of atoms and their relative numbers. It also has some quantitative applications. By using the atomic weights assigned to the elements, we can find the *formula weight* of a compound. If we are sure that the formula represents the actual makeup of one molecule of the substance, the term *molecular weight* or *molar mass* may be used as well. In some cases the formula represents an ionic lattice and no discrete molecule exists, as in the case of table salt, NaCl, or the formula merely represents the simplest ratio of the combined substances and not a molecule of the substance. For example, CH_2 is the simplest ratio of carbon and hydrogen united to form the actual compound ethylene, C_2H_4. This simplest ratio formula is called the *empirical formula*, and the actual formula is the *true formula*. The formula mass is determined by multiplying the atomic mass (in whole numbers) by the subscript for that element in the formula. For example:

$Ca(OH)_2$ (one calcium atomic mass + two hydrogen and two oxygen atomic masses = formula mass).

$$1 \text{ Ca (at. wt.} = 40) = 40$$
$$2 \text{ O (at. wt.} = 16) = 32$$
$$\underline{2 \text{ H (at. wt.} = 1) = 2}$$
$$\text{Formula weight } Ca(OH)_2 = 74$$

Fe_2O_3

$$2 \text{ Fe (at. wt. } = 56) = 112$$

$$\underline{3 \text{ O (at. wt. } = 16) = 48}$$

$$\text{Formula weight } Fe_2O_3 = 160$$

It is sometimes useful to know what percent of the total weight of a compound is made up of a particular element. This is called finding the *percentage composition*. The simple formula for this is:

$$\frac{\text{Total weight of the element in the compound}}{\text{Total formula weight}} \times 100\% = \text{Percentage composition of that element}$$

To find the percent composition of calcium in calcium hydroxide in the example above, we set the formula up as follows:

$$\frac{Ca = 40}{\text{Formula weight } = 74} \times 100\% = 54\% \text{ Calcium}$$

To find the percent composition of oxygen in calcium hydroxide:

$$\frac{O = 32}{\text{Formula weight } = 74} = 100\% = 43\% \text{ Oxygen}$$

To find the percent composition of hydrogen in calcium hydroxide:

$$\frac{H = 2}{\text{Formula weight } = 74} \times 100\% = 2.7\% \text{ Hydrogen}$$

• ANOTHER EXAMPLE:

Find the percent compositions of Cu and H_2O in the compound $CuSO_4 \cdot 5\,H_2O$ (the dot is read "with").

First, we calculate the formula weight:

$$1 \quad Cu = 64$$

$$1 \quad S = 32$$

$$4 \quad O = 64\ (4 \times 16)$$

$$\underline{5\,H_2O = 90\ (5 \times 18)}$$

$$250$$

and then find the percentages:

Percentage Cu:

$$\frac{Cu = 64}{\text{Formula weight } = 250} \times 100\% = 26\%$$

Percentage H_2O:

$$\frac{5\,H_2O = 90}{\text{Formula weight } = 250} \times 100\% = 36\%$$

When you are given the percentage of each element in a compound, you can find the empirical formula as shown with the following example:

> Given that a compound is composed of 60.0% Mg and 40.0% O, find the empirical formula of the compound.

1. It is easiest to think of 100 weight units of this compound. In this case, the 100 weight units are composed of 60 weight units of Mg and 40 weight units of O. Since you know that 1 unit of Mg is 24 weight units (from its atomic weight) and, likewise, 1 unit of O is 16, you can divide 60 by 24 to find the number of units of Mg in the compound and divide 40 by 16 to find the number of units of O in the compound.

$$\begin{array}{cc} \text{Mg} & \text{O} \\ 24)\overline{60} & 16)\overline{40} \\ 2.5 \text{ units Mg} & 2.5 \text{ units O} \end{array}$$

2. Now, since we know formulas are made up of whole number units of the elements, which are expressed as subscripts, we must manipulate these numbers to get whole numbers. This is usually accomplished by dividing these numbers by the smallest quotient. In this case they are equal, so we divide by 2.5.

$$\begin{array}{cc} \text{Mg} & \text{O} \\ 2.5)\overline{2.5} & 2.5)\overline{2.5} \\ 1 & 1 \end{array}$$

- **ANOTHER EXAMPLE:**

  Given: Ba = 58.81%, S = 13.73%, and O = 27.46%.
  Find the empirical formula.

1. Divide each percent by the atomic weight of the element.

$$\begin{array}{ccc} \text{Ba} & \text{S} & \text{O} \\ 137)\overline{58.8} & 32)\overline{13.7} & 16)\overline{27.5} \\ 0.43 & 0.43 & 1.72 \end{array}$$

2. Manipulate numbers to get small whole numbers. Try dividing them all by the smallest first.

$$\begin{array}{ccc} \text{Ba} & \text{S} & \text{O} \\ 0.43)\overline{0.43} & 0.43)\overline{0.43} & 0.43)\overline{1.72} \\ 1 & 1 & 4 \end{array}$$

3. The formula is $BaSO_4$.

In some cases you may be given the true formula weight of the compound. To check if your empirical formula is correct, add up the formula weight of the empirical formula and compare it with the given formula weight. If it is *not* the same, multiply the empirical formula by the small, whole number that gives you the correct formula weight. For example, if your empirical formula is CH_2 (which has a formula weight of 14) and the true formula weight is given as 28, you can see that you must double the empirical formula by doubling all the subscripts. The true formula is C_2H_4.

LAWS OF DEFINITE COMPOSITION AND MULTIPLE PROPORTIONS

In the problems involving percent composition, we have depended on two things: each unit of an element has the same atomic mass, and every time the particular compound forms, it forms in the same percent composition. That this latter statement is true no matter the source of the compound is the *Law of Definite Composition*. There are some compounds formed by the same two elements in which the weight of one element is constant, but the weight of the other varies. In every case, however, the weight of the other element is present in a small-whole-number ratio to the weight of the first element. This is called the *Law of Multiple Proportions*. An example is H_2O and H_2O_2.

In H_2O the proportion of H:O = 2:16 or 1:8
In H_2O_2 the proportion of H:O = 2:32 or 1:16

The ratio of the weight of oxygen in each is 8:16 or 1:2 (a small-whole-number ratio).

WRITING AND BALANCING SIMPLE EQUATIONS

An equation is a simplified way of recording a chemical change. Instead of words, chemical symbols and formulas are used to represent the reactants and the products. Here is an example of how this can be done. The following is the word equation of the reaction of burning hydrogen with oxygen:

Hydrogen + oxygen yields water.

Replacing the words with the chemical formulas, we have

$$H_2 + O_2 \rightarrow H_2O$$

We replaced hydrogen and oxygen with the formulas for their diatomic molecular states and wrote the appropriate formula for water based on the respective oxidation (valence) numbers for hydrogen and oxygen. Note that the word *yields* was replaced with the arrow.

Although the chemical statement tells what happened, it is not an equation because the two sides are not equal. While the left side has two atoms of oxygen, the right side has only one. Knowing that the Law of Conservation of Matter dictates that matter cannot easily be created or destroyed, we must get the number of atoms of each element represented on the left side to equal the number on the right. To do this, we can only use numbers, called *coefficients*, in front of the formulas. It is important to note that in attempting to balance equations THE SUBSCRIPTS IN THE FORMULAS MAY NOT BE CHANGED.

Looking again at the skeleton equation, we notice that if 2 is placed in front of H_2O the numbers of oxygen atoms represented on the two sides of the equation are equal. However, there are now four hydrogens on the right side with only two on the left. This can be corrected by using a coefficient of 2 in front of H_2. Now we have a balanced equation:

$$2H_2 + O_2 \rightarrow 2H_2O$$

This equation tells us more than merely that hydrogen reacts with oxygen to form water. It has quantitative meaning as well. It tells us that two molecular weights of hydrogen react with one molecular weight of oxygen to form two molecular weights of water. Because molecular weights are indirectly related to grams, we may also relate the weights of reactants and products in grams.

$$2H_2 \quad + \quad O_2 \quad \rightarrow \quad 2H_2O$$
$$2(2) \qquad\qquad 32 \qquad\qquad 2(18)$$
$$4 \text{ units} \quad + \quad 32 \text{ units} \quad = \quad 36 \text{ units}$$
$$4 \text{ grams of } H_2 + 32 \text{ grams of } O_2 = 36 \text{ grams of water}$$

This aspect will be important in solving problems related to the weights of substances in a chemical equation.

Here is another, more difficult example: Write the balanced equation for the burning of butane (C_4H_{10}) in oxygen. First, we write the skeleton equation:

$$C_4H_{10} + O_2 \text{ yields } CO_2 + H_2O.$$

Looking at the oxygens, we see that there are an even number on the left but an odd number on the right. This is a good place to start. If we use a coefficient of 2 for H_2O, that will even out the oxygens but introduce four hydrogens on the right while there are ten on the left. A coefficient of 5 will give us the right number of hydrogens but introduces an odd number of oxygens. Therefore we have to go to the next even multiple of 5, which is 10. Ten gives us 20 hydrogen atoms on the right. By placing another coefficient of 2 in front of C_4H_{10}, we also have 20 hydrogen atoms on the left. Now the carbons need to be balanced. By placing an 8 in front of CO_2, we have eight carbons on both sides. The remaining step is to balance the oxygens. We have 26 on the right side, so we need a coefficient of 13 in front of the O_2 on the left to give us 26 oxygens on both sides. Our balanced equation is:

$$2C_4H_{10} + 13O_2 \rightarrow 8CO_2 + 10H_2O$$

For more practice in balancing equations, see page 118. (The balancing of equations using the electron exchange method is introduced at the end of Chapter 12.)

SHOWING PHASES IN CHEMICAL EQUATIONS

Once the equation is balanced, you may choose to give additional information in an equation. This can be done by indicating the phases of substances, telling whether the substance is in the liquid phase (ℓ), the gaseous phase (g), or the solid phase (s). Since many solids will not react to any appreciable extent unless they are dissolved in water, the notation (aq) is used to indicate that the substance exists in a water (aqueous) solution. Information concerning phase is given in parentheses following the formula for each substance. Several illustrations of this notation are given below:

Formula with Phase Notation	Meaning
$Cl_2(g)$	Chlorine gas
$H_2O(\ell)$	Water in the liquid state as opposed to ice or steam
$NaCl(s)$	Sodium chloride as a solid
$NaCl(aq)$	A water solution of dissolved sodium chloride

An example of phase notation in an equation:

$$2HCl(aq) + Zn(s) \rightarrow ZnCl_2(aq) + H_2(g)$$

In words, this says that a water solution of hydrogen chloride (called hydrochloric acid) reacts with solid zinc to produce zinc chloride dissolved in water plus hydrogen gas.

WRITING IONIC EQUATIONS

At times, chemists choose to show only the substances that react in the chemical action. These are called *ionic* equations because they stress the reaction and production of ions. If we look at the preceding equation, we see that the complete cast of "actors" in this equation are:

Reactants

$2HCl(aq)$ releases $2H^+ (aq) + 2Cl^-(aq)$ in solution

$Zn(s)$ stay as $Zn(s)$ particles

Products

$ZnCl_2(aq)$ releases $Zn^{2+}(aq) + 2Cl^-(aq)$ in solution

$H_2(g)$ stay as $H_2(g)$ molecules

Writing the complete reaction using these results, we have:

$$2H^+(aq) + 2Cl^-(aq) + Zn(s) \rightarrow Zn^{2+}(aq) + 2Cl^-(aq) + H_2(g)$$

Notice that nothing happened to the chloride ion. It appears the same on both sides of the equation. It is referred to as a spectator ion. In writing the simple ionic equation, spectator ions are omitted. So, the ionic equation is:

$$2H^+(aq) + Zn(s) \rightarrow Zn^{2+}(aq) + H_2(g)$$

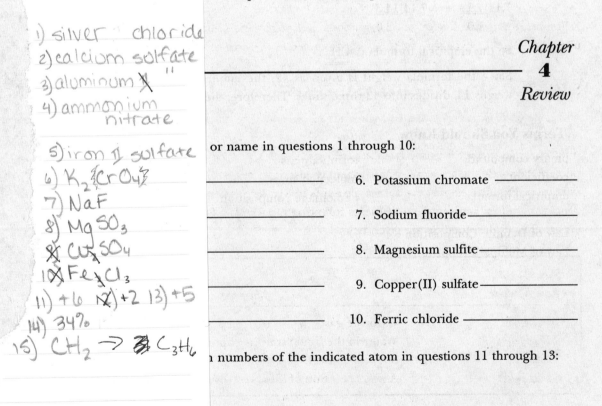

Chapter

4

Review

or name in questions 1 through 10:

6. Potassium chromate —————

7. Sodium fluoride————

8. Magnesium sulfite————

9. Copper(II) sulfate————

10. Ferric chloride ————

numbers of the indicated atom in questions 11 through 13:

ntage of sulfur in H_2SO_4.

empirical formula and the true formula of a compound com-

% C and 14.3% H with a true formula weight of 42?

Handwritten notes:

1) silver chloride
2) calcium sulfate
3) aluminum X ''
4) ammonium nitrate

5) iron II sulfate
6) K_2CrO_4
7) NaF
8) $MgSO_3$
X Cu_2SO_4
1X Fe_2Cl_3
11) +6 X) +2 13) +5
14) 34%
15) $CH_2 \rightarrow$ ~~3~~ C_3H_6

• ANSWERS:

1. Silver chloride
2. Calcium sulfate
3. Aluminum sulfate
4. Ammonium nitrate

5. Iron(II) sulfate or ferrous sulfate
6. K_2CrO_4
7. NaF

8. $MgSO_3$
9. $CuSO_4$
10. $FeCl_3$

11. 6+ because $\overset{1+}{K}$ $\overset{x}{Cr}$ $\overset{2-}{O}$, $2(1+) + 1(x) + 4(2-) = 0$, $x = 6+$

12. 2+ because $\overset{1+}{Na}$ $\overset{x}{S}$ $\overset{2-}{O}$, $2(1+) + 2(x) + 3(2-) = 0$, $x = 2+$

13. 5+ because $\overset{x}{P}$ $\overset{2-}{O}$, $1(x) + 4(2-) = 3-$, $x = 5+$

14. H_2SO_4 is composed of:

   $2H = 4$
   $1S = 32$
   $4O = \underline{64}$
   Total $= 98$

   Percentage of S:

   $\dfrac{S = 32}{\text{Total} = 98} \times 100\% = \underline{\underline{32.65}}$ or $\underline{\underline{33\%}}$

15. $C = 12\overline{)85.7\%\ C}$ $H = 1\overline{)14.3\%\ H}$
   7.14 14.3

   To find the lowest ratio of the whole numbers:

   $7.14\overline{)7.14}$ $7.14\overline{)14.3}$
   1.0 2.0

   So the empirical formula is $\underline{\underline{CH_2}}$.

   Since the formula weight is given as 42, the empirical formula CH_2, which weighs 14, divides into 42 three times. Therefore, the true formula is $\underline{\underline{C_3H_6}}$.

Terms You Should Know

binary compound	formula mass	polyatomic ion
coefficient	molecular mass	ternary compound
empirical formula	percentage composition	true formula

Law of Definite Composition
Law of Multiple Proportions

5

Gases

SOME REPRESENTATIVE GASES

Oxygen

Of the gases that occur in the atmosphere, the most important one to us is oxygen. Although it makes up only approximately 21% of the atmosphere, by volume, the oxygen found on the earth is equal in weight to all the other elements combined. About 50% of the earth's crust (including the waters on the earth, and the air surrounding it) is oxygen. (Note Figure 15.)

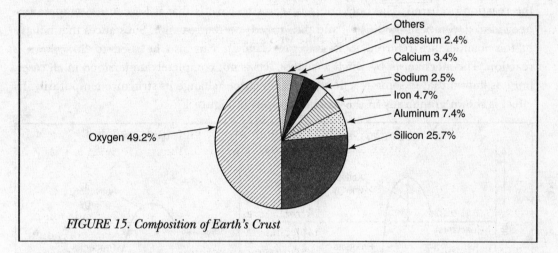

FIGURE 15. *Composition of Earth's Crust*

The composition of air varies slightly from place to place because it is a mixture of gases. The composition by volume is approximately as follows: nitrogen, 78%; oxygen, 21%; argon, 1%. There are also small amounts of carbon dioxide, water vapor, and trace gases.

Preparation of oxygen. In 1774, an English scientist named Joseph Priestley discovered oxygen by heating mercuric oxide in an enclosed container with a magnifying glass. That mercuric oxide decomposes into oxygen and mercury can be expressed in an equation: $2HgO \rightarrow 2Hg + O_2$. After his discovery, Priestley visited one of the greatest of all scientists, Antoine Lavoisier, in Paris. As early as 1773 Lavoisier had carried on experiments concerning burning, and they had caused him to doubt the phlogiston theory (that a substance called phlogiston was released when a substance burned; the theory went through several modifications before it was finally abandoned). By 1775, Lavoisier had demonstrated the true nature of burning and called the resulting gas "oxygen."

Today oxygen is usually prepared in the lab by heating an easily decomposed oxygen compound such as potassium chlorate ($KClO_3$). The equation for this reaction is:

$$2KClO_3 + MnO_2 \rightarrow 2KCl + 3O_2(g) + MnO_2$$

The usual laboratory setup is shown in Figure 16.

FIGURE 16. Laboratory Preparation of Oxygen

In this preparation manganese dioxide (MnO_2) is often used. This compound is not used up in the reaction and can be shown to have the same composition as it had before the reaction occurred. The only effect is seems to have is that it lowers the temperature needed to decompose the $KClO_3$, and thus speeds up the reaction. Substances that behave in this manner are referred to as *catalysts*. Catalysts can also be used to slow down a reaction. The mechanism by which a catalyst acts is not completely understood in all cases, but it is known that in some reactions the catalyst does change its structure temporarily. Its effect is shown graphically in the reaction graphs in Figure 17.

FIGURE 17. Reaction

You notice that in the reaction with the catalyst the activation energy necessary to make the reaction take place is lower so that the reaction can take place at a more rapid rate.

Commercially, oxygen is prepared by either fractional distillation of liquid air or the electrolysis of water. The fractional distillation of liquid air is typical of all fractional distillation processes in that it attempts to separate different components of a mixture in solution by taking advantage of the fact that each component has a distinct boiling point. By raising the temperature slowly and holding it constant when any boiling is taking place, the various components can be separated. With liquid air at –200°C, the temperature can be controlled so that nitrogen will boil off at –196°C. Then the oxygen is boiled off at –183°C.

Electrolysis of water will be more completely explained in Chapter 12, but essentially it consists of passing an electric current through water to cause it to decompose into hydrogen gas (H_2) and oxygen gas (O_2).

Oxidation. When oxygen combines with an element slowly so that *no* noticeable heat and light are given off, the process is referred to as slow oxidation. A common example of this is the rusting of iron. When the combination of oxygen with an element occurs so rapidly that the amount of energy released is visible as light and felt as heat, the process is referred to as rapid oxidation or normal burning. It is not always necessary to have oxygen present for burning or combustion to take place. A jet of hydrogen gas will burn in an atmosphere of chlorine.

Properties of oxygen. That oxygen is a gas under ordinary conditions of temperature and pressure, and that it is a gas which is colorless, odorless, tasteless, and slightly heavier than air, are all physical properties of this element. Oxygen is only slightly soluble in water, thus making it possible to collect the gas over water, as shown in Figure 16. Oxygen at −183°C changes to a pale blue liquid with magnetic properties. This attraction by a magnet (paramagnetism) is believed to be due to an unpaired group of electrons found in the *p* orbitals. The Lewis structure for oxygen is:

$$: \overset{\textstyle .\,.}{\underset{\textstyle .\,.}{\bigcirc}} : \overset{\textstyle .\,.}{\underset{\textstyle .\,.}{\bigcirc}} :$$

Although oxygen will support combustion, as noted above, it will not burn. This is one of its chemical properties. The usual test for oxygen is to lower a glowing splint into the gas and see if the oxidation increases in its rate to reignite the splint. (Note: This is not the only gas that does this. N_2O reacts the same.)

When a compound containing oxygen gives up its oxygen to another substance, the compound is called an *oxidizing agent*. Examples are potassium chlorate, used in making matches, and potassium nitrate, used in the manufacture of gunpowder. This definition will be further expanded in Chapter 12 to include substances that do not contain oxygen.

Ozone. Ozone is another form of oxygen and contains three atoms in its molecular structure (O_3). Since ordinary oxygen and ozone differ in energy content and form, they have slightly different properties. They are called allotropic forms of oxygen. Ozone occurs in small quantities in the upper layers of the earth's atmosphere, and can be formed in the lower atmosphere where high-voltage electricity in lightning passes through the air. This formation of ozone also occurs around machinery using high voltage. The reaction can be shown by this equation:

$$3O_2 + \text{elec.} \rightarrow 2O_3$$

Because of its higher energy content, ozone is more reactive chemically than oxygen.

Hydrogen

Of all the 109 elements in the periodic chart, hydrogen is number 1 by reason of the fact that it is the lightest element known. Its lightness makes it useful in balloons, but care must be taken because of its extreme combustibility. Its ability to burn well makes it an important part of most fuels, and its powerful attraction to oxygen makes it a good reducing agent in some important industrial reactions. Hydrogen can be added to liquid fats, under the proper conditions, to make solid shortenings, or it can be added to certain fuels to improve their burning properties. Its isotopes were used in making the hydrogen bomb.

Preparation of hydrogen. Although there is evidence of the preparation of hydrogen before 1766, Henry Cavandish was the first person to recognize this gas as a separate substance. He observed that, whenever it burned, it produced water. Lavoisier named it *hydrogen*, which means "water former."

Electrolysis of water, which is the process of passing an electric current through water to cause it to decompose, is one method of obtaining hydrogen. This is a widely used commercial method, as well as a laboratory method.

Another method of producing hydrogen is to displace it from the water molecule by using a metal. To choose the metal you must be familiar with its activity with respect to hydrogen. The acitivities of the common metals are shown in Table 8.

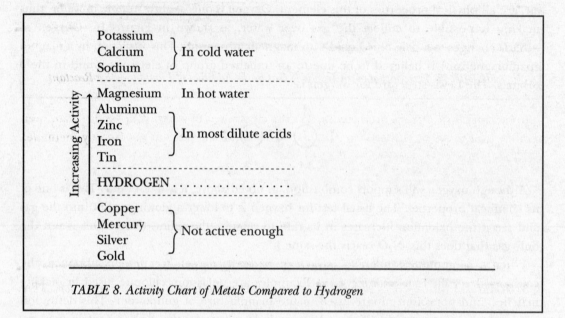

TABLE 8. *Activity Chart of Metals Compared to Hydrogen*

As noted in Table 8, the first three metals will react with water in the following manner:

Very active metal + Water = Hydrogen + Metal hydroxide

Using sodium as an example:

$$2Na + 2HOH \rightarrow H_2(g) + 2NaOH$$

With the metals that react slower, a dilute acid reaction is needed to produce hydrogen in sufficient quantities to collect in the laboratory. This general equation is:

Active metal + Dilute acid $\rightarrow$ Hydrogen + Salt of the acid

An example:

$$Zn + dil. H_2SO_4 \rightarrow H_2(g) + ZnSO_4$$

This equation shows the usual laboratory method of preparing hydrogen. Mossy zinc is used in a setup as shown in Figure 18. The acid is introduced down the thistle tube after the zinc is placed in the reacting bottle. In this sort of setup, you would not begin collecting the gas that bubbles out of the delivery tube for a few minutes so that the air in the system has a chance to be expelled and you can collect a rather pure volume of the gas generated.

FIGURE 18. Preparation of an Insoluble Gas by the Addition of Liquid to Other Reactant

In industry, hydrogen is produced by (1) the electrolysis of water, (2) passing steam over red-hot iron or through hot coke, or (3) by decomposing natural gas (mostly methane, CH_4) with heat ($CH_4 + H_2O \rightarrow CO + 3H_2$).

Properties of hydrogen. Hydrogen has the following important physical properties:

1. It is ordinarily a gas; colorless, odorless, tasteless when pure.

2. It weighs 0.9 gram per liter at 0°C and 1 atmosphere pressure. This is 1/14 as heavy as air.

3. It is slightly soluble in water.

4. It becomes a liquid at a temperature of –240°C and a pressure of 13 atmospheres.

5. It diffuses (moves from place to place in gases) more rapidly than any other gas. This property can be demonstrated as shown in Figure 19.

Here the H_2 in the beaker that is placed over the porous cup diffuses faster through the cup than the air can diffuse out. Consequently, there is a pressure buildup in the cup, which pushes the gas out through the water in the lower beaker.

FIGURE 19. Diffusion of Hydrogen

The chemical properties of hydrogen are:

1. It burns in air or in oxygen, giving off large amounts of heat. Its high heat of combustion makes it a good fuel.

2. It does not support ordinary combustion.

3. It is a good reducing agent in that it withdraws oxygen from many hot metal oxides.

GENERAL CHARACTERISTICS OF GASES

Measuring the Pressure of a Gas

Pressure is defined as force per unit area. With respect to the atmosphere, pressure is the result of the weight of a mixture of gases. This pressure, which is called *atmospheric pressure, air pressure,* or *barometric pressure,* is approximately equal to the weight of a kilogram mass on every square centimeter of surface exposed to it. This weight is about 10 newtons.

The pressure of the atmosphere varies with altitude. At higher altitudes, the weight of the overlying atmosphere is less, so the pressure is less. Air pressure also varies somewhat with weather conditions as low- and high-pressure areas move with weather fronts. On the average, however, the air pressure at sea level can support a column of mercury 760 mm in height. This average sea-level air pressure is known as *normal atmospheric pressure,* also called *standard pressure.*

The instrument most commonly used for measuring air pressure is the *mercury barometer.* The diagram below shows how it operates. Atmospheric pressure is exerted on the mercury in the dish, and this in turn holds the column of mercury up in the tube. This column at standard pressure will measure 760 mm above the level of the mercury in the dish below.

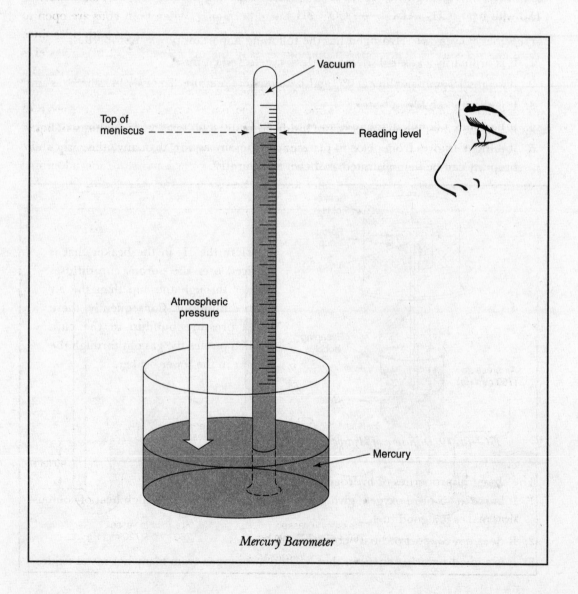

Mercury Barometer

In gas-law problems pressure may be expressed in various units. One standard atmosphere (1 atm) is equal to 760 millimeters of mercury (760 mm Hg) or 760 torr, a unit named for Evangelista Torricelli. In the SI system, the unit of pressure is the pascal (Pa), named in honor of the scientist of the same name. Standard pressure in pascals is 101,325 Pa or 101.325 kPa.

SUMMARY OF UNITS OF PRESSURE

Unit	Abbreviation	Unit Equivalent to 1 atm
Atmosphere	atm	1 atm
Millimeters of Hg	mm Hg	760 mm Hg
Torr	torr	760 torr
Inches of Hg	in. Hg	29.9 in. Hg
Pounds per square inch	lb./in.^{2} (psi)	14.7 lb./in.^{2}
Pascal	Pa	101,325 Pa

A device similar to the barometer can be used to measure the pressure of a gas in a confined container. This apparatus, called a *manometer*, is illustrated below. A manometer is basically a U-tube containing mercury or some other liquid. When both ends are open to the air, as in (1) in the diagram, the level of the liquid will be the same on both sides since the same pressure is being exerted on both ends of the tube. In (2) and (3), a vessel is connected to one end of the U-tube. Now the height of the mercury column serves as a means of reading the pressure inside the vessel if the atmospheric pressure is known. When the pressure inside the vessel is the same as the atmospheric pressure outside, the levels of liquid are the same. When the pressure inside is greater than outside, the column of liquid will be higher on the side that is exposed to the air, as in (2). Conversely, when the pressure inside the vessel is less than the outside atmospheric pressure, the additional pressure will force the liquid to a higher level on the side near the vessel, as in (3).

Manometer

Kinetic-Molecular Theory

By indirect observations, the kinetic-molecular theory has been arrived at to explain the forces between molecules and the energy the molecules possess. There are three basic assumptions to the kinetic-molecular theory:

1. Matter in all its forms (solid, liquid, and gas) is composed of extremely small particles. In many cases these are called molecules. The space occupied by the gas particles themselves is ignored in comparison with the volume of the space they occupy.

2. The particles of matter are in constant motion. In solids, this motion is restricted to a small space. In liquids, the particles have a more random pattern but still are restricted to a kind of rolling over one another. In a gas, the particles are in continuous, random, straight-line motion.

3. When these particles collide with each other or with the walls of the container, there is no loss of energy.

Some Particular Properties of Gases

As the temperature of a gas is increased, its kinetic energy is increased, thereby increasing the random motion. At a particular temperature not all the particles have the same kinetic energy, but the temperature is a measure of the average kinetic energy of the particles. A graph of the various kinetic energies would resemble a normal bell-shaped curve with the average found at the peak of the curve (see Figure 20).

FIGURE 20. Molecular Speed Distribution in a Gas at Different Temperatures

When the temperature is lowered, the gas reaches a point at which the kinetic energy can no longer overcome the attractive forces between the particles (or molecules) and the gas condenses to a liquid. The temperature at which this condensation occurs is related to the type of substance the gas is composed of and the type of bonding in the molecules themselves. This relationship of bond type to condensation point (or boiling point) is pointed out in Chapter 3, "Bonding."

That gases are moving in a random motion so that they may move from one position to another is referred to as *diffusion*. You know that, if a bottle of perfume is opened in one corner of a room, the perfume, that is, its molecules, will move or diffuse to all parts of the room in time. The rate of diffusion is the rate of the mixing of gases.

Effusion is the term used to describe the passage of a gas through a tiny orifice into an evacuated chamber. The rate of effusion measures the speed at which the gas is transferred into the chamber.

GAS LAWS AND RELATED PROBLEMS

Graham's Law

This law relates the rate at which a gas diffuses (or effuses) to the type of molecule in the gas. It can be expressed as follows: The rate of diffusion of a gas is inversely proportional to the square root of its molecular mass. Hydrogen, with the lowest molecular mass, can diffuse more rapidly than other gases under similar conditions. A sample problem:

Compare the rate of diffusion of hydrogen to that of oxygen under similar conditions. The formula is

$$\frac{\text{rate A}}{\text{rate B}} = \frac{\sqrt{\text{Molecular mass of B}}}{\sqrt{\text{Molecular mass of A}}}$$

Let A be H_2 and B be O_2.

$$\frac{\text{rate } H_2}{\text{rate } O_2} = \frac{\sqrt{32}}{\sqrt{2}} = \frac{\sqrt{16}}{\sqrt{1}} = \frac{4}{1}$$

Therefore hydrogen diffuses four times as fast as oxygen.

In dealing with the gas laws, a student must know what is meant by standard conditions of temperature and pressure (abbreviated as STP). The standard pressure is defined as the height of mercury that can be held in an evacuated tube by 1 atmosphere of pressure (14.7 pounds per square inch). This is usually expressed as 760 mm of Hg or 101.3 Pa. Standard temperature is defined as 273 Kelvin or absolute (which corresponds to 0° Celsius).

Charles's Law

Jacques Charles, a French chemist of the early nineteenth century, discovered that, when a gas under constant pressure is heated from 0°C to 1°C, it expands 1/273 of its volume. It contracts this amount when the temperature is dropped 1 degree to –1°C. Charles reasoned that, if a gas at 0°C was cooled to –273°C (actually found to be –273.15°C), its volume would be zero. Actually, all gases are converted into liquids before this temperature is reached. By using the Kelvin scale to rid the problem of negative numbers, we can state Charles's Law as follows:

If the Pressure Remains Constant, the Volume of a Gas Varies Directly as the Absolute Temperature. Then

Initial $\dfrac{V_1}{T_1}$ = Final $\dfrac{V_2}{T_2}$ at constant pressure

Graphic relationship—Charles's Law. The dashed lines represent extrapolation of the data into regions where the gas would become liquid or solid. Extrapolation shows that each gas, if it remained gaseous, would reach zero volume at 0 K or –273°C.

Plots of V versus T for representative gases.

• TYPE PROBLEM:

The volume of a gas at 20°C is 500 mL. Find its volume at standard temperature if pressure is held constant.

Convert temperatures:

$$20°C = 20° + 273° = 293 \text{ K}$$
$$0°C = 0° + 273° = 273 \text{ K}$$

If you know that cooling a gas decreases its volume, then you know that 500 mL will have to be multiplied by a fraction (made up of the Kelvin temperatures) that has a smaller numerator than the denominator. So

$$500 \text{ mL} \times \frac{273}{293} = 465 \text{ mL}$$

Or you can use the formula:

$$\frac{V_1}{T_1} = \frac{V_2}{T_2} \quad \text{SO} \quad \frac{500 \text{ mL}}{293} = \frac{X \text{ mL}}{273}$$
$$X \text{ mL} = \textbf{465 mL}$$

Boyle's Law

Robert Boyle, a seventeenth century English scientist, found that the volume of a gas decreases when the pressure on it is increased, and vice versa, when the temperature is held constant. Boyle's Law can be stated as follows:

If the Temperature Remains Constant, the Volume of a Gas Varies Inversely as the Pressure Changes. Then

$$P_1V_1 = P_2V_2 \text{ at a constant temperature}$$

Graphic relationship—Boyle's Law

• TYPE PROBLEM:

Given the volume of a gas as 200 mL at 800 mm pressure. Calculate the volume of the same gas at 765 mm. Temperature is held constant.

If you know that this *decrease* in pressure will cause an increase in the volume, then you know 200 mL must be multiplied by a fraction (made up of the two pressures) that has a larger numerator than the denominator. So

$$200 \text{ mL} \times \frac{800}{765} = 209 \text{ mL}$$

Or you can use the formula:

$$P_1V_1 = P_2V_2$$

$$\text{So, } V_2 = V_1 \times \frac{P_1}{P_2}$$

$$V_2 = 200 \text{ mL} \times \frac{800}{765} = 209 \text{ mL}$$

Combined Gas Law

This is a combination of the two preceding gas laws and can be expressed as follows:

$$\frac{P_1V_1}{T_1} = \frac{P_2V_2}{T_2}$$

• TYPE PROBLEM:

The volume of a gas at 780 mm pressure and 30°C is 500 mL. What volume would the gas occupy at STP?

You again can use reasoning to determine the kind of fractions the temperatures and pressures must be to arrive at your answer. Since the pressure is going from 780 mm to 760 mm, the volume should increase. The fraction must then be $\frac{780}{760}$. Also, since the temperature is going from 30°C (303 K) to 0°C (273 K), the volume should decrease; this fraction must be $\frac{273}{303}$. So

$$500 \text{ mL} \times \frac{780}{760} \times \frac{273}{303} = 462 \text{ mL}$$

Or you can use the formula:

$$\frac{P_1V_1}{T_1} = \frac{P_2V_2}{T_2}$$

$$\text{Solve for } V_2 = V_1 \times \frac{P_1}{P_2} \times \frac{T_2}{T_1}$$

$$V_2 = 500 \text{ mL} \times \frac{780 \text{ mm Hg}}{760 \text{ mm Hg}} \times \frac{273 \text{ K}}{303 \text{ K}} = 462 \text{ mL}$$

Pressure versus Temperature

At Constant Volume, the Pressure of a Given Mass of Gas Varies Directly with the Absolute Temperature. Then

$$\frac{P_1}{T_1} = \frac{P_2}{T_2} \text{ at constant volume}$$

• TYPE PROBLEM:

A steel tank contains a gas at 27°C and a pressure of 12 atmospheres. Determine the gas pressure when the tank is heated to 100°C.

Reasoning that an increase in temperature will cause an increase in pressure at constant volume, you know the pressure must be multiplied by a fraction that has a larger numerator than denominator. The fraction must be $\frac{373 \text{ K}}{300 \text{ K}}$. So

$$12 \text{ atm.} \times \frac{373 \text{ K}}{300 \text{ K}} = 14.9 \text{ atm.}$$

Or you can use the formula:

$$\frac{P_1}{T_1} = \frac{P_2}{T_2}$$

$$P_2 = P_1 \times \frac{T_2}{T_1}$$

So $P_2 = 12 \text{ atm.} \times \dfrac{373\,\cancel{K}}{300\,\cancel{K}} = 14.9 \text{ atm.}$

Dalton's Law of Partial Pressures

When a gas is made up of a mixture of different gases, the total pressure of the mixture is equal to the sum of the partial pressures of the components, that is, the partial pressure of the gas would be the pressure of the individual gas if it alone occupied the volume.

$$P_{\text{total}} = P_{\text{gas 1}} + P_{\text{gas 2}} + P_{\text{gas 3}} + \cdot \cdot \cdot$$

• TYPE PROBLEM:

A mixture of gases at 760 mm pressure contains 65% nitrogen, 15% oxygen, and 20% carbon dioxide by volume. What is the partial pressure of each gas?

$0.65 \times 760 = 494$ mm pressure (N_2)
$0.15 \times 760 = 114$ mm pressure (O_2)
$0.20 \times 760 = 152$ mm pressure (CO_2)

Corrections of Pressure

Correction of pressure when a gas is collected over water. When a gas is collected over a volatile liquid, such as water, some of the water vapor is present in the gas and contributes to the total pressure. Assuming that the gas is saturated with water vapor for that temperature, you can find the partial pressure due to the water vapor in a table of such water vapor values. (See Chart Ⓝ in the Reference Tables section at the back of the book.) This vapor pressure, which depends only on the temperature, must be subtracted from the total pressure to find the partial pressure of the gas being measured.

Correction of difference in the height of the fluid. When gases are collected in eudiometers (glass tubes closed at one end), it is not always possible to get the level of the liquid inside the tube to equal the level on the outside. This deviation of levels must be taken into account when determining the pressure of the enclosed gas. There are then two possibilities: (1) When the level inside is higher than the level outside the tube, the pressure on the inside is less, by the height of fluid in excess, than the outside pressure. If the fluid is mercury, you simply subtract the differ-

FIGURE 21. *Same Pressure Exerted on Both Liquids*

ence from the outside pressure reading (also in height of mercury and in the same units) to get the corrected pressure of the gas. If the fluid is water, you must first convert the difference to an equivalent height of mercury by dividing the difference by 13.6 (since mercury is 13.6 times as heavy as water, the height expressed in Hg will be 1/13.6 the height of water). Again, care must be taken that this equivalent height of Hg is in the same units as the expression for the outside pressure before it is subtracted. This gives the corrected pressure for the gas in the eudiometer. (2) When the level inside is lower than the level outside the tube, a correction must be added to the outside pressure. If the difference in height between the inside and the outside is expressed in height of water, you must take 1/13.6 of this quantity to correct it to mm of mercury units. This quantity is then added to the expression of the outside pressure, which must also be in mm of mercury. If the tube contains mercury, then the difference between the inside and outside levels is merely added to the outside pressure to get the corrected pressure for the enclosed gas.

• TYPE PROBLEM:

Hydrogen gas was collected in a eudiometer tube over water. It was impossible to level the outside water with that in the tube, so the water level inside the tube was 40.8 mm higher than that outside. The barometric pressure was 730 mm of Hg. The water vapor pressure at the room temperature of 29°C was found in a handbook to be 30.0 mm. What is the pressure of the dry hydrogen?

Step 1: To find the true pressure of the gas, we must first subtract the water level difference expressed in mm of Hg:

$\frac{40.8}{13.6}$ = 3 mm of Hg So 730 mm – 3 mm = 727 mm total pressure of gases in the eudiometer

Step 2: Correcting for the partial pressure due to water vapor in the hydrogen, we subtract the vapor pressure (30.0 mm) from 727 mm and get our answer: 697 mm.

Ideal Gas Law

The previous laws discussed do not include the relationship of number of moles of a gas to the pressure, volume, and temperature of the gas. A law derived from the kinetic-molecular theory relates these variables. It is called the Ideal Gas Law and is expressed as

$$PV = nRT$$

P, V, and T retain their usual meanings, but n stands for the number of moles of the gas and R represents the ideal gas constant.

Boyle's Law and Charles's Law are actually derived from the Ideal Gas Law. Boyle's Law applies when the number of moles and the temperature of the gas are constant. Then in $PV = nRT$, the number of moles, n, is constant; the gas constant (R) remains the same; and by definition T is constant. Therefore, $PV = k$. At the initial set of conditions of a problem, P_1V_1 = a constant (k). At the second set of conditions, the terms on the right side of the equation are equal to the same constant, so $P_1V_1 = P_2V_2$. This matches the Boyle's Law equation introduced earlier.

The same can also be done with Charles's Law, since $PV = nRT$ can be expressed with the variables on the left and the constants on the right.

$$\frac{V}{T} = \frac{nR}{P}$$

In Charles's Law the number of moles and the pressure are constant. Substituting k for the constant term, $\dfrac{nR}{P}$, we have

$$\frac{V}{T} = k.$$

The expression relating two sets of conditions can be written as

$$\frac{V_1}{T_1} = \frac{V_2}{T_2}$$

To use the Ideal Gas Law in the form $PV = nRT$, the gas constant, R, must be determined. This can be done mathematically as shown in the following example.

One mole of oxygen gas was collected in the laboratory at a temperature of 24.0°C and a pressure of exactly 1 atmosphere. The volume was 24.35 liters. Find the value of R.

$$PV = nRT$$

Rearranging the equation to solve for R gives

$$R = \frac{PV}{nT}$$

Substituting the known values on the right, we have

$$R = \frac{1 \text{ atmosphere} \times 24.35 \text{ liters}}{1 \text{ mole} \times 297 \text{ kelvins}}$$

Calculating R, we get

$$R = 0.0820 \frac{\text{L atm}}{\text{mol K}}$$

Once R is known, the Ideal Gas Law can be used to find any of the variables, given the other three.

For example, calculate the pressure, at 16.0°C, of 1.00 gram of hydrogen gas occupying 2.54 liters.

Rearranging the equation to solve for P, we get

$$P = \frac{nRT}{V}$$

The molar mass of hydrogen is 2.00 g/mol, so the number of moles in this problem would be

$$\frac{1.00 \text{ g}}{2.00 \text{ g/mol}} = 0.500 \text{ mol}$$

Substituting the known values, we have

$$P = \frac{(0.500 \text{ mol}) (0.0820 \frac{\text{L atm}}{\text{mol K}}) 289 \text{ K}}{2.54 \text{ L}}$$

Calculating the value, we get

$$P = 4.66 \text{ atm}$$

Another use of the ideal gas law is to find the number of moles of a gas when P, T, and V are known.

For example, how many moles of nitrogen gas are there in 0.38 liter of gas at 0°C and 380 mm Hg pressure?

Rearranging the equation to solve for n gives

$$n = \frac{PV}{RT}$$

Changing temperature to kelvins and pressure to atmospheres gives

$$T = 0° + 273 = 273 \text{ K}$$

$$P = \frac{380 \text{ mm Hg}}{760 \text{ mm Hg/atm}} = 0.5 \text{ atm}$$

Substituting in the equation, we have

$$n = \frac{(0.5 \text{ atm})(0.38 \text{ L})}{(0.220 \frac{\text{L atm}}{\text{mol K}})(273 \text{ K})} = 0.0085 \text{ mole}$$

$$n = 0.0085 \text{ mole of nitrogen gas}$$

Ideal Gas Deviations

In the use of the gas laws, we have assumed that the gases involved were "ideal" gases. This means that the molecules of the gas were not taking up space in the gas volume and that no intermolecular forces of attraction were serving to pull the molecules closer together. You will find that a gas behaves like an ideal gas at low pressures and high temperatures which move the molecules as far as possible from conditions that would cause condensation. In general, pressures below a few atmospheres will cause most gases to exhibit sufficiently ideal properties for the application of the gas laws with a reliability of a few percent or better.

If, however, high pressures are used, the molecules will be forced into closer proximity with each other as the volume decreases until the attractive force between molecules becomes a factor. This factor decreases the volume and, therefore, the PV values at high pressure conditions will be less than that predicted by the ideal gas law, where PV remains a constant.

Examining what occurs at very low temperatures creates a similar situation. Again, the molecules, because they have slowed down at low temperatures, come into closer proximity with each other and begin to feel the attractive force between them. This tends to make the gas volume smaller and, therefore, causes the PV to be lower than that expected in the ideal gas situation. So, under conditions of very high pressures and low temperatures, deviations from the expected results of the ideal gas law will occur.

Chapter

5

Review

• QUESTIONS

1. The most abundant element in the earth's crust is (a) sodium   (b) oxygen (c) silicon   (d) aluminum.

2. A compound that can be decomposed to produce oxygen gas in the lab is (a) MnO_2 (b) NaOH   (c) CO_2 (d) $KClO_3$.

3. In the usual laboratory preparation equation for the reaction in question 2, what is the coefficient of the O_2? (a) 1   (b) 2   (c) 3   (d) 4

4. In the graphic representation of the energy contents of the reactants and the resulting products, which would have a higher energy content in an exothermic reaction? (a) the reactants   (b) the products   (c) both the same (d) neither one

5. The process of separating components of a mixture by making use of the difference in their boiling points is called (a) destructive distillation (b) displacement   (c) fractional distillation   (d) filtration.

6. When oxygen combines with an element to form a compound, the resulting compound is called a(n) (a) salt   (b) oxide   (c) oxidation   (d) oxalate.

7. According to the activity chart of metals, which metal would react most vigorously in a dilute acid solution?   (a) zinc   (b) iron   (c) aluminum (d) magnesium

8. Graham's Law refers to   (a) boiling points of gases   (b) gaseous diffusion (c) gas compression problems   (d) volume changes of gases when the temperature changes.

9. When 200 mL of a gas at constant pressure is heated, its volume (a) increases (b) decreases   (c) remains unchanged.

10. When 200 mL of a gas at constant pressure is heated from 0°C to 100°C, the volume must be multiplied by   (a) 0/100   (b) 100/0   (c) 273/373 (d) 373/273.

11. If you wish to find the corrected volume of a gas that was at 20°C and 760 mm pressure and conditions were changed to 0°C and 700 mm pressure, by what fractions would you multiply the original volume?

(a) $\dfrac{293}{273} \times \dfrac{760}{700}$ (b) $\dfrac{273}{293} \times \dfrac{700}{760}$ (c) $\dfrac{273}{293} \times \dfrac{760}{700}$ (d) $\dfrac{293}{273} \times \dfrac{700}{760}$

12. When the level of mercury inside a gas tube is higher than the level in the reservoir, you find the correct pressure inside the tube by taking the outside pressure reading and _____?_____ the difference in the height of mercury. (a) subtracting   (b) adding   (c) dividing by 13.6   (d) doing both (c) and (a)

13. If water is the liquid in question 12 instead of mercury, you can change the height difference to an equivalent mercury expression by (a) dividing by 13.6 (b) multiplying by 13.6   (c) adding 13.6   (d) subtracting 13.6

14. Standard conditions are (a) 0°C and 14.7 mm   (b) 32°F and 76 cm   (c) 273°C and 760 mm   (d) 4°C and 7.6 m.

15. When a gas is collected over water, the pressure is corrected by (a) adding the vapor pressure of water   (b) multiplying by the vapor pressure of water   (c) subtracting the vapor pressure of water at that temperature   (d) subtracting the temperature of the water from the vapor pressure.

16. At 5 atmospheres pressure and 70°C how many moles are present in 1.5 liters of O_2 gas? (a) 0.036   (b) 0.267   (c) 0.536   (d) 1.60

• **ANSWERS:**

1.	(b)	7.	(d)	12.	(a)
2.	(d)	8.	(b)	13.	(a)
3.	(c)	9.	(a)	14.	(b)
4.	(a)	10.	(d)	15.	(c)
5.	(c)	11.	(c)	16.	(b)
6.	(b)				

Terms You Should Know

atmospheric pressure	oxidation	standard pressure
manometer	ozone	standard temperature
mercury barometer	pascal	torr

Boyle's Law
Charles's Law
Combined Gas Law
Dalton's Law of Partial Pressures
Graham's Law
Ideal Gas Law
Kinetic-Molecular Theory

6

Chemical
Calculations
(Stoichiometry)
and the Mole
Concept

THE MOLE CONCEPT

Providing a name for a quantity of things taken as a whole is common in everyday life. Some examples are a dozen, a gross, and a ream. Each of these represents a specific number of items and is not dependent on the commodity. A dozen eggs, oranges, or bananas will always represent 12 items.

In chemistry we have a unit that decribes a quantity of particles. It is called the *mole* (sometimes abbreviated as *mol*). A mole is 6.02×10^{23} particles. The particles can be atoms, molecules, ions, electrons, and so forth. Because particles are so small in chemistry, the mole is a very convenient unit. The number 6.02×10^{23} is often referred to as *Avogadro's number* in honor of the Italian scientist whose hypothesis led to its determination.

In 1811, Amedeo Avogadro made a far-reaching scientific assumption (hypothesis) that also bears his name. He stated that equal volumes of different gases contain equal numbers of molecules at the same pressure and temperature. The statement was called Avogadro's Hypothesis. It means that, under the same conditions, the number of molecules of hydrogen in a 1 liter container is exactly the same as the number of molecules of carbon dioxide or any other gas in a 1 liter container even though the individual molecules of the different gases have different masses and size. Because of the substantiation of this hypothesis by much data since its inception, it is often referred to as *Avogadro's Law*.

MOLAR MASS AND MOLES

When the formula weight of an ionic compound is determined by the addition of its component relative atomic masses and expressed in grams, it is called the *gram-formula weight* or *molar mass*. An example is $CaCO_3$:

$$1Ca = 40$$
$$1C = 12$$
$$\underline{3O = 48 = (3 \times 16)}$$
$$CaCO_3 = 100 \text{ formula weight}$$

$$100 \text{ g} = \text{gram-formula weight or molar mass}$$

The molar mass is equivalent to the mass of 1 *mole* of that material expressed in grams.

The term *molar mass* or *gram-molecular weight* (gmw) can also be used when it is known that the material is a molecular substance and not an ionic lattice like NaCl or NaOH.

• ANOTHER EXAMPLE:

Find the weight of 1 mole of $KAl(SO_4)_2 \cdot 12H_2O$.

$$1K = 39$$
$$1Al = 27$$
$$2(SO_4) = 192 = 2(32 + 16 \times 4)$$
$$\underline{12H_2O = 216 = 12(2 + 16)}$$
$$1 \text{ mole} = 474 \text{ g}$$

• TRY THESE PROBLEMS:

Find the molecular mass or formula weight of the following:

1. HNO_3 Ans. 63
2. $C_6H_{10}O_5$ Ans. 162
3. H_2SO_4 Ans. 98
4. KCl   Ans. 74.5
5. $C_{12}H_{22}O_{11}$ Ans. 342

In some cases, there may be a question of the mass of 1 mole of an element if you are not told specifically what is referred to—the single atoms or a molecular state. An example of this situation may arise with hydrogen and other elements that form diatomic molecules. If you are asked the mass of 1 mole of hydrogen, you could say it is 1 gram if you are dealing with single atoms of hydrogen. This 1 mole of hydrogen atoms would contain 6.02×10^{23} atoms. In a similar fashion, 1 mole of hydrogen molecules (H_2) would have a molar mass of 2 grams. This 1 mole of hydrogen molecules would contain 6.02×10^{23} molecules and, since each molecule is composed of two atoms, 12.04×10^{23} atoms.

The other elements that exist as diatomic molecules are oxygen, nitrogen, flourine, chlorine, bromine, iodine, and astatine.

Try these questions.

What is the mass of 1 mole of oxygen?

Answer: Since the atomic mass is 16 grams and the molecule is diatomic, 2×16 grams = 32 grams.

What is the mass of one molecule of oxygen?

Answer: There are 6.02×10^{23} molecules in 1 mole of this gas. 32 grams ÷ 6.02×10^{23} molecules = 5.32×10^{-23} gram.

What is the mass of 1 mole of oxygen atoms?

Answer: Without the diatomic molecule structure, the atomic mass of oxygen expressed in grams is the answer. That is 16 grams.

What is the mass of one atom of oxygen?

Answer: The atomic molar mass, 16 grams, is divided by 6.02×10^{23}. This gives 2.66×10^{-23} gram.

If you gave the same answer to the first two questions, you are confused about the mole concept. To say that a molecule of oxygen gas has the same mass as a mole of oxygen gas is the same as saying that an apple has the same mass as a dozen apples.

MOLE RELATIONSHIPS

Since the mole is used often in chemistry to quantify the number of atoms, molecules, and several other items, it is important to know the relationships that exist and how to move from one to another. The following summarizes a number of these relationships:

When dealing with elements—
Moles of an element × molar mass (atomic) = mass of the element
Mass of an element / molar mass (atomic) = moles of the element
When dealing with compounds—
Moles of a compound × molar mass (molecular) = mass of the compound
Mass of a compound / molar mass (molecular) = moles of the compound
When dealing with the molecules of a compound—
Moles of molecules × 6.02×10^{23} = number of molecules
Number of molecules / 6.02×10^{23} = moles of molecules
When dealing with the atoms of elements—
Moles of atoms × 6.02×10^{23} = number of atoms
Number of atoms / 6.02×10^{23} = moles of atoms

GAS VOLUMES AND MOLAR MASS

Because the volume of a gas may vary depending on the conditions of temperature and pressure, a standard is set for comparing gases. The standard conditions of temperature and pressure (abbreviated STP) are 0°C and 760 mm of mercury pressure.

The molecular mass of a gas expressed in grams and under standard conditions occupies 22.4 liters. This is an important relationship to remember! The 22.4 liters is referred to as the gram-molecular volume (gmv) or **molar volume**. Two scientists are associated with this relationship.

Gay-Lussac's Law states that, when only gases are involved in a reaction, the volumes of the reacting gases and the volumes of the gaseous products are in ratio to each other as small whole numbers. This law may be illustrated by the following cases:

1 vol. hydrogen + 1 vol. chlorine = 2 vols. hydrogen chloride
2 vols. hydrogen + 1 vol. oxygen = 2 vols. steam

Avogadro's Law, which explains Gay-Lussac's, states that equal volumes of gases under the same conditions of temperature and pressure contain equal numbers of molecules. This means that 1 mole of any gas at STP occupies 22.4 liters, so:

32 g O_2 at STP occupies 22.4 liters
2 g H_2 at STP occupies 22.4 liters
44 g CO_2 at STP occupies 22.4 liters
$2O_2$ (2 moles O_2) = 64 g = 44.8 liters
$3H_2$ (3 moles H_2) = 6 g = 67.2 liters

DENSITY AND MOLAR MASS

Since the density of a gas is usually given in grams/liter of gas at STP, we can use the molar volume to molar mass relationship to solve the following types of problems.

- **EXAMPLE:**

Find the molar mass of a gas when the density is given as 1.25 g/L.

Since it is known that 1 mole of a gas occupies 22.4 liters at STP, we can solve this problem by multiplying the mass of 1 liter by 22.4 liters/mole.

$$\frac{1.25 \text{ g}}{\text{liter}} \times \frac{22.4 \text{ liters}}{1 \text{ mole}} = 28 \text{ g/mole}$$

Even if the weight given is not for 1 liter, the same setup can be used.

- **PROBLEM:**

If 3 liters of a gas weighs 2 grams, find the molar mass.

Solution:

$$\frac{2 \text{ grams}}{3 \text{ liters}} \times \frac{22.4 \text{ liters}}{1 \text{ mole}} = 14.9 \text{ g/mole}$$

You can also find the density of a gas if you know the molar mass. Since the molar mass occupies 22.4 liters at STP, dividing the molar mass by 22.4 liters will give you the mass per liter, or the density.

- **EXAMPLE:**

Find the density of oxygen at STP.

Solution:

Oxygen is diatomic in its molecular structure.

O_2 = molar mass of 2×16 or 32 g/mole

32 g/mole occupies 22.4 liters

Therefore

$$\frac{32 \text{ g}}{1 \text{ mole}} \div \frac{22.4 \text{ L}}{1 \text{ mole}} = \frac{32 \text{ g}}{1 \text{ mole}} \times \frac{1 \text{ mole}}{22.4 \text{ L}} = 1.43 \text{ g/L}$$

You can find the density of a gas, then, by dividing its molar mass by 22.4 liters.

MASS-VOLUME RELATIONSHIPS

A typical *mass-volume problem*: How many liters of oxygen (STP) can you prepare from the decomposition of 42.6 grams of sodium chlorate?

1st step.   Write the balanced equation for the reaction.

$$2NaClO_3 \xrightarrow{\Delta} 2NaCl + 3O_2 \text{ (g)}$$

2nd step.   Write the given quantity and the unknown quantity above the appropriate substances.

$$\overset{42.6 \text{ g}}{2NaClO_3} \xrightarrow{\Delta} 2NaCl + \overset{x \text{ L}}{3O_2}$$

3rd step.   Calculate the equation mass or volume under the substances that have something indicated above them. Note that the units above and below *must* match.

$$42.6 \text{ g} \qquad\qquad x \text{ L}$$
$$2NaClO_3 \xrightarrow{\Delta} 2NaCl + 3O_2$$
$$213 \text{ g} \qquad\qquad 67.2 \text{ L}$$
$$(2 \times \text{molar mass} \qquad (3 \times 22.4 \text{ L})$$
$$\text{of } NaClO_3)$$

4th step.   Form the proportion.

$$\frac{42.6 \text{ g}}{213 \text{ g}} = \frac{x \text{ L}}{67.2 \text{ L}}$$

5th step.   Solve for x.

$$x = 13.4 \text{ L } O_2$$

This problem can also be solved using methods other than the *proportion method* shown above.

Another method to proceed from Step 3 is called the *factor-label method.* The reasoning is this: Since the equation shows that 213 grams of reactant produces 67.2 liters of the required product, multiplying the given amount by this equality (so that the units of the answer will be correct) will give the same answer as above. So step 4 would be:

4th step. $42.6 \text{ grams } NaClO_3 \times \dfrac{67.2 \text{ L } O_2}{213 \text{ grams } NaClO_3} = 13.4 \text{ L } O_2$

Still another method of solving this problem is called the *mole method.* Steps 1 and 2 are the same. Then step 3 is as follows:

3rd step.   Determine how many moles of substance are given.

$$42.6 \text{ g} \div \frac{106.5 \text{ g}}{1 \text{ mole } NaClO_3} = 0.4 \text{ mole } NaClO_3$$

The equation shows that 2 moles $NaClO_3$ makes 3 moles O_2. So 0.4 mole $NaClO_3$ will yield

$$0.4 \text{ mole } NaClO_3 \times \frac{3 \text{ moles } O_2}{2 \text{ moles } NaClO_3} = 0.6 \text{ mole } O_2$$

4th step.   Convert the moles of O_2 to liters.

$$0.6 \text{ mole } O_2 \times \frac{22.4 \text{ L } O_2}{1 \text{ mole } O_2} = 13.4 \text{ L } O_2$$

• **ANOTHER EXAMPLE:**

Find the weight of $CaCO_3$ needed to produce 11.2 liters of CO_2 when the calcium carbonate is reacted with hydrochloric acid.

1st step.   Write the balanced equation for the reaction.
$$CaCO_3 + 2HCl \rightarrow CaCl_2 + H_2O + CO_2$$

2nd step.   Write the given quantity and the unknown quantity above the appropriate substances.
$$x \text{ g} \qquad\qquad\qquad 11.2 \text{ L}$$
$$CaCO_3 + 2HCl \rightarrow CaCl_2 + H_2O + CO_2$$

3rd step.   Calculate the equation mass or volume under substances that have something indicated above them. Note the units above and below *must* match.

$$x \text{ g} \qquad\qquad\qquad 11.2 \text{ L}$$
$$CaCO_3 + 2HCl \rightarrow CaCl_2 + H_2O + CO_2$$
$$100 \text{ g} \qquad\qquad\qquad 22.4 \text{ L}$$

• PROPORTION METHOD:

4th step.   Form the proportion.

$$\frac{x \text{ g}}{100 \text{ g}} = \frac{11.2 \text{ L}}{22.4 \text{ L}}$$

5th step.   Solve for x.

$$x = 50 \text{ g CaCO}_3$$

• FACTOR-LABEL METHOD:

4th step. $11.2 \; \cancel{L \; CO_2} \times \dfrac{100 \text{ g CaCO}_3}{22.4 \; \cancel{L \; CO_2}} = 50 \text{ g CaCO}_3$

• MOLE METHOD:

$$11.2 \text{ L CO}_2 \div \frac{22.4 \text{ L}}{1 \text{ mole}} = 0.5 \text{ mole CO}_2$$

Equation shows 1 mole $CaCO_3$ yields 1 mole CO_2, so

$$0.5 \; \cancel{\text{mole CO}_2} \times \frac{1 \text{ mole CaCO}_3}{1 \; \cancel{\text{mole CO}_2}} = 0.5 \text{ mole CaCO}_3$$

Converting moles to grams:

$$0.5 \; \cancel{\text{mole CaCO}_3} \times \frac{100 \text{ g CaCO}_3}{1 \; \cancel{\text{mole CaCO}_3}} = 50 \text{ g CaCO}_3$$

• TRY THESE PROBLEMS:

1. What mass of water must be electrolyzed to obtain 20 liters of oxygen?

Ans. 32.2 g*

2. How many grams of aluminum will be completely oxidized by 44.8 liters of oxygen?

Ans. 72 g Al*

MASS-MASS PROBLEMS

A typical problem concerning just mass relationships is as follows:

What mass of oxygen can be obtained from heating 100 grams of potassium chlorate?

1st step.   Write the balanced equation for the reaction.

$$2KClO_3 \quad \rightarrow \quad 2KCl + 3O_2$$

2nd step.   Write the given quantity and the unknown quantity above the appropriate substances.

$$100 \text{ g} \qquad\qquad x \text{ g}$$
$$2KClO_3 \rightarrow 2KCl + 3O_2$$

*Answer explained at the end of this chapter, page 116.

3rd step. Calculate the equation mass under the substances that have something indicated above them. Note the units above and below *must* match.

$$\begin{array}{cc} 100\text{ g} & x\text{ g} \\ 2KClO_3 \rightarrow 2KCl + 3O_2 \\ 245\text{ g} & 96\text{ g} \end{array}$$

• USING THE PROPORTION METHOD:

4th step. Form the proportion.

$$\frac{100\text{ g}}{245\text{ g}} = \frac{x\text{ g}}{96\text{ g}}$$

5th step. Solve for x.

$$x = 39.3\text{ g of }O_2$$

• USING THE FACTOR-LABEL METHOD:

From the 3rd step on you would proceed as follows:

4th step. The equation indicates that 245 g $KClO_3$ yields 96 g of O_2. Therefore multiplying the given quantity by a factor made up of these two quantities arranged appropriately so that the units of the answer remain uncanceled will give the answer to the problem.

$$100\text{ g }KClO_3 \times \frac{96\text{ g }O_2}{245\text{ g }KClO_3} = 39.3\text{ g }O_2$$

• USING THE MOLE METHOD:

To solve this problem you would proceed as follows from the 2nd step:

3rd step. Determine how many moles of substance are given.

$$100\text{ g }KClO_3 \times \frac{1\text{ mole }KClO_3}{122.5\text{ g }KClO_3} = 0.815\text{ mole }KClO_3$$

The equation shows that 2 moles of $KClO_3$ yields 3 moles of O_2. So 0.815 mole $KClO_3$ will yield

$$0.815\text{ mole }KClO_3 \times \frac{3\text{ moles }O_2}{2\text{ moles }KClO_3} = 1.22\text{ moles }O_2$$

4th step. Convert the moles of O_2 to grams of O_2.

$$1.22\text{ moles }O_2 \times \frac{32\text{ g }O_2}{1\text{ mole }O_2} = 39\text{ g }O_2$$

• ANOTHER EXAMPLE:

What mass of potassium hydroxide is needed to neutralize 20 grams of sulfuric acid?

1st step. $2KOH + H_2SO_4 \rightarrow K_2SO_4 + 2H_2O$

2nd step. $\begin{array}{cc} x\text{ g} & 20\text{ g} \\ 2KOH + H_2SO_4 \rightarrow K_2SO_4 + 2H_2O \end{array}$

3rd step.
$$x \text{ g} \qquad 20 \text{ g}$$
$$2KOH + H_2SO_4 \rightarrow K_2SO_4 + 2H_2O$$
$$112 \text{ g} \qquad 98 \text{ g}$$

• USING THE PROPORTION METHOD:

4th step. $\dfrac{x \text{ g}}{112 \text{ g}} = \dfrac{20 \text{ g}}{98 \text{ g}}$

5th step. $x = 22.8$ g KOH

• USING THE FACTOR-LABEL METHOD:

4th step. $20 \text{ g } H_2SO_4 \times \dfrac{112 \text{ g KOH}}{98 \text{ g } H_2SO_4} = 22.8$ g KOH

• USING THE MOLE METHOD:

3rd step. $20 \text{ g } H_2SO_4 \times \dfrac{1 \text{ mole } H_2SO_4}{98 \text{ g } H_2SO_4} = 0.204$ mole H_2SO_4

$0.204 \text{ mole } H_2SO_4 \times \dfrac{2 \text{ mole KOH}}{1 \text{ mole } H_2SO_4} = 0.408$ mole KOH

4th step. $0.408 \text{ mole KOH} \times \dfrac{56 \text{ g KOH}}{1 \text{ mole KOH}} = 22.8$ g KOH

• TRY THESE PROBLEMS:

3. What mass of manganese dioxide is needed to react with an excess of hydrochloric acid so that 100 g of chlorine is liberated?   Ans.: 122.5 g*

4. A 20-g sample of Mg is burned in 20 g of O_2. How much MgO is formed?
   (Hint: Determine which is in excess.)   Ans. 33.3 g MgO*

VOLUME-VOLUME PROBLEMS

This type of problem involves only volume units and therefore can make use of Gay-Lussac's Law: "When gases combine, they combine in simple whole number ratios." These simple numbers are the coefficients of the equation.

A typical problem concerning just volume relationships is as follows:

What volume of NH_3 is produced when 22.4 L of nitrogen are made to combine with a sufficient quantity of hydrogen under the appropriate conditions?

1st step.   Write the balanced equation for the reaction.

$$N_2 + 3H_2 \rightarrow 2NH_3$$

2nd step.   Write the given quantity and the unknown quantity above the respective substances.

$$22.4 \text{ L} \qquad x \text{ L}$$
$$N_2 + 3H_2 \rightarrow 2NH_3$$

3rd step.   Set up a proportion using the coefficient of the substances that have something indicated above them as denominators.

$$\dfrac{22.4 \text{ L}}{1 \text{ L}} = \dfrac{x \text{ L}}{2 \text{ L}}$$

*Answer explained at the end of this chapter, pages 116–117.

4th step.   Solve for x.

   $x = 44.8$ L

• USING THE FACTOR-LABEL METHOD:

From the 2nd step on you would proceed as follows:

3rd step.   The equation shows that 1 volume of N_2 will yield 2 volumes of NH_3. Therefore multiplying the given quantity by a factor made up of these two quantities appropriately arranged so that the units of the answer remain uncanceled will solve the problem.

$$22.4 \text{ } \cancel{L N_2} \times \frac{2 \text{ L } NH_3}{1 \text{ } \cancel{L N_2}} = 44.8 \text{ L } NH_3$$

• USING THE MOLE METHOD:

To solve this problem you would proceed as follows:

3rd step.   The given quantity is converted to moles.

$$\cancel{22.4 \text{ L } N_2} \times \frac{1 \text{ mole } N_2}{\cancel{22.4 \text{ L } N_2}} = 1 \text{ mole } N_2$$

The equation shows that 1 mole of N_2 will yield 2 moles of NH_3. Therefore using this relationship will yield

$$\cancel{1 \text{ mole } N_2} \times \frac{2 \text{ moles } NH_3}{\cancel{1 \text{ mole } N_2}} = 2 \text{ moles } NH_3$$

Convert the moles of NH_3 to liters of NH_3

$$2 \text{ } \cancel{\text{moles } NH_3} \times \frac{22.4 \text{ L } NH_3}{1 \text{ } \cancel{\text{mole } NH_3}} = 44.8 \text{ L } NH_3$$

• ANOTHER EXAMPLE:

What volume of SO_2 will result from the complete burning of pure sulfur in 8 liters of oxygen until all the oxygen is used?

1st step. $S + O_2 \rightarrow SO_2$

$$\begin{array}{ccc} & 8 \text{ L} & x \text{ L} \\ \end{array}$$
2nd step. $S + O_2 \rightarrow SO_2$

• USING THE PROPORTION METHOD:

3rd step. $\dfrac{8 \text{ L}}{1 \text{ L}} = \dfrac{x \text{ L}}{1 \text{ L}}$

4th step. $x = 8$L of SO_2

• USING THE FACTOR-LABEL METHOD:

3rd step. $8 \text{ } \cancel{L O_2} \times \dfrac{1 \text{ L } SO_2}{1 \text{ } \cancel{L O_2}} = 8 \text{ L } SO_2$

• USING THE MOLE METHOD (NOT REALLY PRACTICAL IN THIS CASE):

3rd step. $8 \text{ } \cancel{L O_2} \times \dfrac{1 \text{ mole } O_2}{22.4 \text{ } \cancel{L O_2}} = 0.358 \text{ mole } O_2$

$$0.358 \text{ } \cancel{\text{mole } O_2} \times \frac{1 \text{ mole } SO_2}{1 \text{ } \cancel{\text{mole } O_2}} = 0.358 \text{ mole } SO_2$$

4th step. $0.358 \text{ mole SO}_2 \times \dfrac{22.4 \text{ L SO}_2}{1 \text{ mole SO}_2} = 8 \text{ L SO}_2$

• TRY THESE PROBLEMS:

5. How many liters of hydrogen are necessary to react with sufficient chlorine to produce 12 liters of hydrogen chloride?   Ans. 6 L*

6. How many liters of oxygen will be needed to burn 100 liters of carbon monoxide?   Ans. 50 L*

PROBLEMS WITH AN EXCESS OF ONE REACTANT

It will not always be true that the amounts given in a particular problem are exactly in the proportion required for the reaction to use up all of the reactants. In other words, at times some of one reactant will be left over after the other has been used up. This is similar to the situation in which two eggs are required to mix with one cup of flour in a particular recipe, and you have four eggs and four cups of flour. Since two eggs require only one cup of flour, four eggs can use only two cups of flour and two cups of flour will be left over.

A chemical equation is very much like a recipe. Consider the following equation:

$$Zn + 2HCl \rightarrow ZnCl_2 + H_2$$

If you are given 65 grams of zinc and 65 grams of HCl, how many grams of hydrogen gas can be produced? Which reactant will be left over? How many grams of this reactant will not be consumed?

1st step.   Set up the problem.

$$
\begin{array}{cccc}
65\text{ g} & 65\text{ g} & & x\text{ g} \\
Zn + & 2HCl & \rightarrow ZnCl_2 + & H_2 \\
65\text{ g} & 73\text{ g} & & 2\text{ g}
\end{array}
$$

The given quantities are above the equation, and the equation masses are given beneath. To solve for the grams of hydrogen gas, the limiting reactant must be determined.

2nd step. Compare the given quantities with the equation requirements.

Knowing that it takes 65 grams of zinc to react with 73 grams of hydrochloric acid, it is reasonable to surmise that, since there are only 65 grams of hydrochloric acid, not all of the 65 grams of zinc can be used. The limiting factor will be the amount of hydrochloric acid.

3rd step.   Solve for the quantity of product that will be produced using the amount of the limiting factor.

Now that we know that the limiting factor is the amount of hydrochloric acid (65 g), the equation can be solved. This means that the proportion is set up ignoring the 65 grams of zinc.

$$\frac{65\text{ g}}{73\text{ g}} = \frac{x\text{ g}}{2\text{ g}}$$

Solving for x, we get $x = 1.78$ g or 1.8 g of hydrogen produced.

*Answer explained at the end of this chapter, page 117.

4th step.   To find the number of grams of zinc that were not consumed, a separate problem is necessary. The equation would be set up as follows:

$$y\,g \quad 65\,g$$
$$Zn + 2HCl \rightarrow ZnCl_2 + H_2$$
$$65\,g \quad 73\,g$$

This gives the proportion:

$$\frac{y\,g}{65\,g} = \frac{65\,g}{73\,g}$$

$$y = \frac{65 \times 65}{73} = 57.8 \text{ or } 58 \text{ g}$$

Subtracting the 58 g of zinc that will be used from the original 65 g leaves 7 g of zinc that will not be consumed.

Answers to Chapter Problems

After each type of problem in this chapter there are several problems to try to solve; therefore a list of review problems is not included here. The answers to these problems are:

Page 111   **1.**   Equation:
$$x\,g \qquad\qquad 20\,L$$
$$2H_2O \rightarrow 2H_2 + O_2$$

Calculate the equation mass and volume so that the units match the ones above the equation.

$$x\,g \qquad\qquad 20\,L$$
$$2H_2O \rightarrow 2H_2 + O_2$$
(2 × molecular mass of water)   (1 × gram-molecular volume)
$$36\,g \qquad\qquad 22.4\,L$$

Then

$$\frac{x\,g}{36\,g} = \frac{20\,L}{22.4\,L}$$

$$\underline{x\,g = 32.2\,g}$$

Page 111   **2.**
$$x\,g \qquad 44.8\,L$$
$$4Al + 3O_2 \rightarrow 2Al_2O_3$$

Calculate the equation mass or volume so that the units match the ones above the equation.

$$x\,g \qquad\qquad 44.8\,L$$
$$4Al \quad + \quad 3O_2 \quad \rightarrow 2Al_2O_3$$
$$(4 \times 27\,g) \quad (3 \times 22.4\,L)$$
$$108\,g \qquad\quad 67.2\,L$$

Then $\dfrac{x\,g}{108\,g} = \dfrac{44.8\,L}{67.2\,L}$

$$x\,g = \underline{\underline{72\,g}}$$

Page 113   **3.**   The equation:
$$x\,g \qquad\qquad\qquad 100\,g$$
$$MnO_2 + 4HCl \rightarrow MnCl_2 + Cl_2 + 2H_2O$$

Calculate the equation masses for the substances involved.

$$x \text{ g} \qquad\qquad\qquad 100 \text{ g}$$
$$MnO_2 \ + 4HCl \rightarrow MnCl_2 + \ Cl_2 \ + 2H_2O$$
$$(1 \times 87 \text{ g}) \qquad\qquad\qquad (1 \times 71 \text{ g})$$
$$87 \text{g} \qquad\qquad\qquad 71 \text{ g}$$

Then $\quad \dfrac{x \text{ g}}{87 \text{ g}} = \dfrac{100 \text{ g}}{71 \text{ g}}$

$$\underline{\underline{x \text{ g} = 122.5 \text{ g}}}$$

Page 113   **4.** The equation:

$$20 \text{ g} \quad\ 20 \text{ g} \qquad x \text{ g}$$
$$2Mg \ + \ O_2 \ \rightarrow \ 2MgO$$
$$48 \text{ g} \quad\ 32 \text{ g} \qquad 80 \text{ g}$$

Since the equation masses indicate you must have more Mg than O_2, there is an excess of oxygen.

$$\dfrac{20 \text{ g oxygen}}{48 \text{ g oxygen}} = \dfrac{x \text{ g MgO}}{80 \text{ g MgO}}$$

$$\underline{\underline{x = 33.3 \text{ g MgO}}}$$

Page 115   **5.** The equation:

$$x \text{ liters} \qquad\quad 12 \text{ liters}$$
$$H_2 \ + \ Cl_2 \ \rightarrow \ 2 \text{ HCl}$$

Set up the proportion using the coefficients of the substances that have something indicated above them as denominators.

$$\dfrac{x \text{ liters}}{1 \text{ liter}} = \dfrac{12 \text{ liters}}{2 \text{ liters}}$$

$$\underline{\underline{x = 6 \text{ liters}}}$$

Page 115   **6.** The equation:

$$100 \text{ liters} \quad\ x \text{ liters}$$
$$2CO \quad + \quad O_2 \quad \rightarrow 2CO_2$$

Set up the proportion using the coefficients of the substances that have something indicated above them as denominators.

$$\dfrac{100 \text{ liters}}{2 \text{ liters}} = \dfrac{x \text{ liters}}{1 \text{ liter}}$$

$$\underline{\underline{x = 50 \text{ liters}}}$$

• **QUESTIONS**

Often there are some equations on a chemistry test to either complete, or complete and balance. The following list of representative equations are some you should be capable of writing and balancing. Below the word equations are the complete balanced equations.

1. zinc + sulfuric acid →
2. iron(III) chloride + sodium hydroxide → iron(III) hydroxide(s) + sodium chloride
3. aluminum hydroxide + sulfuric acid → aluminum sulfate + water
4. potassium + water →
5. magnesium + oxygen →
6. silver nitrate + copper → copper(II) nitrate + silver(s)
7. magnesium bromide + chlorine → magnesium chloride + bromine

• **ANSWERS**

1. zinc + sulfuric acid → zinc sulfate + hydrogen(g)
   $Zn + H_2SO_4 → ZnSO_4 + H_2(g)$
2. $FeCl_3 + 3NaOH → Fe(OH)_3 (s) + 3NaCl$
3. $2Al(OH)_3 + 3H_2SO_4 → Al_2(SO_4)_3 + 6H_2O$
4. potassium + water → potassium hydroxide + hydrogen(g)
   $2K + 2H_2O → 2KOH + H_2(g)$
5. magnesium + oxygen → magnesium oxide
   $2Mg + O_2 → 2MgO$
6. $2AgNO_3 + Cu → Cu(NO_3)_2 + 2Ag(s)$
7. $MgBr_2 + Cl_2 → MgCl_2 + Br_2$

Terms You Should Know

molar mass	mole
molar volume	STP

Avogadro's Law
Avogadro's number
Gay-Lussac's Law

Liquids, Solids, and Phase Changes

LIQUIDS

Importance of Intermolecular Interaction

In a liquid, the volume of the molecules and the intermolecular forces between them are much more important than in a gas. When you consider that in a gas the molecules constitute far less than 1% of the total volume, while in the liquid state the molecules constitute 70% of the total volume, it is clear that in a liquid the forces between molecules are more important. Because of this decreased volume and increased intermolecular interaction, a liquid expands and contracts only very slightly with a change in temperature and lacks the compressibility typical of gases.

Kinetics of Liquids

Even though the volume of space between molecules has decreased in a liquid and the mutual attraction forces between neighboring molecules can have great effects on the molecules, they are still in motion. This motion can be verified under a microscope when colloidal particles are suspended in a liquid. The particles' zigzag path, called *Brownian movement*, indicates molecular motion and supports the *Kinetic Molecular Theory*.

Increases in temperature increase the average kinetic energy of molecules and the rapidity of their movement. This is shown graphically in Figure 22. The molecules in the sample of cold liquid have, on the average, less kinetic energy than those in the warmer sample. Hence the temperature reading T_1 will be less than the temperature reading T_2. If a particular molecule gains enough kinetic energy when it is near the surface of a liquid, it can overcome the attractive forces of the liquid phase and escape into the gaseous phase. This is called a *change of phase*. When fast-moving molecules with high kinetic energy escape, the average energy of the remaining molecules is lower; hence the temperature is lowered.

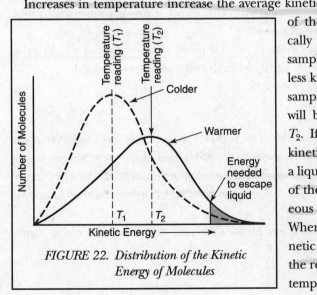

FIGURE 22. *Distribution of the Kinetic Energy of Molecules*

PHASE EQUILIBRIUM

Figure 23 shows water in a container enclosed by a bell jar. Observation of this closed system would show an initial small drop in the water level, but after some time the level would become constant. The explanation is that, at first, more energetic molecules near the surface are escaping into the gaseous phase faster than some of the gaseous water molecules are returning to the surface and possibly being caught by the attractive forces that will retain them in the liquid phase. After some time the rates of evaporation and condensation equalize. This is known as *phase equilibrium.*

FIGURE 23. Closed System in Dynamic Equilibrium

Note: Arrows' lengths show extent of evaporation (↑) and condensation (↓).

In a closed system like this, when opposing changes are taking place at equal rates, the system is said to have *dynamic equilibrium.* At higher temperatures, since the number of molecules at higher energies increases, the number of molecules in the liquid phase will be reduced and the number of molecules in the gaseous phase will be increased. The rates of evaporation and condensation, however, will again become equal.

The behavior of the system described above illustrates what is known as *Le Châtelier's Principle.* It is stated as follows: When a system at equilibrium is disturbed by the application of a stress (change in temperature, pressure, or concentration) it reacts so as to minimize the stress, and attain a new equilibrium position.

In the discussion above, if the 20°C system is heated to 30°C, the number of gas molecules will be increased while the number of liquid molecules will be decreased:

$$\text{Heat} + H_2O(\ell) \rightleftharpoons H_2O(g)$$

The equation shifts to the right (any similar system that is endothermic shifts to the right when temperature is increased) until equilibrium is reestablished at the new temperature.

The molecules in the vapor that are in equilibrium with the liquid at a given temperature exert a constant pressure. This is called the *equilibrium vapor pressure* at that temperature.

Chart (N) in the Reference Tables section at the back of this book gives the vapor pressure of water at various temperatures.

BOILING POINT

The vapor pressure-temperature relation can be plotted on a graph for a closed system. (See Figure 24.) When a liquid is heated in an open container, the liquid and vapor are not in equilibrium and the vapor pressure increases until it becomes equal to the pressure above the liquid. At this point the average kinetic energy of the molecules is such that they are rapidly converted from the liquid to the vapor phase within the liquid as well as at the surface. The temperature at which this occurs is known as the *boiling point.*

FIGURE 24. *Vapor Pressure-Temperature Relationship for Carbon Tetrachloride and Water*

CRITICAL TEMPERATURE AND PRESSURE

There are conditions for particular substances when it is impossible for the liquid or gaseous phase to exist. Since the kinetic energy of a molecular system is directly proportional to the Kelvin temperature, it is logical to assume that there is a temperature at which the kinetic energy of the molecules is so great that the attractive forces between molecules are insufficient for the liquid phase to remain. The temperature above which the liquid phase of a substance cannot exist is called its *critical temperature.* Above its critical temperature, no gas can be liquefied regardless of the pressure applied. The minimum pressure required to liquefy a gas at its critical temperature is called its *critical pressure.*

SOLIDS

Whereas particles in gases have the highest degree of disorder, the solid state has the most ordered system. Particles are fixed in a rather definite position and maintain a definite shape. Particles in solids do vibrate in position, however, and may even diffuse through the solid. (Example: Gold clamped to lead shows diffusion of some gold atoms into the lead over long periods of time.) Other solids do not show diffusion because of strong ionic or covalent bonds in network solids. (Examples: NaCl and diamond, respectively.)

When heated at certain pressures, some solids vaporize directly without passing through the liquid phase. This is called *sublimation.* Solids like solid carbon dioxide and solid iodine exhibit this property because of unusually high vapor pressure.

The temperature at which atomic or molecular vibrations of a solid become so great that the particles break free from fixed positions and begin to slide freely over each other in a liquid state is called the *melting point.* The amount of energy required at the melting point temperature to cause the change of phase to occur is called the *heat of fusion.* The amount of this energy depends on the nature of the solid and the type of bonds present.

PHASE DIAGRAMS

The simplest way to discuss a phase diagram is by an example, such as Figure 25.

A phase diagram ties together the effects of temperature and also pressure on the phase changes of a substance. In Figure 25 the line *BD* is essentially the vapor-pressure curve for the liquid phase. Notice that at 760 mm (the pressure for 1 atmosphere) the water will boil (change to the vapor phase) at 100°C (point *F*). However, if the pressure is raised, the boiling point temperature increases; and, if the pressure is less than 760 mm, the boiling point decreases along the *BD* curve down to point *B*.

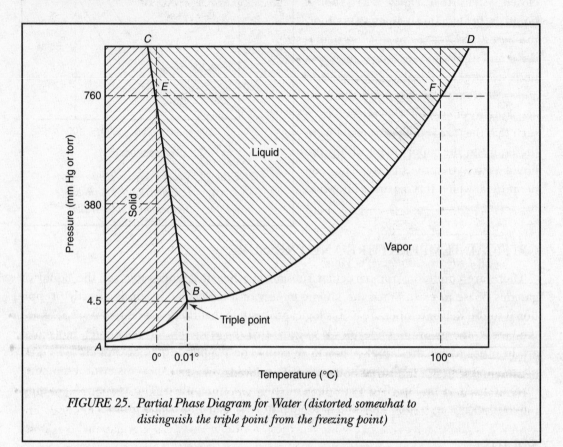

FIGURE 25. *Partial Phase Diagram for Water (distorted somewhat to distinguish the triple point from the freezing point)*

At 0°C the freezing point of water is found along the line *BC* at point *E* for pressure at 1 atmosphere or 760 mm. Again, this point is affected by pressure along the line *BC* so that, if pressure is decreased, the freezing point is slightly higher up to point *B* or 0.01°C.

Point *B* represents the point at which the solid, liquid, and vapor phases may all exist at equilibrium. This point is known as the *triple point*. It is the only temperature and pressure at which three phases of a pure substance can exist in equilibrium with one another in a system containing only the pure substance.

WATER

Purification of Water

Water is so often involved in chemistry that it is important to have a rather complete understanding of this compound and its properties. Pure water has become a matter of national concern. Although commercial methods of purification will not be discussed here, the usual laboratory method of obtaining pure water, distillation, will be covered.

The process of distillation involves the evaporation and condensation of the water molecules. The usual apparatus for the distillation of any liquid is shown in Figure 26.

This method of purification will remove any substance that has a boiling point higher than that of water. It cannot remove dissolved gases or liquids that boil off before water. These substances will be carried over into the condenser and subsequently into the distillate.

FIGURE 26. Distillation of Water

When water contains bicarbonates of Ca, Mg, or Fe it is said to have carbonate hardness, formerly called "temporary" hardness. Such hardness may be removed by boiling since heat causes the bicarbonate to decompose to a carbonate precipitate. This precipitate can cause detrimental scales in industrial boilers if not removed.

Natural waters are also apt to contain varying amounts of calcium sulfate, magnesium sulfate, calcium chloride, and magnesium chloride. Water in which these chemicals are dissolved is said to have noncarbonate hardness, formerly known as "permanent" hardness. One way of softening water is to pass it through an artificial zeolite that replaces the "hard" positive ions (called cations) with "soft" ions of sodium. It is the "hard" cations that are responsible for "bathtub ring" by reacting with the soap to form an insoluble compound. Another effective method that gives pure water like that obtained by distillation is the use of organic deionizers (e.g., Dowex, Amberlite, and zeolite). These organic resins exchange H ions for the cations, and OH ions for the anions.

Composition of Water

Water can be analyzed, that is, broken into its components, by electrolysis, which is shown on page 185. This process shows that its composition by volume is 2 parts of hydrogen to 1 part of oxygen. Water composition can also be arrived at by synthesis. Synthesis is the formation of a compound by uniting its components. Water can be made by mixing hydrogen and oxygen in a eudiometer over mercury and passing an igniting spark through the mixture. Again the ratio of combination is found to be 2 parts hydrogen: 1 part oxygen. In a steam-jacketed eudiometer, which keeps the water formed in the gas phase, 2 volumes of hydrogen combine with 1 volume of oxygen to form 2 volumes of steam. Another interesting method is the Dumas experiment pictured in Figure 27.

Data obtained would show that the ratio of hydrogen to oxygen combined to form water would be 1 : 8 by mass. This means that 1 gram of hydrogen combines with 8 grams of oxygen to form 9 grams of water.

$$Cu\,O + H_2 \rightarrow Cu + H_2O$$

FIGURE 27. Synthesis of Water

Some sample problems involving the composition of water are shown below.

• TYPE PROBLEM (BY MASS):

An electric spark is passed through a mixture of 12 grams of hydrogen and 24 grams of oxygen in the eudiometer setup shown. Find the number of grams of water formed and the number of grams of gas left uncombined.

Since water forms in a ratio of 1 : 8 by mass, to use up the oxygen (which by inspection will be the limiting factor since it has enough hydrogen present to react completely) we need only 3 grams of hydrogen.

3 g of hydrogen + 24 g of oxygen = 27 g of water

This leaves 12 g – 3 g = 9 g of hydrogen uncombined.

• TYPE PROBLEM (BY VOLUME):

A mixture of 8 mL of hydrogen and 200 mL of air is placed in a steam-jacketed eudiometer, and a spark is passed through the mixture. What will be the total volume of gases in the eudiometer?

Since this is a combination by volume, 8 mL of hydrogen require 4 mL of oxygen. (Ratio $H_2 : O_2$ by volume = 2 : 1.)

The 200 mL of air is approximately 21% oxygen. This will more than supply the needed oxygen and leave 196 mL of the air uncombined.

The 8 mL of hydrogen and 4 mL of oxygen will form 8 mL of steam since the eudiometer is steam-jacketed and keeps the water formed in the gaseous state.

(Ratio by volume of hydrogen : oxygen : steam = 2 : 1 : 2)
TOTAL VOLUME = 196 mL of air + 8 mL of steam = 204 mL

Heavy water. A small portion of water is called "heavy" water because it contains an isotope of hydrogen, deuterium (symbol D), rather than ordinary hydrogen nuclei. Deuterium has a nucleus of one proton and one neutron rather than just one proton. Another isotope of hydrogen is tritium. Its nucleus is composed of two neutrons and one proton. Both of these isotopes have had use in the nuclear energy field.

Hydrogen peroxide. The prefix *per-* indicates that this compound contains more than the usual oxide. Its formula is H_2O_2. It is a well-known bleaching and oxidizing agent. Its electron-dot formula is shown in Figure 28.

FIGURE 28. *Hydrogen Peroxide*

Properties and Uses of Water

Water has been used in the definition of various standards.

1. For mass—1 mL (or 1cc) of water at 4°C is 1 gram
2. For heat—(a) the heat needed to raise one gram of water one degree on the Celsius scale = 1 calorie (cal) 1000 calories = 1 kilocalorie (kcal)
   (b) the heat needed to raise one pound of water one degree on the Fahrenheit scale = 1 British thermal unit (BTU)
3. Degree of heat—the freezing point of water = 0°C, 32°F, 273K
   —the boiling point of water = 100°C, 212°F, 373K

Water Calorimetry Problems

A calorimeter is a container well insulated from outside sources of heat or cold so that most of its heat is contained in the vessel. If a very hot object were placed in a calorimeter containing some ice crystals, we could find the final temperature of the mixture mathematically and check it experimentally. To do this, certain behaviors must be understood. Ice changing to water and then to steam is not a continuous and constant change of temperature as time progresses. In fact, the chart would look as shown in Figure 29.

FIGURE 29. *Changing Ice to Steam*

From this graph, you see that heat is being used at 0°C and 100°C to change the state of water, but not its temperature. One gram of ice at 0°C needs 80 calories or 3.34×10^2 J to change to water at 0°C. This is called its *heat of fusion.* Likewise, energy is being used at 100°C to change water to steam, not to change the temperature. One gram of water at 100°C absorbs 540 calories or 2.26×10^3 J of heat to change to 1 gram of steam at 100°C. This is called its *heat of vaporization.* This energy being absorbed at the plateaus in the curve is being used to break up the bonding forces between molecules by increasing their energy content so that a specific change of state can occur.

- **TYPE OF PROBLEM:**

What quantity of ice at $0°C$ can be melted by 6,800 calories of heat?

Heat to fuse (melt) a substance = heat of fusion of the substance × mass of the substance
So,

$$6,800 \text{ cal} = 80 \text{ cal/g} \times \text{mass of substance}$$

Solving this equation for the mass of the substance, we get

$$\text{Mass of substance} = \frac{6,800 \text{ cal}}{80 \text{ cal/g}} = 85 \text{ g of ice melted}$$

Using the factor-label method, we have

$$6,800 \text{ cal} \times \frac{1 \text{ g}}{80 \text{ cal}} = 85 \text{ g}$$

- **ANOTHER EXAMPLE:**

How much heat is needed to change 100 grams of ice at 0°C to steam at 100°C? To melt 100 g of ice at 0°C:

$$100 \text{ g} \times \frac{80 \text{ cal}}{1 \text{ g}} = 8,000 \text{ cal} = 8 \text{ kcal}$$

To heat 100 g of water from 0°C to 100°C:

$$100 \text{ g} \times (\overbrace{100° - 0°}^{\text{temperature change}})°C \times \frac{1 \text{ cal}}{1 \text{ g} \times 1°C} = 10,000 \text{ cal} = 10 \text{ kcal}$$

To vaporize 100 g of water at 100°C to steam at 100°C:

$$100 \text{ g} \times \frac{540 \text{ cal}}{1 \text{ g}} = 54,000 \text{ cal} = 54 \text{ kcal}$$

Total heat = 8 + 10 + 54 = 72 kcal

Using the factor-label system to express the answer in joules, we have

$$72 \text{ kcal} \times \frac{1,000 \text{ cal}}{1 \text{ kcal}} \times \frac{4.18 \text{ J}}{1 \text{ cal}} = 3.01 \times 10^5 \text{ J}$$

Water's Reactions with Anhydrides

Anhydrides are certain oxides that react with water to form two classes of compounds—acids and bases.

Many metal oxides react with water to form bases such as sodium hydroxide, potassium hydroxide, and calcium hydroxide. For this reason, they are called *basic anhydrides*. The common bases are water solutions which contain the hydroxyl (OH^-) ion. Some common examples are:

$$Na_2O + H_2O \rightarrow 2NaOH, \text{ sodium hydroxide}$$
$$CaO + H_2O \rightarrow Ca(OH)_2, \text{ calcium hydroxide}$$

In general then: Metal oxide + H_2O → Metal hydroxide

In a similar manner, nonmetallic oxides react with water to form an acid such as sulfuric acid, carbonic acid, or phosphoric acid. For this reason, they are referred to as *acid anhy-*

drides. The common acids are water solutions containing hydrogen ions (H^+). Some common examples are:

$$CO_2 + H_2O \rightarrow H_2CO_3, \text{ carbonic acid}$$
$$SO_3 + H_2O \rightarrow H_2SO_4, \text{ sulfuric acid}$$
$$P_2O_5 + 3H_2O \rightarrow 2H_3PO_4, \text{ phosphoric acid}$$

In general, then: Nonmetallic oxide + H_2O $\rightarrow$ Acid

POLARITY AND HYDROGEN BONDING

Water is different from most liquids in that it reaches its greatest density at 4°C and then begins to expand from 4°C to 0°C (which is its freezing point). When it freezes at 0°C, its volume expands by about 9 percent. Most liquids contract as they cool and change state to a solid because their molecules have less energy, move more slowly, and are closer together. This abnormal behavior can be explained as follows. X-ray studies of ice crystals show that H_2O molecules are bound into large molecules in which each oxygen atom is connected through *hydrogen bonds* to four other oxygen atoms as shown in Figure 30.

FIGURE 30. *Study of Ice Crystal*

This is a rather wide open structure that accounts for the low density of ice. As heat is applied and melting begins, this structure begins to collapse but not all the hydrogen bonds are broken. The collapsing increases the density but the remaining bonds keep the structure from completely collapsing. As the heat is absorbed, the kinetic energy of the molecules breaks more of these bonds as the temperature goes from 0° to 4°C. At the same time this added kinetic energy is tending to distribute the molecules farther apart. At 4°C these opposing forces are in balance—thus the greatest density. Above 4°C the increasing molecular motion again causes a decrease in density since it is the dominate force and offsets the breaking of any more hydrogen bonds.

This behavior of water can be explained by studying the water molecule itself. The water molecule is composed of two hydrogen atoms bonded by a polar covalent bond to one oxygen atom.

Because of the polar nature of the bond, the molecule exhibits the charges shown in the above drawing. It is this polar charge that causes the polar bonding discussed in Chapter 3 as the hydrogen bond. This bonding is stronger than the usual molecular attraction called van der Waals forces.

SOLUBILITY

Water is often referred to as "the universal solvent" because of the number of common substances that dissolve in water. When substances are dissolved in water to the extent that no more will dissolve at that temperature, the solution is said to be *saturated*. The substance dissolved is called the *solute* and the dissolving medium is called a *solvent*. To give an accurate statement of a substance's solubility, three conditions are mentioned: the amount of solute, the amount of solvent, and the temperature of the solution. Since the solubility varies for each substance and for different temperatures, a student must be acquainted with the use of solubility curves such as those shown in Figure 31 and Table B in the Tables for Reference section.

These curves show the number of grams of solute that will dissolve in 100 grams (milliliters) of water over a temperature range of 0°C to 100°C. Take, for example, the very lowest curve at 0°C. This curve shows the number of grams of $KClO_3$ that will dissolve in 100 g (ml) of water over a temperature range of 0°C to 100°C. To find the solubility at any particular temperature, for example at 50°C, you follow the vertical line up from 50°C until it crosses the curve. At that point you place a ruler horizontally across the page and take the reading on the vertical axis. This point happens to be slightly below the 20 gram mark, or 18 grams. This means that 18 grams of $KClO_3$ will dissolve in 100 grams (milliliters) of water at 50°C.

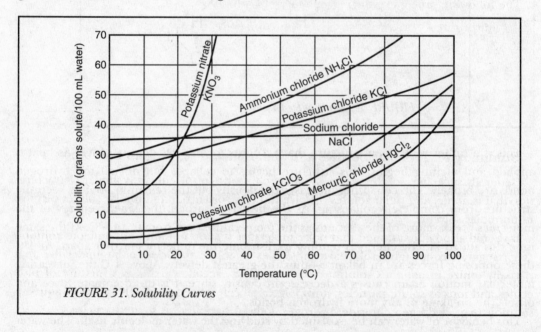

FIGURE 31. Solubility Curves

• TYPE PROBLEM USING THE SOLUBILITY CURVE:

A solution contains 20 grams of $KClO_3$ in 200 grams of H_2O at 80°C. How many more grams of $KClO_3$ can be dissolved to saturate the solution at 90°C?

Reading the graph at 90° and up to the graph line for $KClO_3$, you find that 100 grams of H_2O can dissolve 48 grams. So 200 grams could hold (2×48) grams or 96 grams. Therefore 96 g – 20 g = 76 g $KClO_3$ can be added to the solution.

General Rules of Solubility

All nitrates, acetates, and chlorates are soluble.

All common compounds of sodium, potassium, and ammonium are soluble.

All chlorides are soluble except those of silver, mercury(ious), and lead. (Lead chloride is noticeably soluble in hot water.)

All sulfates are soluble except those of lead, barium, strontium, and calcium. (Calcium sulfate is slightly soluble.)

The normal carbonates, phosphates, silicates, and sulfides are insoluble except those of sodium, potassium, and ammonium.

All hydroxides are insoluble except those of sodium, potassium, ammonium, calcium, barium, and strontium.

Some general trends of solubility are shown in the chart below.

	Temperature Effect	*Pressure Effect*
Solid	Solubility usually increases with temperature increase	Little effect
Gas	Solubility usually decreases with temperature increase	Solubility varies in direct proportion to the pressure applied to it. *Henry's Law*

Factors That Affect Rate of Solubility

The following procedures increase the rate of solubility

Pulverizing	increases surface exposed to solvent
Stirring	brings more solvent that is unsaturated into contact with solute
Heating	increases molecular action and gives rise to mixing by convection currents. (This heating affects the solubility as well as the rate of solubility.)

Summary of Types of Solutes and Relationship of Type to Solubility

Generally speaking, solutes are most likely to dissolve in solvents with similar characteristics, that is, ionic, and polar solutes dissolve in polar solvents and nonpolar solutes dissolve in nonpolar solvents.

It should also be mentioned that polar molecules that do not ionize in aqueous solution (e.g., sugar, alcohol, glycerol) have molecules as solute particles; polar molecules that partially ionize in aqueous solution (e.g., ammonia, acetic acid) have a mixture of molecules and ions as solute particles; and polar molecules that completely ionize in aqueous solution (e.g., hydrogen chloride, hydrogen iodide) have ions as solute particles.

WATER SOLUTIONS

In order to make molecules or ions of another substance go into solution, water molecules must overcome the forces that hold these molecules or ions together. The mechanism of the actual process is complex. To make sugar molecules go into solution, the water molecules cluster around the sugar molecules, pull them off, and disperse, forming the solution.

For an ionic crystal such as salt, the water molecules orient themselves around the ions (which are charged particles) and again must overcome the forces holding the ions together. Since the water molecule is polar, this orientation around the ion is an attraction of the polar ends of the water molecule. For example:

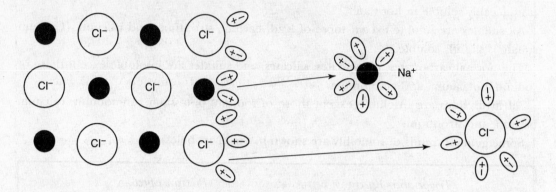

Once surrounded, the ion is insulated to an extent from other ions in solution because of the *dipole* property of water. The water molecules that surround the ion differ in number for various ions, and the whole group is called a *hydrated ion*. In general, as stated in the preceding section, polar substances and ions dissolve in polar solvents and nonpolar substances such as fats dissolve in nonpolar solvents such as gasoline. The process of going into solution is *exothermic* if energy is released in the process, and *endothermic* if energy from the water is used up to a greater extent than energy is released in freeing the particle.

When two liquids are mixed and they dissolve in each other, they are said to be completely *miscible*. If they separate and do not mix, they are said to be *immiscible*.

Two molten metals may be mixed and allowed to cool. This gives a "solid solution" called an *alloy*.

CONTINUUM OF WATER MIXTURES

Figure 32 shows the general sizes of the particles found in a water mixture.

FIGURE 32. Size of Particles in Water Mixture

The basic difference between a colloid and a suspension is the diameter of the particles dispersed. All the boundaries marked are only the general ranges in which the distinctions between solutions, colloids, and suspensions are usually made.

The characteristics of water mixtures are as follows:

Solutions	Colloids	Suspensions
.... 1 mμ .... 100 mμ .... 1,000 mμ .... 10,000 mμ		
Clear; may have color		Cloudy; opaque color
Particles do not settle		Particles settle on standing
Particles pass through ordinary filter paper		Do not pass through ordinary filter paper
Particles pass through membranes	*Do not pass through semipermeable membranes like animal bladders, cellophane, and parchment, which have very small pores	
Particles are not visible	Visible in ultramicroscope	Visible with microscope or naked eye
	Show Brownian movement	No Brownian movement

*Separation of a solution from a colloidal dispersion through a semipermeable membrane is called *dialysis*.

When a bright light is directed at right angles to the stage of an ultramicroscope, the individual reflections of colloidal particles can be seen. They can be observed to be following a random zigzag path. This is explained as follows: The molecules in the dispersing medium are in motion and continuously bumping into the colloidal particles, causing them to change direction in a random fashion. This motion is called *Brownian movement* after the scientist who first observed it.

EXPRESSIONS OF CONCENTRATION

There are general terms and very specific terms used to express the concentration of a solution. The general terms and their definitions are:

Dilute = small amount of solute is dispersed in the solvent.

Concentrated = large amount of solute is dissolved in the solvent.

Saturated = the solution is holding all the solute possible at that temperature. This is not a static condition; that is, some solute particles are exchanging places with some of the undissolved particles, but the total solute in solution remains the same. This is an example of equilibrium.

Unsaturated = more solute can go into solution at that temperature. The solvent has further capacity to hold more solute.

Supersaturated = sometimes a saturated solution at a higher temperature can be carefully cooled so that the solute does not get a chance to come out of solution. At a lower temperature, then, the solution will be holding more solute in solution than it should for saturation and is said to be supersaturated. As soon as the solute particles are jarred or a "seed" particle is added to the solution to act as a nucleus, they rapidly come out of solution so that the solution reverts to the saturated state.

It is interesting to note that the words *saturated* and *concentrated* are NOT synonymous. In fact, it is possible to have a saturated dilute solution when the solute is only slightly soluble and a small amount of it makes the solution saturated, but its concentration is still dilute.

The more specific terms used to describe concentration are mathematically calculated. **Percentage concentration** is based on the percent of solute in the solution by mass. The general formula is:

$$\frac{\text{No. of grams of solute}}{\text{No. of grams of solution}} \times 100\% = \% \text{ concentration}$$

• TYPE PROBLEM:

How many grams of NaCl are needed to prepare 200 grams of a 10% salt solution?

10% of 200 grams = 20 grams of salt

You could also solve the problem using the above formula and solving for the unknown quantity.

$$\frac{x \text{ g solute}}{200 \text{ g solution}} \times 100\% = 10\%$$

$$x \text{ g solute} = \frac{10}{100} \times 200 = 20 \text{ grams of solute}$$

Specific gravity can also be used to describe the concentration.
Specific gravity is the ratio of the mass of a substance to the mass of an equal volume of water. Often chemistry handbooks give the specific gravity of solutions. These, in turn, can be used to determine the mass of a solute in solution. Specific gravity has no units. However, since the density of water is 1.0 g/mL, specific gravity is numerically equal to the density in grams per milliliter. For example, 7.9 is the specific gravity of iron. This means that the density of iron is 7.9 g/mL.

This relationship can be used in solving problems of concentration that use percentage composition and a known specific gravity.

• TYPE PROBLEM:

Suppose that a solution of hydrogen chloride gas dissolved in water (hydrochloric acid) has a concentration of 30% HCl by mass. How much solute is there in 100 mL of this solution if the specific gravity of the solution is 1.15? (This means that its density is 1.15 g/mL.)

First, determine the mass of 100 mL of the solution:

mass = density × volume
mass = 1.15 g/mL × 100 mL = 115 g

Then, determine what mass of HCl is dissolved in 115 g of the solution:

mass HCl = 30% × mass of solution
mass HCl = 0.30 × 115 g = 34.5g

So, 34.5 g of HCl gas are dissolved in 100 mL of this solution.

The next two expressions depend on the fact that, if the formula mass of a substance is expressed in grams, it is called a *gram-formula mass* (gfw), *molar mass*, or 1 *mole. Gram-molecular mass* can be used in place of gram-formula mass when the substance is really of molecular composition, and not ionic like NaCl or NaOH. The definitions and examples are:

Molarity (abbreviated M) is defined as the number of moles of a substance dissolved in 1 liter of solution.

- **EXAMPLE:**

    A 1 molar H_2SO_4 solution would have 98 grams of H_2SO_4 (its gram-formula mass) in 1 liter of the solution.

This can be expressed as a formula so that

$$\text{Molarity} = \frac{\text{No. of moles of solute}}{1 \text{ liter of solution}}$$

If the molarity (M) and the volume of a solution are known, the mass of the solute can be determined. First, use the above equation and solve for the number of moles of solute. Then, multiply this number by the molar mass.

- **TYPE PROBLEM**

How many grams of NaOH are dissolved in 200 mL of solution if its concentration is 1.50 M?

$$\text{Since } M = \frac{\text{No. of moles of solute}}{1 \text{ liter of solution}}$$

By solving for no. of moles, we have:

No. of moles of solute = M × volume in liters of solution
No. of moles of NaOH = 1.50 M × volume in liters of solution

$$\text{No. of moles of NaOH} = 1.50 \frac{\text{mole}}{\text{liter}} \times 0.200 \text{ liter} = 0.30 \text{ mole of NaOH}$$

The molar mass of NaOH = 23 + 16 + 1 = 40 g of NaOH

$$0.30 \text{ mole NaOH} \times \frac{40 \text{ g of NaOH}}{1 \text{ mole NaOH}} = 12 \text{ g of NaOH}$$

Molality (abbreviated m) is defined as the number of moles of the solute dissolved in 1,000 grams of solvent.

- **EXAMPLE:**

    A 1 molal solution of H_2SO_4 would have 98 grams of H_2SO_4 dissolved in 1,000 grams of water. This, you will notice, gives a total volume greater than 1 liter, whereas the molar solution had 98 grams in 1 liter of solution.

$$\text{Molality (m)} = \frac{\text{Moles of solute}}{1,000 \text{ grams of solvent}}$$

- **TYPE PROBLEM:**

Suppose that 0.25 moles of sugar are dissolved in 500 grams of water. What is the molality of this solution?

$$m = \frac{\text{Moles of solute}}{1,000 \text{ grams of solvent}}$$

If 0.25 moles are in 500 grams of H_2O, then 0.50 moles are in 1,000 grams of H_2O. So,

$$m = \frac{0.50 \text{ moles of sugar}}{1,000 \text{ grams of } H_2O} \text{ or}$$

$$m = 0.50$$

Normality (abbreviated N) is used less frequently to express concentration and depends on knowledge of what a *gram-equivalent mass* is. This can be defined as the amount of a substance that reacts with or displaces 1 gram of hydrogen or 8 grams of oxygen. A simple method of determining the number of equivalents in a formula is to count the number of hydrogens or find the total positive oxidation numbers since each +1 charge can be replaced by a hydrogen.

In H_2SO_4 (gram-formula mass = 98 g), there are 2 hydrogens so the gram-equivalent mass will be:

$$\frac{98 \text{ g}}{2} = 49 \text{ g}$$

In Al_2SO_4 (molar mass = 150 g), there are 2 aluminums, each of which has a +3 oxidation number, making a total of +6. The gram-equivalent mass will be:

$$\frac{150 \text{ g}}{6} = 49 \text{ g}$$

This idea of gram-equivalent mass is used in the definition of this means of expressing concentration, namely, the normality of a solution.

Normality is defined as the number of gram-equivalent masses of solute in 1 liter of solution.

- **EXAMPLE:**

    If 49 grams of H_2SO_4 is mixed with enough H_2O to make 500 mL of solution, what is the normality?

    Since normality is expressed for a liter, we must double the expression to 98 g of H_2SO_4 in 1,000 mL of solution. To find the normality we find the number of gram-equivalents in a liter of solution. So

    $$\frac{98 \text{ g (no. of grams in 1 liter of solution)}}{49 \text{ g (gram-equivalent mass)}} = 2 \text{ N (normal)}$$

DILUTION

Since the expression of molarity gives the quantity of solute per volume of solution, the amount of solute dissolved in a given volume of solution is equal to the product of the concentration times the volume. Hence 0.5 liter of 2 M solution contains

$$M \times V = \text{amount of solute (in moles)}$$

$$\frac{2 \text{ moles}}{\text{liter}} \times 0.5 \text{ liter} = 1 \text{ mole (of solute in 0.5 liter)}$$

Notice that volume units must be identical.

If you dilute a solution with water, the amount or number of moles of solute present remains the same, but the concentration changes. So you can use the expression:

Before   After
$$M_1 V_1 = M_2 V_2$$

This expression is useful in solving problems involving dilution.

• TYPE PROBLEM:

If you wish to make 1 liter of solution that is 6 M into 3 M solution, how much water must be added?

$$M_1 V_1 = M_2 V_2$$
$$6 \text{ M} \times 1 \text{ L} = 3 \text{ M} \times ? \text{ L}$$

Solving this expression:

? L = 2 liters. This is the total volume of the solution after dilution and means that 1 liter of water had to be added to the original volume of 1 liter to get a total of 2 liters for the dilute solution volume.

An important use of the molarity concept is in the solution of *titration* problems, which are covered in Chapter 11, along with pH expressions of concentration for acids.

COLLIGATIVE PROPERTIES OF SOLUTIONS

Colligative properties are properties that depend primarily on the concentration of particles and not the type of particle. There is usually a direct relationship between the concentration of particles and the effect recorded.

The vapor pressure of an aqueous solution is always lowered by the addition of more solute. From the molecular standpoint, it is easy to see that there are fewer molecules of water per unit volume in the liquid, and therefore fewer molecules of water in the vapor phase are required to maintain equilibrium. The concentration in the vapor drops and so does the pressure that molecules exert. This is shown graphically on the next page.

Notice that the effects of this change in vapor pressure are registered in the freezing point and the boiling point. The freezing point is lowered, and the boiling point is raised, in direct proportion to the number of particles of solute present. For water solutions, the concentration expression that expresses this relationship is molality (m), that is, the number of moles of solute per kilogram of solvent. For molecules that do not dissociate, it has been found that a 1 m solution freezes at –1.86°C and boils at 100.51°C. A 2 m solution would then freeze at twice this lowering, or –3.72°C, and boil at twice the 1 molal increase of 0.51°C, or 101.02°C.

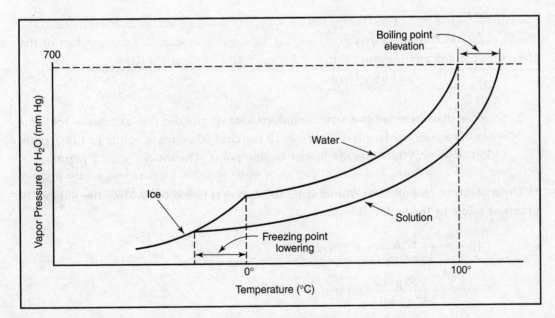

The chart below summarizes the colligative effect for aqueous solutions.

Type	Concen- tration (m)	Examples	Moles of Particles	Freezing Point (°C)	Boiling Point (°C)
Molecular nonionizing	1	Sugar, urea	1	–1.86	100.51
Molecular completely ionized	1	HCl	2	–3.72	101.02
Ionic completely dissociated	1	NaCl	2	–3.72	101.02
Ionic completely dissociated	1	$CaCl_2$ $Cu(NO_3)_2$	3	–5.58	101.53

• SAMPLE PROBLEMS:

1. A 1.50-g sample of urea is dissolved in 105.0 g of water and produces a solution that boils at 100.12°C. From the data, what is the molecular mass of urea?

   Since this property is related to the molality, then

   $$\frac{1.5 \text{ g}}{105.0 \text{ g}} = \frac{x \text{ g}}{1,000 \text{ g}}$$

   $x = 14.28$ g in 1,000 g of water

   The boiling point change is 0.12°C, and since each mole of particles causes a 0.51° increase, then

   $$0.12°\cancel{C} \div \frac{0.51°\cancel{C}}{\text{mol}} = 0.235 \text{ mol}$$

   Then 14.28 g = 0.235 mol

   and

   $$\frac{14.28 \text{ g}}{0.235 \text{ mol}} = \frac{x \text{ g}}{1 \text{ mol}}$$

   $x = 60.8$ g/mol

2. Suppose that there are two water solutions, one of glucose (molar mass = 180), the other of sucrose (molar mass = 342). Each contains 50 grams of solute in 1,000 grams (1 kg) of water. Which has the higher boiling point? The lower freezing point?

   The molality of each of these nonionizing substances is found by dividing the number of grams of solute by the molecular mass.

   Glucose: $\dfrac{50 \text{ g}}{180 \text{ g/mol}} = 0.278 \text{ mol}$

   Sucrose: $\dfrac{50 \text{ g}}{342 \text{ g/mol}} = 0.146 \text{ mol}$

Therefore their respective molalities are 0.278 m and 0.146 m. Since the freezing point and boiling point are colligative properties, the effect depends only on the concentration. Because the glucose has a higher concentration, it will have a higher boiling point and a lower freezing point. The respective boiling points would be:

$$0.278 \; m \times \frac{0.51°C \text{ rise}}{1 \; m} = 0.14°C \text{ rise or } 100.14°C \text{ boiling point for glucose.}$$

$$0.146 \; m \times \frac{0.51°C \text{ rise}}{1 \; m} = 0.07°C \text{ rise in boiling point or } 100.07°C \text{ for sucrose.}$$

The lowering of the freezing point would be:

$$0.278 \; m \times \frac{-1.86°C \text{ drop}}{1 \; m} = -0.52°C \text{ drop below } 0°C \text{ for glucose.}$$

$$0.146 \; m \times \frac{-1.86°C \text{ drop}}{1 \; m} = -0.27°C \text{ drop below } 0°C \text{ for sucrose.}$$

Using a solute that is an ionic solid and that completely ionizes in an aqueous solution introduces the consideration of the number of particles present in the solution. Notice in the previous chart that a 1 molal solution of NaCl yields a solution with 2 moles of particles. This is because:

NaCl(aq)		Na^+		Cl^-
1 mole of ionic	=	1 mole of	+	1 mole of
sodium chloride salt		Na^+ ions		Cl^- ions

So a 1 molal solution of NaCl has 2 moles of ion particles in 1,000 grams of solvent. The colligative property of lowering the freezing point and raising the boiling point depends primarily on the concentration of particles and not the type of particles.

In a 1 molal solution of NaCl, 1 mole of sodium chloride salt would be dissolved in 1,000 grams of water so that a total of 2 moles of ions were released—1 mole of Na^+ ions and 1 mole of Cl^- ions. Because this provides 2 moles of particles, it will cause a $2 \times -1.86°C$ (drop caused by 1 mole) $= -3.72°C$ drop in the freezing point. Likewise, it will cause a $2 \times 0.51°C$ (rise caused by 1 mole) $= +1.02°C$ rise in the boiling point or a boiling point of $101.02°C$.

The chart also shows that $CaCl_2$ releases 3 moles of particles from 1 mole of $CaCl_2$ dissolved in 1,000 g of water. Note its effect is to lower the freezing point to $3 \times -1.86°C$ or to $-5.58°C$. The boiling point rise is also 3 times the molal rise of $0.51°C$ and gives a boiling point of $101.53°C$.

This explains the use of salt on icy roads in the winter and the increased effectiveness of calcium chloride per mole of solute. The use of glycols in antifreeze solutions in automobile radiators is also based on this same concept.

CRYSTALLIZATION

Many substances form a repeated pattern structure as they come out of solution. The structure is bounded by plane surfaces that make definite angles with each other to produce a geometric form called a *crystal*. The smallest portion of the crystal lattice that is repeated throughout the crystal is called the *unit cell*. The kinds of unit cells are shown in Figure 33.

The crystal structure can also be classified by its internal axis, as shown in Figure 34.

A substance that holds a definite proportion of water in its crystal structure is called a *hydrate*. The formulas of hydrates show this water in the following manner: $CuSO_4 \cdot 5 H_2O$; $CaSO_4 \cdot 2 H_2O$; and $Na_2CO_3 \cdot 10 H_2O$. (The · is read as "with.") When these crystals are heated gently, the water of hydration can be forced out of the crystal and the structure collapses into an anhydrous (without water) powder. The dehydration of hydrated $CuSO_4$ serves as a good example since the hydrated crystals are deep blue because of the water molecules present with the copper ions. When this water is removed, the structure crumbles into the anhydrous white powder. Some hydrated crystals, such as magnesium sulfate (Epsom salt), lose the water of hydration on exposure to air at ordinary temperatures. They are said to be *efflorescent*. Other hydrates, such as magnesium chloride and calcium chloride, absorb water from the air and become wet. They are said to be *deliquescent*, or *hydroscopic*. This property explains why calcium chloride is often used as a drying agent in laboratory experiments.

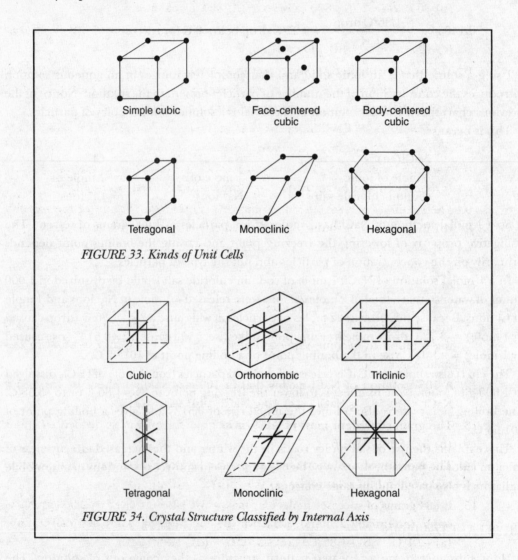

FIGURE 33. *Kinds of Unit Cells*

FIGURE 34. *Crystal Structure Classified by Internal Axis*

Chapter

7

Review

• QUESTIONS

1. Distillation of water cannot remove (a) volatile liquids like alcohol (b) dissolved salts (c) suspensions (d) precipitates.

2. The ratio in water of hydrogen to oxygen by mass is (a) 1 : 9 (b) 2 : 1 (c) 1 : 2 (d) 1 : 8.

3. Decomposing water by an electric current will give a ratio of hydrogen to oxygen by volume equal to (a) 1 : 9 (b) 2 : 1 (c) 1 : 2 (d) 1 : 8.

4. If 10 g of ice melts at 0°C, the total quantity of heat absorbed is (a) 10 calories (b) 80 calories (c) 800 calories (d) 800 kilocalories.

5. To heat 10 g of water from 4°C to 14°C requires (a) 10 calories (b) 14 calories (c) 100 calories (d) 140 calories.

6. The abnormally high boiling point of water in comparison to similar compounds is due to (a) van der Waals forces (b) polar covalent bonding (c) dipole insulation (d) hydrogen bonding.

7. A metallic oxide placed in water would most likely yield a(n) (a) acid (b) base (c) metallic anhydride (d) basic anhydride.

8. A solution can be *both* (a) dilute and concentrated (b) saturated and dilute (c) saturated and unsaturated (d) supersaturated and saturated.

9. The solubility of a solute must indicate (a) the temperature of the solution (b) the quantity of solute (c) the quantity of solvent (d) all of these.

10. A foam is an example of (a) a gas dispersed in a liquid (b) a liquid dispersed in a gas (c) a solid dispersed in a liquid (d) a liquid dispersed in a liquid.

11. When another crystal was added to a water solution of the same substance, the crystal seemed to remain unchanged. Its particles were (a) going into an unsaturated solution (b) exchanging places with others in the solution (c) causing the solution to become supersaturated (d) not going into solution in this static condition.

12. A 10% solution of NaCl means that in 100 g of solution there is (a) 5.85 g NaCl (b) 58.5 g NaCl (c) 10 g NaCl (d) 94 g of H_2O.

13. The gram-equivalent mass of $AlCl_3$ is its gram-formula mass divided by (a) 1 (b) 3 (c) 4 (d) 6.

14. The molarity of a solution made by placing 98 g of H_2SO_4 in sufficient water to make 500 mL of solution is (a) 0.5 (b) 1 (c) 2 (d) 3.

15. If 684 grams of sucrose (molecular mass = 342 g) is dissolved in 2,000 grams of water (essentially 2 liters), what will be the freezing point of this solution? (a) –.51°C (b) –1.86°C (c) –3.72°C (d) –6.58°C

• ANSWERS:

1. (a)	4. (c)	7. (b)	10. (a)	13. (b)
2. (d)	5. (c)	8. (b)	11. (b)	14. (c)
3. (b)	6. (d)	9. (d)	12. (c)	15. (b)

• SOLUTIONS TO 4 AND 5

4. Each gram of ice that dissolves at 0°C absorbs 80 calories.

    So,   10 g × 80 calories/gram = <u>800 calories</u>

5. By definition, 1 calorie is the amount of heat needed to heat 1 gram of water 1° on the Celsius scale.

    To heat 10 g from 4° to 14° would require

$$\text{Temperature difference}$$
$$10 \text{ g} \times \overbrace{(14° - 4°)} = \underline{100 \text{ calories}}$$

• SOLUTIONS TO 12, 14, AND 15

12. A 10% solution contains 10 g of solute/100 g of solution because

$$10\% \times 100 \text{ g} = \underline{\underline{10 \text{ g}}}$$

14. 98 grams of H_2SO_4 = 1 mole H_2SO_4

    500 mL = 0.5 liter

    So 1 mole in 0.5 liter would be 2 moles in 1 liter or 2 molar solution.

15. 684 grams of sucrose = $\dfrac{684}{342 \text{ g/mole}}$ = 2 moles of sucrose

$$\dfrac{2 \text{ moles}}{2,000 \text{ g } H_2O} = 1 \text{ molal solution}$$

    So freezing point is lowered only −1.86°C.

Terms You Should Know

anhydride	exothermic	normality
boiling point	heat of fusion	phase equilibrium
Brownian movement	heat of vaporization	polarity
colligative property	"heavy" water	solute
critical pressure	hydrate	solvent
critical temperature	Le Châtelier's Principle	specific gravity
crystal	melting point	sublimation
dynamic equilibrium	molality	van der Waals forces
endothermic	molarity	

Chemical Reactions

TYPES OF REACTIONS

The many kinds of reactions you may encounter can be placed into four basic categories: combination, decomposition, single replacement, and double replacement.

The first type, *combination*, can also be called *synthesis*. This means the formation of a compound from the union of its elements. Some examples of this type are:

$$Zn + S \rightarrow ZnS$$
$$2H_2 + O_2 \rightarrow 2H_2O$$
$$C + O_2 \rightarrow CO_2$$

The second type of reaction, *decomposition*, can also be referred to as *analysis*. This means the breakdown of a compound to release its components as individual elements or other compounds. Some examples of this type are:

$$2H_2O \rightarrow 2H_2 + O_2 \text{ (electrolysis of water)}$$
$$C_{12}H_{22}O_{11} \rightarrow 12C + 11H_2O$$
$$2HgO \rightarrow 2Hg + O_2$$

The third type of reaction is called *single replacement* or *single displacement*. This type can best be shown by examples in which one substance is displacing another. Some examples are:

$$Fe + CuSO_4 \rightarrow FeSO_4 + Cu$$
$$Zn + H_2SO_4 \rightarrow ZnSO_4 + H_2(g)$$
$$Cu + 2\,AgNO_3 \rightarrow Cu(NO_3)_2 + 2Ag$$

The last type of reaction is called *double replacement* or *double displacement* because there is an actual exchange of "partners" to form new compounds. Some examples of this are:

$$AgNO_3 + NaCl \rightarrow AgCl + NaNO_3$$
$$H_2SO_4 + 2NaOH \rightarrow Na_2SO_4 + 2H_2O \text{ (neutralization)}$$
$$CaCO_3 + 2HCl \rightarrow H_2CO_3 + CaCl_2$$
$$\downarrow \text{(unstable)}$$
$$\hookrightarrow H_2O + CO_2(g)$$

PREDICTING REACTIONS

One of the most important topics of chemistry deals with the reasons why reactions take place. Taking each of the above types of reactions, let us see how a prediction can be made concerning how the reaction gets the driving force to make it occur.

1. Combination

The best source of information to predict a chemical combination is the heat of formation table. A *heat of formation* table gives the number of calories evolved or absorbed when a mole (gram-formula mass) of the compound in question is formed by the direct union of its elements. In this book a positive number indicates that heat is absorbed and a negative number that heat is evolved. It makes some difference whether the compounds formed are in the solid, liquid, or gaseous state. Unless otherwise indicated (g = gas, ℓ = liquid), the compounds are in the solid state. The values given are in kilocalories (1 kilocalorie = 1,000 calories). One calorie is the amount of heat needed to raise the temperature of 1 gram of water 1 degree on the Celsius scale. The symbol ΔH is used to indicate the heat of formation.

If the heat of formation is a large number preceded by a minus sign, the combination is likely to occur spontaneously and the reaction is exothermic. If, on the other hand, the number is small and negative or is positive, heat will be needed to get the reaction to go at any noticeable rate. Some examples are:

$$Zn + S \rightarrow ZnS + 48.5 \text{ kcal} \qquad \Delta H = -48.5 \text{ kcal}$$

This means that 1 mole of zinc (65 grams) reacts with 1 mole of sulfur (32 grams) to form 1 mole of zinc sulfide (97 grams) and releases 48.5 kilocalories of heat.

• SIMILARLY:

$$Mg + \tfrac{1}{2}O_2 \rightarrow MgO + 143.84 \text{ kcal} \qquad \Delta H = -143.84 \text{ kcal}$$

indicates that the formation of 1 mole of magnesium oxide requires 1 mole of magnesium and $\tfrac{1}{2}$ mole of oxygen with the release of 143.84 kilocalories of heat. Notice the use of the fractional coefficient for oxygen. If the equation had been written with the usual whole-number coefficients, 2 moles of magnesium oxide would have been released.

$$2Mg + O_2 \rightarrow 2MgO + 2(+143.84) \text{ kcal}$$

Since, by definition, the heat of formation is given for the formation of *one* mole, this latter thermal equation shows $2 \times (-143.84)$ kcal released.

• ANOTHER EXAMPLE:

$$H_2(g) + \tfrac{1}{2}O_2(g) \rightarrow H_2O(1) + 68.32 \text{ kcal} \qquad \Delta H = -68.32 \text{ kcal}$$

In combustion reactions the heat evolved when 1 mole of a substance is completely oxidized is called the *heat of combustion* of the substance. So, in this equation:

$$C(g) + O_2(g) \rightarrow CO_2(g) + 94.05 \text{ kcal} \qquad \Delta H = -94.05 \text{ kcal}$$

ΔH is the heat of combustion of carbon. Because the energy of a system is conserved during chemical activity, the same equation could be arrived at by adding the following equations:

$$C(s) + \tfrac{1}{2}O_2(g) \rightarrow CO(g) \qquad \Delta H = -26.41 \text{ kcal}$$

$$CO(g) + \tfrac{1}{2}O_2(g) \rightarrow CO_2(g) \qquad \Delta H = -67.64 \text{ kcal}$$

$$\overline{C(s) + O_2(g) \rightarrow CO_2(g) \qquad \Delta H = -94.05 \text{ kcal}}$$

2. Decomposition

The prediction of decomposition reactions uses the same source of information, the heat of formation table. If the heat of formation is a high exothermic (ΔH is negative) value, the compound will be difficult to decompose since this same quantity of energy must be returned to the compound. A low heat of formation indicates decomposition would not be difficult, such as the decomposition of mercuric oxide with $\Delta H = -21.68$ kcal/mole:

$$2HgO \rightarrow 2Hg + O_2 \text{ (Priestley's method of preparation)}$$

A high positive heat of formation indicates extreme instability of a compound, and it can explosively decompose.

3. Single Replacement

A prediction of the feasibility of this type of reaction can be based on a comparison of the heat of formation of the original compound and that of the compound to be formed. For example, in a reaction of zinc with hydrochloric acid, the 2 moles of HCl have $\Delta H = 2 \times -22.06$ kcal.

$$Zn + 2HCl \rightarrow ZnCl_2 + H_2(g) \quad \textit{Note: } \Delta H = 0 \text{ for elements}$$
$$2 \times -22.06$$
$$-44.12 \text{ kcal} \neq -99.40 \text{ kcal}$$

and the zinc chloride has $\Delta H = -99.40$. This comparison leaves an excess of 55.28 kcal of heat given off so the reaction would occur.

In this next example, $-220.5 - (-184.0) = -36.5$ excess kcal to be given off as the reaction occurs:

$$Fe + \quad CuSO_4 \quad \rightarrow \quad FeSO_4 \quad + Cu$$
$$-184.0 \text{ kcal} \qquad -220.5 \text{ kcal}$$

Another simple way of predicting single replacement reactions is to check the relative positions of the two elements in the electromotive chart (Table 10, p. 181). If the element that is to replace the other in the compound is higher on the chart, the reaction will occur. If it is below, there will be no reaction.

Some simple examples of this are the following reactions.

In predicting the replacement of hydrogen by zinc in hydrochloric acid, reference to the electromotive chart shows that zinc is above hydrogen. This reaction would occur:

$$Zn + 2HCl \quad \rightarrow \quad ZnCl_2 + H_2(g)$$

In fact, all the metals above hydrogen in the chart on p. 181 would replace hydrogen in an acid solution. If a metal such as copper were chosen, no reaction would occur.

$$Ca + HCl \quad \rightarrow \quad \text{no reaction}$$

The determination of these replacements using a quantative method is covered in Chapter 12.

4. Double Replacement

For double replacement reactions to go to completion, that is, proceed until the supply of one of the reactants is exhausted, one of the following conditions must be present: an insoluble precipitate is formed, a nonionizing substance is formed, or a gaseous product is given off.

To predict the formation of an insoluble precipitate, you should have some knowledge of the solubilities of compounds. Table 9 gives some general solubility rules.

SOLUBLE   *EXCEPT*

Na^+
NH_4^+ } compounds
K^+

Acetates
Bicarbonates
Chlorates
Chlorides . . . . Ag^+, Hg^+, Pb ($PbCl_2$, sol. in hot water)
Nitrates
Sulfates . . . . Ba, Ca (slight), Pb

INSOLUBLE

Carbonates, phosphates . . . . Na, NH_4, K compounds
Sulfides, hydroxides . . . . Na, NH_4, K, Ba, Ca

---

TABLE 9. *Solubilities of Compounds*

(A table of solubilities could also be used as reference.)

An example of this type of reaction is given in its complete ionic form.

$$(K^+ + Cl^-) + (Ag^+ + NO_3^-) \rightarrow AgCl(s) + (K^+ + NO_3^-)$$

The silver ions combine with the chloride ions to form an insoluble precipitate, silver chloride. If the reaction had been like this:

$$(K^+ + Cl^-) + (Na^+ + NO_3^-) \rightarrow K^+ + NO_3^- + Na^+ + Cl^-$$

merely a mixture of the ions would have been shown in the final solution.

Another reason for a reaction of this type to go to completion is the formation of a nonionizing product such as water. This weak electrolyte keeps its component ions in molecular form and thus removes the possibility of reversing the reaction. All neutralization reactions are of this type.

$$(H^+ + Cl^-) + (Na^+ + OH^-) \rightarrow H_2O + Na^+ + Cl^-$$

This example shows the ions of the reactants, hydrochloric acid and sodium hydroxide, and the nonelectrolyte product water with sodium and chloride ions in solution. Since the water does not ionize to any extent, the reverse reaction cannot occur.

The third reason for double displacement to occur is the evolution of a gaseous product. An example of this is calcium carbonate reacting with hydrochloric acid:

$$CaCO_3 + 2HCl \rightarrow CaCl_2 + H_2O + CO_2(g)$$

Another example of a compound that evolves a gas in sodium sulfite with an acid:

$$Na_2SO_3 + 2HCl \rightarrow 2NaCl + H_2O + SO_2(g)$$

In general, acids with carbonates or sulfites are good examples of this type of equation.

Hydrolysis Reactions

Hydrolysis reactions are the opposite of neutralization reactions. In hydrolysis the salt and water react to form an acid and a base. For example, if sodium chloride is placed in solution, this reaction occurs to some degree:

$$(Na^+ + Cl^-) + H_2O \rightarrow (Na^+ + OH^-) + (H^+ + Cl^-)$$

In this hydrolysis reaction the same number of hydrogen ions and of hydroxide ions is released so that the solution is neutral. This occurs because sodium hydroxide is a strong base and hydrochloric acid is a strong acid. (There are charts of relative acid and base strengths on pages 169 and 170 to use as a reference.) Because they are both classified as strong, sodium hydroxide and hydrochloric acid essentially exist as ions in solutions. Therefore, the solution has neither an excess of hydrogen nor hydroxide ions, and it will test neutral. So, the salt of a strong acid and strong base forms a neutral solution when dissolved in water. However, if Na_2CO_3 is dissolved, we have:

$$(2Na^+ + CO_3^{2-}) + 2H_2O \rightarrow (2Na^+ + 2OH^-) + H_2CO_3$$

The H_2CO_3 is written together because it is a slightly ionized acid or, in other words, a weak acid. Since the hydroxide ions are free in the solution, the solution is basic. Notice that here it was the salt of a strong base and a weak acid that formed a basic solution. This generalization is true for this type of salt.

If we use $ZnCl_2$, which is the salt of a strong acid and a weak base, the reaction would be:

$$(Zn^{2+} + 2Cl^-) + H_2O \rightarrow (2H^+ + 2Cl^-) + Zn(OH)_2$$

In this case the hydroxide ions are held in the weakly ionizing compound while the hydrogen ions are free to make the solution acidic. In general then, the salt of a strong acid and a weak base forms an acid solution by hydrolysis.

The fourth possibility is that of a salt of a weak acid and weak base dissolving in water. An example would be ammonium carbonate, $(NH_4)_2CO_3$, which is the salt of a weak base and a weak acid. The hydrolysis reaction would be:

$$(NH_4)_2 CO_3 + 2 H_2O \rightarrow 2 NH_4OH + H_2CO_3$$

Both the ammonium hydroxide, NH_4OH, and the carbonic acid, H_2CO_3, are written as nonionized compounds because they are classified as a weak base and a weak acid, respectively. Therefore, a salt of a weak acid and a weak base forms a neutral solution since neither hydrogen ion nor hydroxide ion will be present in excess.

Entropy

In many of the preceding predictions of reactions, we used the concept that reactions will occur when they result in the lowest possible energy state.

There is, however, another driving force to reactions that relates to their state of disorder or of randomness. This state of disorder is called *entropy*. A reaction is also driven, then, by a need for a greater degree of disorder. An example is the intermixing of gases in two connected flasks when a valve is opened to allow the two previously isolated gases to travel between the two flasks. Since temperature remains constant throughout the process, the total heat content cannot have changed to a lower energy level, and yet the gases will become evenly distributed in the two flasks. The system has thus reached a higher degree of disorder or entropy. A quantitative treatment of entropy is given on page 166.

ORIGIN OF CHEMICAL ENERGY

In general, all chemical reactions either liberate or absorb heat. The origin of chemical energy lies in the position and motion of atoms, molecules, and subatomic particles. The total energy possessed by a molecule is the sum of all the forms of potential and kinetic energy associated with it.

The energy changes in a reaction are due, to a large extent, to the changes in potential energy that accompany the breaking of chemical bonds in reactants to form new bonds in products.

The molecule may also have rotational, vibrational, and translational energy, along with some nuclear energy sources. All these make up the total energy of molecules. In beginning chemistry, the greatest concern in reactions is the electronic energy involved in the making and breaking of chemical bonds.

Since it is virtually impossible to measure the total energy of molecules, the energy change is usually the experimental data that we deal with in reactions. This change in quantity of energy is known as the change in *enthalpy* (heat content) of the chemical system and is symbolized by ΔH.

CHANGES IN ENTHALPY

Changes in enthalpy for exothermic and endothermic reactions can be shown graphically.

Course of Endothermic Reaction   *Course of Exothermic Reaction*

Notice that the ΔH for an endothermic reaction is positive, while that for an exothermic reaction is negative. It should be noted also that changes in enthalpy are always independent of the path taken to change a system from the initial state to the final state.

Since the quantity of heat absorbed or liberated during a reaction varies with the temperature, scientists have adopted 25°C and 1 atmosphere pressure as the *standard state* condition for reporting heat data. A superscript zero on ΔH (i.e., ΔH^0) indicates that the corresponding process was carried out under standard conditions. The standard enthalpy of formation (ΔH_f^0) of a compound is defined as the change in enthalpy that accompanies the formation of 1 mole of a compound from its elements with all substances in their standard states at 25°C.

To calculate the enthalpy of a reaction, it is necessary to write an equation for the reaction. The standard enthalpy change, ΔH, for a given reaction is usually expressed in kilocalories and depends on how the equation is written. For example, here is the equation for the reaction of hydrogen with oxygen in two ways:

$$H_2(g) + \tfrac{1}{2}O_2(g) \rightarrow \quad H_2O(g) \qquad \Delta H_f^0 = -57.8 \text{ kcal}$$

$$2H_2(g) + O_2(g) \rightarrow \quad 2H_2O(g) \qquad \Delta H_f^0 = -115.6 \text{ kcal}$$

Experimentally, ΔH_f^0 for the formation of 1 mole of $H_2O(g)$ is –57.8 kcal. Since the second equation represents the formation of 2 moles of $H_2O(g)$, the quantity is twice –57.8 or –115.6 kcal. It is assumed that the initial and final states are measured at 25°C and 1 atmosphere, although the reaction occurs at a higher temperature.

• PROBLEM:

How much heat is liberated when 40.0 g of $H_2(g)$ reacts with excess $O(g)$? The reaction equation is:

$$H_2(g) + \tfrac{1}{2}O_2(g) \rightarrow H_2O(g) \qquad \Delta H_f^0 = -57.8 \text{ kcal}$$

This represents 1 mole or 2 g of $H(g)$ forming 1 mole of $H_2O(g)$:

$$40 \text{ g} \times \frac{1 \text{ mole}}{2 \text{ g}} = 20 \text{ moles of hydrogen}$$

Since each mole gives off –57.8 kcal, then

$$20 \text{ moles} \times \frac{-57.8 \text{ kcal}}{1 \text{ mole}} = -1,156 \text{ kcal}$$

Notice that the physical state of each participant must be given since the phase changes involve energy changes.

ADDITIVITY OF REACTION HEATS

Chemical equations and ΔH^0 values may be manipulated algebraically. Finding the ΔH for the formation of vapor from liquid water shows how this can be done.

$$H_2(g) + \tfrac{1}{2}O_2(g) \rightarrow H_2O(g) \qquad \Delta H_f^0 = -57.8 \text{ kcal}$$

$$H_2(g) + \tfrac{1}{2}O_2(g) \rightarrow H_2O(\ell) \qquad \Delta H_f^0 = -68.3 \text{ kcal}$$

Since we want the equation for $H_2O(\ell) \rightarrow H_2O(g)$, we can reverse the second equation. This changes the sign of ΔH.

$$H_2O(\ell) \rightarrow H_2O(g) + \tfrac{1}{2}O_2(g) \qquad \Delta H_f^0 = 68.3 \text{ kcal}$$

Adding

$$H_2(g) + \tfrac{1}{2}O_2(g) \rightarrow H_2O(g) \qquad \Delta H_f^0 = -57.8 \text{ kcal}$$

yields

$$H_2O(\ell) + H_2(g) + \tfrac{1}{2}O_2(g) \rightarrow$$

$$H_2(g) + \tfrac{1}{2}O_2(g) + H_2O(g) \qquad \Delta H_f^0 = 10.5 \text{ kcal}$$

Simplification gives a net equation of

$$H_2O(\ell) \rightarrow H_2O(g) \qquad \Delta H_f^0 = 10.5 \text{ kcal}$$

The principle underlying the preceding calculations is known as *Hess's Law of Heat Summation*. This principle states that, when a reaction can be expressed as the algebraic sum of two or more other reactions, the heat of the reaction is the algebraic sum of the heats of these reactions. This is based upon the *First Law of Thermodynamics*, which, simply stated, says that the total energy of the universe is constant and cannot be created or destroyed.

This allows calculations of ΔH's that cannot be easily determined experimentally. An example is the determination of the ΔH of CO from the ΔH_f^0 of CO_2.

$$C(s) + O_2(g) \quad \rightarrow CO_2(g) \qquad \Delta H_f^0 = -94.0 \text{ kcal}$$

$$CO(g) + \tfrac{1}{2}O_2(g) \rightarrow CO_2(g) \qquad \Delta H_f^0 = -67.6 \text{ kcal}$$

The equation wanted is

$$C(s) + \tfrac{1}{2}O_2(g) \rightarrow CO(g)$$

To get this, we reverse the second equation and add it to the first:

$$C(s) + O_2(g) \rightarrow CO_2(g) \qquad \Delta H_f^0 = -94.0 \text{ kcal}$$

$$CO_2(g) \rightarrow \tfrac{1}{2}O_2(g) + CO(g) \qquad \Delta H_f^0 = +67.6 \text{ kcal}$$

Addition yields

$$C(s) + \tfrac{1}{2}O_2(g) \rightarrow CO(g) \qquad \Delta H_f^0 = -26.4 \text{ kcal}$$

This relationship can be shown schematically as done below:

Some commonly used standard heats of formation (enthalpy), designated as ΔH_f^0, are listed in the Table (E) in the Reference Tables section.

An alternative (and easier) method of calculating enthalpies is based on the concept that $\Delta H_{\text{reaction}}^0$ is equal to the difference between the total enthalpy of the reactants and that of the products. This can be expressed as follows:

$$\Delta H_{\text{reaction}}^0 = \Sigma \text{ (sum of) } \Delta H_f^0 \text{ (products)} - \Sigma \text{ (sum of) } \Delta H_f^0 \text{ (reactants)}$$

• **PROBLEM:**

Calculate $\Delta H^0_{reaction}$ for the decomposition of sodium chlorate.

$$NaClO_3(s) \rightarrow NaCl(s) + \tfrac{3}{2}O_2(g)$$

Step 1. Obtain ΔH^0_f for all substances.

$NaClO_3(s) = -85.7$ kcal/mole

$NaCl(s) \quad = -98.2$ kcal/mole

$O_2(g) \quad = \quad 0$ kcal/mole (all elements = 0)

Step 2. Substitute these values in the equation.

$$\Delta H^0_{reaction} = \Sigma \Delta H^0_f(products) - \Sigma \Delta H^0_f(reactants)$$

$$\Delta H^0_{reaction} = -98.2 - (-85.7) \text{ kcal}$$

$$\Delta H^0_{reaction} = -12.5 \text{ kcal}$$

• **ANOTHER EXAMPLE:**

Calculate $\Delta H_{reaction}$ for this oxidation of ammonia:

$$4NH_3(g) + 5O_2(g) \rightarrow 6H_2O(g) + 4NO(g)$$

The individual ΔH^0_f's are

$4NH_3 \quad = 4(-11 \text{ kcal}) \quad = -44.9 \text{ kcal}$

$5O_2(g) \quad = 0$

$6H_2O(g) = 6(-57.8 \text{ kcal}) = -346.8 \text{ kcal}$

$4NO(g) = 4(21.6 \text{ kcal}) = \quad 86.4 \text{ kcal}$

Substituting these in the $\Delta H_{reaction}$ equation gives

$$\Delta H_{reaction} = [-346.8 \text{ (6H}_2\text{O)} + 86.4 \text{ (4NO)}] - [-44.0 \text{ (4NH}_3\text{)} + 0 \text{ (5O}_2\text{)}]$$

$$\Delta H_{reaction} = -216.4 \text{ kcal}$$

BOND DISSOCIATION ENERGY

The same principle of additivity applies to bond energies. Experimentation has found average bond energies for particular bonds. Some of the more common are shown in the chart.

Bond	Energy (kcal/mole)	Bond	Energy (kcal/mole)
H—H	104	O=O	119
C—H	99	O—H	111
C—C	83	C—O	86
C=C	146	C=O	177
C≡C	200	H—F	135
O—O	35	H—Cl	103
N—N	39	H—Br	87
Cl—Cl	58	H—I	71

• **SAMPLE CALCULATION:**

Find the bond energy, H_{be}, for ethane.

$$\text{Ethane} = \quad H-\overset{\overset{\displaystyle H}{|}}{\underset{\underset{\displaystyle H}{|}}{C}}-\overset{\overset{\displaystyle H}{|}}{\underset{\underset{\displaystyle H}{|}}{C}}-H \quad \text{and has 6 C-H bonds and 1 C-C bond.}$$

$$H_{be} = 6\ E_{C-H} + 1\ E_{C-C} = 6(99\text{ kcal}) + 1(83\text{ kcal}) = 677\text{ kcal}$$

ENTHALPY FROM BOND ENERGIES

A reaction's enthalpy can be approximated through the summation of bond energies. An example is this reaction:

$$H_2(g) + Cl_2(g) \rightarrow 2HCl(g)$$

Bonds Broken	Heat Energy Absorbed (kcal)	Bonds Formed	Heat Energy Evolved (kcal)
H—H	104.2		
Cl—Cl	57.8	2 H—Cl	206.0
Totals	162.0 kcal		206.0 kcal

The difference between heat evolved and heat absorbed is 206.0 – 162.0 = 44.0 kcal. This is for 2 moles of HCl, so –44 kcal divided by 2 = –22 kcal/mole. This answer is very close to the experimentally determined ΔH_f^0.

Further discussions of enthalpy (ΔH) and entropy (ΔS) are found on pages 166–167.

Chapter

8

Review

• **QUESTIONS**

1. A synthesis reaction will occur spontaneously if the heat of formation of the product is (a) large and negative (b) small and negative (c) large and positive (d) small and positive.

2. The reaction of aluminum with dilute H_2SO_4 can be classified as (a) synthesis (b) decomposition (c) single replacement (d) double replacement.

3. For a metal atom to replace another kind of metallic ion in a solution, the metal atom must be (a) a good oxidizing agent (b) higher in the electromotive chart than the metal in solution (c) lower in the electromotive chart than the metal in solution (d) equal in activity to the metal in solution.

4. One reason for a double displacement reaction to go to completion is that (a) a product is soluble (b) a product is given off as a gas (c) the products can react with each other (d) the products are miscible.

5. Hydrolysis will give an acid reaction when which of these is placed in solution with water? (a) Na_2SO_4 (b) K_2SO_4 (c) $NaNO_3$ (d) $Cu(NO_3)_2$

6. A salt derived from a strong base and a weak acid will undergo hydrolysis and give a solution that will be (a) basic   (b) acid   (c) neutral   (d) volatile.

7. Enthalpy is an expression for the (a) heat content (b) energy state (c) reaction rate   (d) activation energy.

8. The ΔH_f^0 of a reaction is recorded for (a) 0°C   (b) 25°C   (c) 100°C   (d) 200°C.

9. The property of being able to add enthalpies is based on the (a) Law of Conservation of Heat   (b) First Law of Thermodynamics   (c) Law of Constants   (d) Law of $E = mc^2$.

10. If $\Delta H_{\text{reaction}}$ is –100 kcal/mole, it indicates the reaction is (a) endothermic   (b) unstable   (c) in need of a catalyst   (d) exothermic.

• ANSWERS:

1. (a)	3. (b)	5. (d)	7. (a)	9. (b)
2. (c)	4. (b)	6. (a)	8. (b)	10. (d)

Terms You Should Know

combination reaction

decomposition reaction

double replacement reaction

enthalpy

entropy

heat of combustion

hydrolysis reaction

reaction mechanism

single replacement reaction

First Law of Thermodynamics

Hess's Law of Heat Summation

Rates of Chemical Reactions

MEASUREMENTS OF REACTION RATES

The measurement of reaction rate is based on the rate of appearance of a product or disappearance of a reactant. It is usually expressed in terms of a change in concentration of one of the participants per unit time.

Experiments have shown that for most reactions the concentrations of all participants change most rapidly at the beginning of the reaction, that is, the concentration of the product shows the greatest rate of increase, and the concentrations of the reactants the highest rate of decrease, at this point. This means that the rate of a reaction changes with time. Therefore a rate must be identified with a specific time.

FACTORS AFFECTING REACTION RATES

Five important factors control the rate of a chemical reaction. These are summarized below.

1. The nature of the reactants. In chemical reactions, some bonds break and others form. Therefore, the rates of chemical reactions should be affected by the nature of the bonds in the reacting substances. For example, reactions between ions in an aqueous solution may take place in a fraction of a second. Thus, the reaction between silver nitrate and sodium chloride is very fast. The white silver chloride precipitate appears immediately. In reactions where many covalent bonds must be broken, reaction usually takes place slowly at room temperatures. The decomposition of hydrogen peroxide into water and oxygen happens slowly at room temperatures. In fact, it takes about 17 minutes for half the peroxide in a 0.50 M solution to decompose.

2. The surface area exposed. Since most reactions depend on the reactants coming into contact, the surface exposed proportionally affects the rate of the reaction. Consider what happens to the surface area exposed when a cube is sliced in half.

   Slicing increases the surface area by $33^1/_3\%$, surface area that can now become involved in a chemical reaction. This is why reactants that depend on surface reactions are often introduced as powders into reaction vessels. It also explains the danger of a dust explosion if the reactant is flammable.

Six surface areas of equal size

New cut

Exposes 2 additional surface areas
or a total of 8 surface areas of equal area.

3. The concentrations. The reaction rate is usually proportional to the concentrations of the reactants. The usual dependence of the reaction rate on the concentration of the reactants can simply be explained by theorizing that if more molecules or ions of the reactant are in the reaction area, then there is a greater chance that more reactions will occur. This idea is further developed in the collision theory below.

4. The temperature. A temperature increase of 10°C above room temperature usually causes the reaction rate to double or triple. The basis for this generality is that as the temperature increases, the average kinetic energy of the particles involved increases. This means that the particles move faster and have a greater probability of hitting other reactant particles. Because the particles have more energy, they can cause an effective collision, resulting in the chemical reaction that forms the product substance.

5. The catalyst. It is a substance that increases the rate of a chemical reaction without itself undergoing any permanent chemical change. The catalyst provides an alternative pathway by which the reaction can proceed and in which the activation energy is lower. It thus increases the rate at which the reaction comes to completion or equilibrium. Generally, the term is used for a substance that increases reaction rate (a positive catalyst). Some reactions can be slowed down by negative catalysts.

COLLISION THEORY OF REACTION RATES

This theory makes the assumption that, for a reaction to occur, there must be collisions between the reacting species. This means that the rate of reaction depends on two factors: the number of collisions per unit time, and the fraction of these collisions that are successful because enough energy is involved.

There is a definite relationship between the concentrations of the reactants and the number of collisions. Graphically, the reaction of

$$A + B \rightarrow AB$$

can be examined as the concentrations are changed.

Single — One possible collision

Double Concentration — Four possible collisions

Triple Concentration of A and Double Concentration of B — Six possible collisions

This shows that the number of collisions, and consequently the rate of reaction, are proportional to the product of the concentrations. Simply stated, then, the rate of reaction is directly proportional to the concentrations.

ACTIVATION ENERGY

Often a reaction rate may be increased or decreased by affecting the activation energy, that is, the energy necessary to cause a reaction to occur. This is shown graphically below.

A *catalyst* is a substance that is introduced into a reaction to either speed up or slow down the reaction. This is accomplished by changing the amount of activation energy needed. The effect of a catalyst used to speed up a reaction can be shown like this:

REACTION RATE LAW

The relationship between the rate of a reaction and the masses (expressed as concentrations) of the reacting substances is summarized in the *Law of Mass Action*. It states that the rate of a chemical reaction is proportional to the product of the concentrations of the reactants. For a general reaction between A and B, represented by

$$aA + bB \rightarrow \cdots$$

the rate law expression is

$$r \propto [A]^a[B]^b$$

or, inserting a constant of proportionality that mathematically changes the expression to an equality, we have

$$r = k[A]^a[B]^b$$

Here k is called the *specific rate* constant for the reaction at the temperature of the reaction.

The exponents a and b may be added to give the total reaction order. For example:

$$H_2(g) + I_2(g) \rightarrow 2HI(g)$$
$$r = k[H_2]^1[I_2]^1$$

The sum of the exponents is $1 + 1 = 2$, and therefore we have a second-order reaction.

Reaction Mechanism and Rates of Reaction

The beginning of this chapter stated that the reaction rate is *usually* proportional to the concentration of the reactants. This occurs because some reactions do not directly occur between the reactants but may go through intermediate steps to get to the final product. The series of steps by which the reacting particles rearrange themselves to form the products of a chemical reaction is called the *reaction mechanism*. For example:

Step 1	A + B	$\rightarrow$	I_1 (fast)
Step 2	A + I_1	$\rightarrow$	I_2 (slow)
Step 3	C + I_2	$\rightarrow$	D (fast)
Net equation	2A + B + C	$\rightarrow$	D

Notice that the reactions of steps 1 and 3 occur relatively fast compared with the reaction of step 2. Now suppose that we increase the concentration of C. This will make the reaction of step 3 go faster, but it will have little effect on the speed of the overall reaction since step 2 is the rate-determining step. If, however, the concentration of A is increased, then the overall reaction rate would increase because it would speed up step 2. Knowing the reaction mechanism provides the basis for predicting the effect of a concentration change of a reactant on the overall rate of reaction. Another way of determining the effect of concentration changes is to do actual experimentation.

Chapter

9

Review

• **QUESTIONS**

1. List the five factors that affect the rate of a reaction:

   1.   4.

   2.   5.

   3.

2. The addition of a catalyst to a reaction   (a) changes the enthalpy   (b) changes the entropy   (c) changes the nature of the products   (d) changes the activation energy.

3. An increase in concentration (a) is related to the number of collisions directly (b) is related to the number of collisions inversely (c) has no effect on the number of collisions.

4. At the beginning of a reaction, the reaction rate for the reactants is (a) largest, then decreasing (b) largest and remains constant (c) smallest, then increasing (d) smallest and remains constant.

5. The reaction rate law applied to the reaction $aA + bB \rightarrow AB$ gives the expression (a) $r \propto [A]^b[B]^a$ (b) $r \propto [AB]^a[A]^b$ (c) $r \propto [B]^a[AB]^b$ (d) $r \propto [A]^a[B]^b$

• **ANSWERS:**

1. (1) Nature of the reactants   (4) Temperature
   (2) Surface area exposed   (5) Presence of a catalyst
   (3) Concentrations

2. (d)   3. (a)   4. (a)   5. (d)

Terms You Should Know

catalyst   Law of Mass Action
collision theory

Chemical Equilibrium

REVERSIBLE REACTIONS AND EQUILIBRIUM

In some reactions no product is formed to allow the reaction to go to completion, that is, the reactants and products can still interact in both directions. This can be shown as follows:

$$A + B \rightleftharpoons C + D$$

The double arrow indicates that C and D can react to form A and B, while A and B react to form C and D.

The reaction is said to have reached *equilibrium* when the forward reaction rate is equal to the reverse reaction rate. Notice that this is a dynamic condition, NOT a static one, although in appearance the reaction *seems* to have stopped. An example of an equilibrium is a crystal of copper sulfate in a saturated solution of copper sulfate. Although to the observer the crystal seems to remain unchanged, there is actually an equal exchange of crystal material with the copper sulfate in solution. As some solute comes out of solution, an equal amount is going into solution.

To express the rate of reaction in numerical terms, we can use the *Law of Mass Action*, discussed in Chapter 9, which states: The rate of a chemical reaction is proportional to the product of the concentrations of the reacting substances. The concentrations are expressed in moles of gas per liter of volume or moles of solute per liter of solution. Suppose, for example, that 1 mole/liter of gas A_2 (diatomic molecule) is allowed to react with 1 mole/liter of another diatomic gas, B_2, and they form gas AB; let R be the rate for the forward reaction forming AB. The bracketed symbols $[A_2]$ and $[B_2]$ represent the concentrations in moles per liter for these diatomic molecules. Then $A_2 + B_2 \rightarrow 2AB$ has the rate expression

$$R \propto [A_2] \times [B_2]$$

where $\propto$ is the symbol for "proportional to." When $[A_2]$ and $[B_2]$ are both 1 mole/liter, the reaction rate is a certain constant value (k_1) at a fixed temperature.

$$R = k_1 \ (k_1 \text{ is called the rate constant})$$

For any concentrations of A and B, the reaction rate is

$$R = k_1 \times [A_2] \times [B_2]$$

If $[A_2]$ is 3 moles/liter and $[B_2]$ is 2 moles/liter, the equation becomes

$$R = k_1 \times 3 \times 2 = 6\ k_1$$

The reaction rate is six times the value for a 1 mole/liter concentration of both reactants.

At the fixed temperature of the forward reaction, AB molecules are also decomposing. If we designate this reverse reaction as R', then, since

$$2\ AB\ (\text{or AB} + \text{AB}) \rightarrow A_2 + B_2$$

two molecules of AB must decompose to form one molecule of A_2 and one of B_2. Thus the reverse reaction in this equation is proportional to the square of the molecular concentration of AB.

$$R' \propto\ [AB] \times [AB]$$
$$\text{or } R' \propto\ [AB]^2$$
$$\text{and } R' \propto\ k_2 \times [AB]^2$$

where k_2 represents the rate of decomposition of AB at the fixed temperature. Both reactions can be shown in this manner:

$$A_2 + B_2 \rightleftharpoons 2\ AB \quad \text{(note double arrows)}$$

When the first reaction begins to produce AB, some AB is available for the reverse reaction. If the initial condition is only the presence of A_2 and B_2 gases, then the forward reaction will occur rapidly to produce AB. As the concentration of AB increases, the reverse reaction will increase. At the same time, the concentrations of A_2 and B_2 will be decreasing and consequently the forward reaction rate will decrease. Eventually the two rates will be equal, that is, $R = R'$. At this point, equilibrium has been established, and

$$k_1[A_2] \times [B_2] = k_2[AB]^2$$

or

$$\frac{k_1}{k_2} = \frac{[AB]^2}{[A_2] \times [B_2]} = K_e$$

The convention is that k_1 (forward reaction) is placed over k_2 (reverse reaction) to get this expression. Then k_1/k_2 can be replaced by K_e, which is called the *equilibrium constant* for this reaction under the particular conditions.

In another general example, like this:

$$aA + bB \rightleftharpoons cC + dD$$

the reaction rates can be expressed as

$$R = k_1[A]^a \times [B]^b$$
$$R' = k_2[C]^c \times [D]^d$$

Note that the values of k_1 and k_2 are different, but that each is a constant for the conditions of the reaction. At the start of the reaction, [A] and [B] will be at their greatest values, and R will be large; [C], [D], and R' will be zero. Gradually R will decrease and R' will become equal. At this point the reverse reaction is forming the original reactants just as rapidly as they are being used up by the forward reaction. Therefore no further change in R, R', or any of the concentrations will occur.

If we set R' equal to R, we have:

$$k_2 \times [C]^c \times [D]^d = k_1 \times [A]^a \times [B]^b$$

or

$$\frac{[C]^c \times [D]^d}{[A]^a \times [B]^b} = \frac{k_1}{k_2} = K_e$$

We see that, for the given reaction and the given conditions, K_e is a constant, called the *equilibrium constant*. If K_e is large, this means that equilibrium does not occur until the concentrations of the original reactants are small and those of the products large. A small value of K_e means that equilibrium occurs almost at once and relatively little product is produced.

The equilibrium constant, K_e, has been determined experimentally for many reactions, and the values are printed in chemical handbooks. (See Tables Ⓙ and Ⓚ in the Reference Tables section.)

Suppose we find the K_e for reacting H_2 and I_2 at 490°C to be equal to 45.9. Then the equilibrium constant for this reaction

$$H_2 + I_2 \rightleftharpoons 2HI \text{ at } 490°C$$

is

$$K_e = \frac{[HI]^2}{[H_2][I_2]} = 45.9$$

• SAMPLE PROBLEM:

Three moles of H_2 and 3 moles of I_2 are introduced into a 1-liter box at a temperature of 490°C. Find the concentration of each substance in the box when equilibrium is established.

Initial conditions:

$[H_2]$ = 3 moles/liter
$[I_2]$ = 3 moles/liter
$[HI]$ = 0 mole/liter

The reaction proceeds to equilibrium and

$$K_e = \frac{[HI]^2}{[H_2][I_2]} = 45.9$$

At equilibrium, then,

$[H_2]$= $(3 - x)$ moles/liter (where x is the number of moles of H_2 that are in the form of HI at equilibrium)

$[I_2]$ = $(3 - x)$ moles/liter (the same x is used since 1 mole of H_2 requires 1 mole of I_2 to react to form 2 moles of HI)

$[HI]$ = $2x$ moles/liter

so

$$K_e = \frac{(2x)^2}{(3-x)(3-x)} = 45.9$$

If

$$\frac{(2x)^2}{(3-x)^2} = 45.9$$

then taking the square root of each side gives

$$\frac{2x}{3-x} = 6.77$$

Solving for x:

$$x = 2.32$$

Substituting this x value into the concentration expressions at equilibrium we have

$$[H_2] = (3 - x) = 0.68 \text{ mole/liter}$$
$$[I_2] = (3 - x) = 0.68 \text{ mole/liter}$$
$$[HI] = 2x = 4.64 \text{ moles/liter}$$

The crucial step in this type of problem is setting up your concentration expressions from your knowledge of the equation. Suppose that this problem had been:

Find the concentrations at equilibrium for the same conditions as in the preceding example except that only 2 moles of HI are injected into the box.

$$[H_2] = 0 \text{ mole/liter}$$
$$[I_2] = 0 \text{ mole/liter}$$
$$[HI] = 2 \text{ moles/liter}$$

At equilibrium

$$[HI] = (2 - x) \text{ moles/liter}$$

(For every mole of HI that decomposes, only 0.5 mole of H_2 and 0.5 mole of I_2 are formed.)

$$[H_2] = 0.5x$$
$$[I_2] = 0.5x$$
$$K_e = \frac{(2-x)^2}{(x/2)^2} = 45.9$$

Solving for x gives:

$$x = 0.456$$

Then, substituting into the equilibrium conditions,

$$[HI] = 2 - x = 1.54 \text{ moles/liter}$$
$$[I_2] = \tfrac{1}{2}x = 0.228 \text{ mole/liter}$$
$$[H_2] = \tfrac{1}{2}x = 0.228 \text{ mole/liter}$$

LE CHÂTELIER'S PRINCIPLE

A general law, *Le Châtelier's Principle*, can be used to explain the results of applying any change of condition (stress) on a system in equilibrium. It states that if a stress is placed upon a system in equilibrium, the equilibrium is displaced in the direction which counter-acts the effect of the stress. An increase in concentration of a substance favors the reaction that uses up that substance and lowers its concentration. A rise in temperature favors the reaction that absorbs heat and so tends to lower the temperature. These ideas are further developed below.

EFFECTS OF CHANGING CONDITIONS

Effect of Changing the Concentrations

When a system at equilibrium is disturbed by adding or removing one of the substances (thus changing its concentration), all the concentrations will change until a new equilib-rium point is reached with the same value of K_e.

If the concentration of a reactant in the forward action is increased, the equilibrium is displaced to the right, favoring the forward reaction. If the concentration of a reactant in the reverse reaction is increased, the equilibrium is displaced to the left. Decreases in concentration will produce effects opposite to those produced by increases.

Effect of Temperature on Equilibrium

If the temperature of a given equilibrium reaction is changed, the reaction will shift to a new equilibrium point. If the temperature of a system in equilibrium is raised, the equilibrium is shifted in the direction that absorbs heat. Note that the shift in equilibrium as a result of temperature change is actually a change in the value of the equilibrium constant. This is different from the effect of changing the concentration of a reactant; when concentrations are changed, the equilibrium shifts to a condition that maintains the same equilibrium constant.

Effect of Pressure on Equilibrium

A change in pressure affects only equilibria in which a gas or gases are reactants or products. Le Châtelier's Law can be used to predict the direction of displacement. If it is assumed that the total space in which the reaction occurs is constant, the pressure will depend on the total number of molecules in that space. An increase in the number of molecules will increase pressure; a decrease in the number of molecules will decrease pressure. If the pressure is increased, the reaction that will be favored is the one that will lower the pressure, that is, decrease the number of molecules.

An example of the application of these principles is the Haber process of making ammo-nia. The reaction is

$$N_2 + 3H_2 \rightleftharpoons 2NH_3 + heat \text{ (at equilibrium)}$$

If the concentrations of the nitrogen and hydrogen are increased, the forward reaction is increased. At the same time, if the ammonia produced is removed by dissolving it into water, the forward reaction is again favored.

Since the reaction is exothermic, the addition of heat must be considered with care. Increasing the temperature causes an increase in molecular motion and collisions, thus allowing the product to form more readily. At the same time, the equilibrium equation shows that the reverse reaction is favored by the increased temperature, so a compromise temperature of about 500°C is used to get the best yield.

An increase in pressure will cause the forward reaction to be favored since the equation shows that four molecules of reactants are forming two molecules of products. This effect tends to reduce the increase in pressure by the formation of more ammonia.

EQUILIBRIA IN HETEROGENEOUS SYSTEMS

The examples so far have been of systems made up of only gaseous substances. The expression of the K_e of systems is changed with the presence of other phases.

Equilibrium Constant for Systems Involving Solids

If the experimental data for this reaction are studied:

$$CaCO_3(s) \rightleftharpoons CaO(s) + CO_2(g)$$

it is found that at a given temperature an equilibrium is established in which the concentration of CO_2 is constant. It is also true that the concentrations of the solids have no effect on the CO_2 concentration as long as both solids are present. Therefore the K_e, which would conventionally be written like this:

$$K_e = \frac{[CaO][CO_2]}{[CaCO_3]}$$

can be modified by incorporating the concentrations of the two solids. This can be done since the concentration of solids is fixed by the density. The K_e becomes a new constant, K:

$$K = [CO_2]$$

Any heterogeneous reaction involving gases does not include the concentrations of pure solids. As another example, K for the reaction

$$NH_4Cl(s) \rightleftharpoons NH_3(g) + HCl(g)$$

is

$$K = [NH_3][HCl]$$

Ionization Constants

When a substance that does not ionize completely is placed in solution, an equilibrium is reached between the substance and its ions. The mass action expression can be used to derive an equilibrium constant, called the *ionization constant*, for this condition. For example, an acetic acid solution ionizing is shown as

$$HC_2H_3O_2 + H_2O \rightleftharpoons H_3O^+ + C_2H_3O_2^-$$

$$K = \frac{[H_3O^+][C_2H_3O_2^-]}{[HC_2H_3O_2][H_2O]}$$

The concentration of water in moles/liter is found by dividing the mass of 1 liter of water (which is 1,000 grams at 4°C) by its gram-molecular mass, 18 g, giving H_2O a value of 55.6 moles/liter. Since this number is so large compared with the other numbers involved in the equilibrium constant, it is practically constant and is incorporated into a new equilibrium constant, designated as K_i. Then the new expression is

$$K_i = \frac{[H_3O^+][C_2H_3O_2^-]}{[HC_2H_3O_2]}$$

Ionization constants have been found experimentally for many substances and listed in chemical tables. The ionization constants of ammonia and acetic acid are about 1.8×10^{-5}. For boric acid $K_i = 5.8 \times 10^{-10}$ and for carbonic acid $K_i = 4.3 \times 10^{-7}$.

If the concentrations of the ions present in the solution of a weak electrolyte are known, the value of the ionization constant can be calculated. Also, if the value of K_i is known, the concentrations of the ions can be calculated.

A small value for K_i means that the concentration of the un-ionized molecule must be relatively large compared with the ion concentrations. Conversely, a large value for K_i means that the concentrations of ions are relatively high. Therefore the smaller the ionization constant of an acid, the weaker the acid. Thus, for the three acids referred to above, the ionization constants show that the weakest of these is boric acid, and the strongest, acetic acid. It should be remembered that, in all cases where ionization constants are used, the electrolytes must be weak in order to be involved in ionic equilibria.

Ionization Constant of Water

Since water is a very weak electrolyte, its ionization constant can be expressed as follows:

$$2H_2O \rightleftharpoons H_3O^+ + OH^-$$

(Equilibrium constant) $K = \dfrac{[H_3O^+][OH^-]}{[H_2O]^2}$ (since $[H_2O]^2$ remains relatively constant, it is incorporated into K_w)

(Ionization constant) $K_w = [H_3O^+][OH^-] = 1 \times 10^{-14}$ at 25°C

From this expression, we see that for distilled water $[H_3O^+] = [OH^-] = 1 \times 10^{-7}$. Therefore the pH, which is $-\log [H_3O^+]$, is

$$pH = -\log [1 \times 10^{-7}]$$
$$pH = -[-7] = 7 \text{ for a neutral solution}$$

The pH range of 1 to 6 is acid, and the pH range of 8 to 14 is basic.

• SAMPLE PROBLEM:

(This sample incorporates the entire discussion of ionization constants, including finding the pH.)

Calculate (a) the $[H_3O^+]$, (b) the pH, and (c) the percentage dissociation for 0.100 M acetic acid at 25°C. The symbol K_a is used for the ionization of acids. K_a for $HC_2H_3O_2$ is 1.8×10^{-5}.

(a) For this reaction

$$H_2O(\ell) + HC_2H_3O_2(\ell) \rightleftharpoons H_3O^+(aq) + C_2H_3O_2^-(aq)$$

and

$$K_a = \frac{[H_3O^+][C_2H_3O_2^-]}{[HC_2H_3O_2]} = 1.8 \times 10^{-5}$$

Let x = number of moles/liter of $HC_2H_3O_2$ that dissociate and reach equilibrium. Then

$$[H_3O^+] = x, \ [C_2H_3O_2^-] = x, \ [HC_2H_3O_2] = 0.1 - x$$

Substituting in the expression for K_a gives

$$K_a = 1.8 \times 10^{-5} = \frac{(x)(x)}{0.10 - x}$$

Since weak acids, like acetic, at concentrations of 0.01 M or greater dissociate very little, the equilibrium concentration of the acid is very nearly equal to the original concentration, that is,

$$0.10 - x \cong 0.10$$

Therefore, the expression can be changed to

$$1.8 \times 10^{-5} = \frac{(x)(x)}{0.10}$$

$$x^2 = 1.8 \times 10^{-6}$$

$$x = 1.3 \times 10^{-3} = [H_3O^+]$$

(b) Substituting this result in the pH expression gives

$$pH = -\log [H_3O^+] = -\log [1.3 \times 10^{-3}]$$
$$pH = 3 - \log 1.3$$
$$pH = 2.9$$

(c) The percentage of dissociation of the original acid may be expressed as

$$\% \text{ dissociation} = \frac{\text{moles/liter that dissociate}}{\text{original concentration}} (100)$$

$$\% \text{ dissociation} \times \frac{1.3 \times 10^{-3}}{1.0 \times 10^{-1}} (100) = 1.3\%$$

Solubility Products

A saturated solution of a substance has been defined as an equilibrium condition between the solute and its ions. For example:

$$AgCl \rightleftharpoons Ag^+ + Cl^-$$

The equilibrium constant would be:

$$K = \frac{[Ag^+][Cl^-]}{[AgCl]}$$

Since the concentration of the solute remains constant for that temperature, the [AgCl] is incorporated into the K to give the K_{sp}, called the *solubility constant*:

$$K_{sp} = [Ag^+][Cl^-] = 1.2 \times 10^{-10} \text{ at } 25°C$$

This setup can be used to solve problems in which the ionic concentrations are given and the K_{sp} is to be found or the K_{sp} is given and the ionic concentrations are to be determined.

• SAMPLE PROBLEM: Finding the K_{sp}.

By experimentation it is found that a saturated solution of $BaSO_4$ at 25°C contains 3.9×10^{-5} mole/liter of Ba^{2+} ions. Find the K_{sp} of this salt.

Since $BaSO_4$ ionizes into equal numbers of Ba^{2+} and SO_4^{2-}, the barium ion concentration will equal the sulfate ion concentration. So the solution is

$$BaSO_4 \rightleftharpoons Ba^{2+} + SO_4^{2-}$$
$$K_{sp} = [Ba^{2+}][SO_4^{2-}]$$

Therefore

$$K_{sp} = (3.9 \times 10^{-5})(3.9 \times 10^{-5}) = 1.5 \times 10^{-9}$$

• **ANOTHER TYPE PROBLEM: Finding the solubility.**

If the K_{sp} of radium sulfate, $RaSO_4$, is 4×10^{-11}, calculate its solubility in pure water. Let x = moles of $RaSO_4$ that dissolve per liter of water. Then, in the saturated solution,

$$[Ra^{2+}] = x \text{ moles/liter}$$
$$[SO_4^{2-}] = x \text{ moles/liter}$$
$$RaSO_4(s) \rightleftharpoons Ra^{2+} + SO_4^{2-}$$
$$[Ra^{2+}][SO_4^{2-}] = K_{sp} = 4 \times 10^{-11}$$

Let $x = [Ra^{2+}]$ and $[SO_4^{2-}]$. Then

$$(x)(x) = 4 \times 10^{-11} = 40 \times 10^{-12}$$
$$x = 6 \times 10^{-6} \text{ mole/liter}$$

Thus the solubility of $RaSO_4$ is 6×10^{-6} mole per liter of water, giving a solution 6×10^{-6} M in Ra^{2+} and 6×10^{-6} M in SO_4^{2-}.

• **ANOTHER TYPE PROBLEM: Predicting the formation of a precipitate.**

In some cases, the solubility products of solutions can be used to predict the formation of a precipitate.

Suppose we have two solutions. One solution contains 1.00×10^{-3} mole of silver nitrate, $AgNO_3$, per liter. The other solution contains 1.00×10^{-2} mole of sodium chloride, NaCl, per liter. If 1 liter of each of these solutions are mixed to make a 2-liter mixture, will a precipitate of AgCl form?

In the $AgNO_3$ solution, the concentrations are:

$$[Ag^+] = 1.00 \times 10^{-3} \text{ mol/L and } [NO_3^-] = 1.00 \times 10^{-3} \text{ mol/L}$$

In the NaCl solution, the concentrations are:

$$[Na^+] = 1.00 \times 10^{-2} \text{ mol/L and } [Cl^-] = 1.00 \times 10^{-2} \text{ mol/L}$$

When 1 liter of each of these solutions are mixed to form a total volume of 2 liters, the concentrations will be halved.

In the mixture then, the initial concentrations will be:

$$[Ag^+] = 0.50 \times 10^{-3} \quad \text{or} \quad 5.0 \times 10^{-4} \text{ mol/L}$$
$$[Cl^-] = 0.50 \times 10^{-2} \quad \text{or} \quad 5.0 \times 10^{-3} \text{ mol/L}$$

In the K_{sp} of AgCl,

$$[Ag^+][Cl^-] = [5.0 \times 10^{-4}][5.0 \times 10^{-3}]$$
$$[Ag^+][Cl^-] = 25 \times 10^{-7} \quad \text{or} \quad 2.5 \times 10^{-6}$$

This is far greater than 1.7×10^{-10}, which is the K_{sp} of AgCl. These concentrations cannot exist, and Ag^+ and Cl^- will combine to form solid AgCl precipitate. Only enough Ag^+ ions and Cl^- ions will remain to make the product of the respective ion concentrations equal 1.7×10^{-10}.

COMMON ION EFFECT

When a reaction has reached equilibrium, and an outside source adds more of one of the ions that is already in solution, the result is to cause the reverse reaction to occur at a faster rate and reestablish the equilibrium. This is called the *common ion effect*. For example, in this equilibrium reaction:

$$NaCl(s) \rightleftharpoons Na^+ + Cl^-$$

the addition of concentrated HCl (12 M) adds H^+ and Cl^- both at a concentration of 12 M. This increases the concentration of the Cl^- and disturbs the equilibrium. The reaction will shift to the left and cause some solid NaCl to come out of solution.

The "common" ion is the one already present in an equilibrium before a substance is added that increases the concentration of that ion. The effect is to reverse the solution reaction and to decrease the solubility of the original substance, as shown in the above example.

FACTORS RELATED TO THE MAGNITUDE OF K_e

Relation of Minimum Energy (Enthalpy) to Maximum Disorder (Entropy)

Some reactions are said to go to completion because the equilibrium condition is achieved when practically all the reactants have been converted to products. At the other extreme, some reactions reach equilibrium immediately with very little product being formed. These two examples are representative of very large K_e values and very small K_e values, respectively. There are essentially two driving forces that control the extent of a reaction and determine when equilibrium will be established. These are the drive to the lowest heat content, or *enthalpy*, and the drive to the greatest randomness or disorder, which is called *entropy*. Reactions with negative ΔH's (enthalpy or heat content) are exothermic, and reactions with positive ΔS's (entropy or randomness) are proceeding to greater randomness.

The Second Law of Thermodynamics states that the entropy of the universe increases for any spontaneous process. This means that the entropy of a system may increase or decrease but that, if it decreases, then the entropy of the surroundings must increase to a greater extent so that the overall change in the universe is positive. In other words,

$$\boxed{\Delta S_{universe} = \Delta S_{system} + \Delta S_{surroundings}}$$

The following is a list of conditions in which ΔS is positive for the system:

1. When a gas is formed from a solid, for example,

    $$CaCO_3(s) \rightarrow CaO(s) + CO_2(g).$$

2. When a gas is evolved from a solution, for example,

    $$Zn(s) + 2H^+(aq) \rightarrow H_2(g) + Zn^{2+}(aq).$$

3. When the number of moles of gaseous product exceeds the moles of gaseous reactant, for example,

    $$2C_2H_6(g) + 7O_2(g) \rightarrow 4CO_2(g) + 6H_2O(g).$$

4. When crystals dissolve in water, for example,

    $$NaCl(s) \rightarrow Na^+(aq) + Cl^-(aq).$$

Looking at specific examples, we find that in some cases endothermic reactions occur when the products provide greater randomness or positive entropy. This reaction is an example:

$$CaCO_3(s) \rightleftharpoons CaO(s) + CO_2(g)$$

The production of the gas and thus greater entropy might be expected to take this reaction almost to completion. However, this is not true because another force is hampering this reaction. It is the absorption of energy, and thus the increase in enthalpy, as the $CaCO_3$ is heated.

The equilibrium condition, then, at a particular temperature, is a compromise between the increase in entropy and the increase in enthalpy of the system.

The Haber process of making ammonia is another example of this compromise of driving forces that affect the establishment of an equilibrium. In this reaction

$$N_2(g) + 3H_2(g) \rightleftharpoons 2NH_3(g) + heat$$

the forward reaction to reach the lowest heat content and thus release energy cannot go to completion because the force to maximum randomness is driving the reverse reaction.

Change in Free Energy of a System

These factors can be combined in an equation that summarizes the change of *free energy* in a system. This is designated as ΔG. The relationship is

$$\Delta G = \Delta H - T \Delta S \qquad (T \text{ is temperature in kelvins})$$

and is called the *Gibbs free energy equation.*

The sign of ΔG can be used to predict the spontaneity of a reaction at constant temperature and pressure. If ΔG is negative, the reaction is (probably) spontaneous; if ΔG is positive, the reaction is improbable; and if ΔG is 0, the system is at equilibrium and there is no net reaction.

The ways in which the factors in the equation affect ΔG are shown in this table:

ΔH	ΔS	Will It Happen?	Comment
Exothermic (−)	+	Yes	No exceptions
Exothermic (−)	−	Probably	At low temperature
Endothermic (+)	+	Probably	At high temperature
Endothermic (+)	−	No	No exceptions

This drive to achieve a minimum of free energy may be interpreted as the driving force of a chemical reaction.

Chapter

10

Review

• QUESTIONS

1. For the reaction $A + B \rightleftharpoons C + D$, the equilibrium constant can be expressed as:

(a) $K_e = \dfrac{[A][B]}{[C][D]}$ (b) $K_e = \dfrac{[C][B]}{[A][D]}$ (c) $K_e = \dfrac{[C][D]}{[A][B]}$ (d) $K_e = \dfrac{C \cdot D}{A \cdot B}$

2. The concentrations in an expression of the equilibrium constant are given in
(a) moles/mL   (b) g/L   (c) gram-equivalents/L   (d) moles/L.

3. In the equilibrium expression for the reaction of

$$BaSO_4 \rightleftharpoons Ba^{2+} + SO_4^{2-}$$

K_{sp} is equal to (a) $[Ba^{2+}][SO_4^{2-}]$ (b) $\dfrac{[Ba^{2+}][SO_4^{2-}]}{BaSO_4}$

(c) $\dfrac{[Ba^{2+}][SO_4^{2-}]}{[BaSO_4]}$ (d) $\dfrac{[BaSO_4]}{[Ba^{2+}][SO_4^{2-}]}$.

4. The K_w of water is equal to (a) 1×10^{-7} (b) 1×10^{-17} (c) 1×10^{-14} (d) 1×10^{-1}.

5. The pH of a solution that has a hydrogen ion concentration of 1×10^{-4} mole/liter is (a) 4   (b) –4   (c) 10   (d) –10.

6. The pH of a solution that has a hydroxide ion concentration of 1×10^{-4} mole/liter is (a) 4   (b) –4   (c) 10   (d) –10.

7. A small value for K_e, the equilibrium constant, indicates that (a) the concentration of the un-ionized molecules must be relatively small compared with the ion concentrations   (b) the concentration of the ionized molecules must be larger than the ion concentrations   (c) the substance ionizes to a large degree   (d) the concentration of the un-ionized molecules must be relatively large compared with the ion concentrations.

8. In the Haber process for making ammonia, an increase in pressure favors (a) the forward reaction   (b) the reverse reaction   (c) neither reaction.

9. A change in which of these conditions will change the K of an equilibrium? (a) temperature (b) pressure (c) concentration of reactants (d) concentration of products.

10. If $Ca(OH)_2$ is dissolved in a solution of NaOH, its solubility, compared with that in pure water, is (a) increased   (b) decreased   (c) unaffected.

• ANSWERS:

1. (c)	5. (a)	8. (a)
2. (d)	6. (c)	9. (a)
3. (a)	7. (d)	10. (b)
4. (c)		

Terms You Should Know

common ion effect	equilibrium	ionization constant
enthalpy	equilibrium constant	solubility product constant
entropy	Gibbs free energy equation	

Le Châtelier's Principle

Acids, Bases, and Salts

DEFINITIONS AND PROPERTIES

Acids

There are some characteristic properties by which an acid may be defined. The most important are:

1. *Water solutions of acids conduct electricity.* This conduction depends on their degree of ionization. A few acids ionize almost completely, while others ionize only to a slight degree. The table below indicates some common acids and their degrees of ionization.

DEGREES OF IONIZATION OF COMMON ACIDS		
Completely or Nearly Completely Ionized	*Moderately Ionized*	*Slightly Ionized*
Nitric	Oxalic	Hydrofluoric
Hydrochloric	Phosphoric	Acetic
Sulfuric	Sulfurous	Carbonic
Hydriodic		Hydrosulfuric
Hydrobromic		(Most others)

2. *Acids will react with metals that are more active than hydrogen ions (see chart on page 92) to liberate hydrogen.* (Some acids are also strong oxidizing agents and will not release hydrogen. Somewhat concentrated nitric acid is such an acid.)

3. *Acids have the ability to change the color of indicators.* Some common indicators are litmus and phenolphthalein. Litmus is a dyestuff obtained from plant life. When litmus is added to an acidic solution, or paper impregnated with litmus is dipped into an acid, the neutral purple color changes to pink-red. Phenolphthalein is red in a basic solution and becomes colorless in a neutral or acid solution.

4. *Acids react with bases so that the properties of both are lost to form water and a salt.* This is called *neutralization.* The general equation is:

    Acid + Base → Salt + Water

An example is:

$$Mg(OH)_2 + H_2SO_4 \rightarrow MgSO_4 + 2H_2O$$

5. *If an acid is known to be a weak solution, you might taste it and note the sour taste.*

6. *Acids react with carbonates to release carbon dioxide.* An example:

$$CaCO_3 + 2HCl \rightarrow CaCl_2 + H_2CO_3 \text{ (unstable and decomposes)}$$
$$\rightarrow H_2O + CO_2(g)$$

The most common theory used in first-year chemistry is the *Arrhenius Theory*, which states that an acid is a substance that yields hydrogen ions in an aqueous solution. Although we speak of these hydrogen ions in the solution, they are really not separate ions but become attached to the oxygen of the polar water molecule to form the H_3O^+ ion (the hydronium ion). So it is really this hydronium ion we are concerned with in an acid solution.

The general reaction for the dissociation of an acid, HX, is commonly written:

$$HX \rightleftharpoons H^+ + X^-$$

To show the formation of the hydronium ion, H_3O^+, the complete equation is:

$$HX + H_2O \rightleftharpoons H_3O^+ + X^-$$

A list of common acids and their formulas is given in Chapter 4, Table 7, with an explanation of the naming procedures for acids preceding Table 7.

Bases

Bases may also be defined by some operational definitions that are based on experimental observations. Some of the important ones are as follows:

1. *Bases are conductors of electricity in an aqueous solution.* Their degrees of conduction depend on their degrees of ionization. See the table of common bases below and their degrees of ionization.

DEGREES OF IONIZATION OF COMMON BASES	
Completely or Nearly Completely Ionized	*Slightly Ionized*
Potassium hydroxide	Ammonium hydroxide
Sodium hydroxide	(All others)
Barium hydroxide	
Strontium hydroxide	
Calcium hydroxide	

2. *Bases cause a color change in indicators.* Litmus changes from red to blue in a basic solution, and phenolphthalein turns pink from its colorless state.

3. *Bases react with acids to neutralize each other and form a salt and water.*

4. *Bases react with fats to form a class of compounds called soaps.* The earlier generations used this method to make their own soap.

5. *Aqueous solutions of bases feel slippery and the stronger bases are very caustic to the skin.*

The Arrhenius Theory defines a base as a substance that yields hydroxide ions (OH^-) in an aqueous solution.

Some of our common bases have common names. The list below gives some of these:

Sodium hydroxide   = lye, caustic soda
Potassium hydroxide   = caustic potash
Calcium hydroxide   = slaked lime, hydrated lime, limewater
Ammonium hydroxide = ammonia water, household ammonia

Much of the sodium hydroxide produced today comes from the Hooker cell electrolysis apparatus. We have already discussed the electrolysis apparatus for the decomposition of water. When an electric current is passed through a saltwater solution, hydrogen, chlorine, and sodium hydroxide are the products. The formula for this equation is:

$$2NaCl + 2HOH \xrightarrow[\text{energy}]{\text{electrical}} H_2(g) + Cl_2(g) + 2NaOH$$

Broader Acid-Base Theories

Besides the common Arrhenius Theory of acids and bases discussed for aqueous solutions, two other theories, the Brønsted Theory and the Lewis Theory, are widely used.

The Brønsted Theory (1923) defines acids as proton donors and bases as proton acceptors. This definition agrees with the aqueous solution definition of an acid giving up hydrogen ions in solution, but goes beyond to other cases as well.

An example of this is dry HCl gas reacting with ammonia gas to form the white solid NH_4Cl.

$$HCl(g) + NH_3(g) \rightarrow NH_4^+ + Cl^-(s)$$

The HCl is the proton donor or acid and the ammonia is a Brønsted base that accepts the proton.

Conjugate Acids and Bases

In an acid-base reaction, the original acid gives up its proton to become a conjugate base. In other words, after losing its proton, the original acid is capable of gaining a proton, thus qualifying as a base. The original base accepts a proton, so it now is classified as a conjugate acid since it can release this newly acquired proton and thus behave like an acid.

Some examples of this are given below:

The Lewis Theory (1916) of acids and bases is defined in terms of the electron-pair concept. It is probably the most generally useful concept of acids and bases. According to the Lewis definition, an acid is an electron-pair acceptor; and a base is an electron-pair donor. An example of this is the formation of ammonium ions from ammonia gas and hydrogen ions.

Notice that the hydrogen ion is in fact accepting the electron pair of the ammonia, so it is a Lewis acid. The ammonia is donating its electron pair, so it is a Lewis base.

Acid Concentration Expressed as pH

Frequently, acid and base concentrations are expressed by means of the pH system. The pH can be defined as $-\log [H^+]$, where $[H^+]$ is the concentration of hydrogen ions expressed in moles per liter. The logarithm is the exponent of 10 when the number is written in the base 10. For example:

$$100 = 10^2 \text{ so logarithm of } 100, \text{ base } 10 = 2$$
$$10,000 = 10^4 \text{ so logarithm of } 10,000, \text{ base } 10 = 4$$
$$0.01 = 10^{-2} \text{ so logarithm of } 0.01, \text{ base } 10 = -2$$

The logarithms of more complex numbers can be found in a logarithm table. An example of a pH problem is:

Find the pH of a 0.1 molar solution of HCl.

1st step. Since HCl ionizes almost completely into H^+ and Cl^-, $[H^+] = 0.1$ mole/liter.
2nd step. By definition
$$pH = -\log [H^+]$$
So
$$pH = -\log [10^{-1}]$$
3rd step. The logarithm of 10^{-1} is -1
So
$$pH = -(-1)$$
4th step. The pH then is $= 1$.

Since water has a normal H^+ concentration of 10^{-7} mole/liter because of the slight ionization of water molecules, the water pH is 7 when it is neither acid nor base. The normal pH range is from 1 to 14.

	Acid		Neutral		Base	
1	$\longleftrightarrow$		7	$\longleftrightarrow$		14

The pOH is the negative logarithm of the hydroxide ion concentration:

$$pOH = -\log [OH^-]$$

If the concentration of the hydroxide ion is 10^{-9} M, then the pOH of the solution is $+9$. From the equation

$$[H^+][OH^-] = 1.0 \times 10^{-14} \text{ at } 25°C$$

the following relationship can be derived:

$$pH + pOH = 14.00$$

In other words, the sum of the pH and pOH of an aqueous solution at 25°C must always equal 14.00. For example, if the pOH of a solution is 9.00, then its pH must be 5.00.

• SAMPLE PROBLEM:

What is the pOH of a solution whose pH is 3.0?
Substituting the 3.0 for pH in the expression

$$pH + pOH = 14.0$$
$$3.0 + pOH = 14.0$$
$$pOH = 14.0 - 3.0$$
$$pOH = 11.0$$

INDICATORS

Some indicators can be used to determine pH because of their color changes somewhere along this pH scale. Some common indicators and their respective color changes are given below.

Indicator	pH Range of Color Change	Color below Lower pH	Color above Higher pH
Methyl orange	3.1–4.4	Red	Yellow
Bromthymol blue	6.0–7.6	Yellow	Blue
Litmus	4.5–8.3	Red	Blue
Phenolphthalein	8.3–10.0	Colorless	Red

An example of how to read this chart is: At pH values below 4.5, litmus is red; above 8.3, it is blue. Between these values, it is a mixture of the two colors.

VOLUMETRIC ANALYSIS

Knowledge of the concentrations of solutions and the reactions they take part in can be used to determine the concentrations of "unknown" solutions or solids. The use of volume measurement in solving these problems is called *volumetric analysis.*

A common example of a volumetric analysis uses acid-base reactions. If you are given a base of known concentration, let us say 0.10 M NaOH, and you want to determine the concentration of an HCl solution, you could *titrate* the solutions in the following manner.

First, introduce a measured quantity, 25.0 mL, of the NaOH into a flask by using a pipet or burette. Next, introduce 2 drops of a suitable indicator. Since NaOH and HCl

Burette Setup for Titration

are considered a strong base and a strong acid, respectively, an indicator that changes color in the middle pH range would be appropriate. Litmus solution would be one choice. It is blue in basic solution but changes to red when the solution becomes acidic. Slowly introduce the HCl until the color change occurs. This is called the *end point.*

Suppose 21.5 mL was needed to produce the color change; the reaction that occured was

$$H^+(aq) + OH^-(aq) \rightarrow H_2O$$

until all the OH^- were neutralized, and then the excess H^+ caused the litmus paper to change color. To solve the question of the concentration of the NaOH this equation is used:

$$M_{acid} \times V_{acid} = M_{base} \times V_{base}$$

Substituting the known amounts in this equation gives

$$xM_{acid} \times 21.5 \text{ mL} = 0.1M \times 25.0 \text{ mL}$$
$$x = 0.116 \text{ M}$$

In choosing an indicator for a titration, we need to consider if the solution formed when the end point is reached has a pH of 7. Depending on the type of acid and base used, the resulting hydrolysis of the salt formed may cause the solution to be slightly acidic, slightly basic, or neutral. If titrating a strong acid and strong base, the end point will be at pH 7, and practically any indicator can be used. This is because adding one drop of either reagent will change the pH at the end point by about six units. For titrations of strong acids and weak bases, we need an indicator, such as methyl orange, that changes color between 3.1 and 4.4 in the acid region. When titrating a weak acid and strong base, we should use an indicator that changes in the basic range. Phenolphthalein is the suitable choice for this type of titration since it changes color in the pH 8.3 to 10.0 range.

• SAMPLE PROBLEMS:

1. Find the concentration of acetic acid in vinegar if 21.6 mL of 0.20 M NaOH is needed to titrate a 25-mL sample of the vinegar.

   Solution
   Using the equation $M_{acid} \times V_{acid} = M_{base} \times V_{base}$ we have

   $$xM \times 25 \text{ mL} = 0.2 \times 21.6 \text{ mL}$$
   $$x = 0.173 \text{ M}_{acid}$$

   Another type of titration problem involves a solid and a titrated solution.

2. A solid mixture contains NaOH and NaCl. If a 0.100-g sample of this mixture required 10 mL of 0.100 M HCl to titrate the sample to its end point, what is the percent of NaOH in the sample?

   Solution
   Since

   $$Molarity = \frac{No. \text{ of moles}}{liter \text{ of solution}}$$

   then

   $$M \times V = \frac{No. \text{ of moles}}{liter \text{ of solution}} \times liter \text{ of solution} = No. \text{ of moles}$$

Substituting the HCl information in the equation, we have

$$\frac{0.100 \text{ mole}}{\text{liter of solution}} \times 0.010 \text{ liter of solution} = 0.001 \text{ mole}$$
(Note : this is 10mL expressed in liters)

Since 1 mole of HCl neutralizes 1 mole of NaOH, 0.001 mole of NaOH must be present in the mixture.

1 mole NaOH = 40.0 g

So

0.001 mole × 40.0 g/mole = 0.04 g NaOH

Therefore 0.04 g of NaOH was in the 0.1-g sample of the solid mixture. The percent is 0.04 g/0.1 g × 100 = 40%.

In the explanations given to this point, the reactions that took place were between monoprotic acids (single hydrogen ions) and monobasic bases (one hydroxide ion per base). This means that each mole of acid had 1 mole of hydrogen ions available, and each mole of base had 1 mole of hydroxide ions available, to interact in the following reaction until the end point was reached:

$$H^+(aq) + OH^-(aq) \rightarrow 2H_2O$$

This is not always the case, however, and it is important to know how to deal with acids and bases that have more than one hydrogen ion and more than one hydroxide ion per formula. The following is an example of such a problem.

If 20.0 mL of an aqueous solution of calcium hydroxide, $Ca(OH)_2$, is used in a titration, and an appropriate indicator is added to show the neutralization point (end point), the few drops of indicator that are added can be ignored in the volume considerations. Therefore, if 25.0 mL of standard 0.050 M HCl is required to reach the end point, what was the original concentration of the $Ca(OH)_2$ solution?

The balanced equation for the reaction gives the relationship of the moles of acid reacting to the moles of base:

$$\underbrace{2HCl}_{2 \text{ moles}} + \underbrace{Ca(OH)_2}_{1 \text{ mole}} \rightarrow CaCl_2 + 2H_2O$$

The mole relationship here is that the number of moles of acid is twice the number of moles of base:

the number of moles of acid = 2 times the number of moles of base
↑ mole factor

Since the molar concentration of the acid times the volume of the acid gives the number of moles of acid:

$$M_a \times V_a = \text{moles of acid}$$

and the molar concentration of the base times the volume of the base gives the number of moles of base:

$$M_b \times V_b = \text{moles of base}$$

then, substituting these products into the mole relationship, we get

$$M_a V_a = 2 M_b V_b$$

Solving for M_b gives

$$M_b = \frac{M_a V_a}{2 V_b}$$

Substituting values, we get

$$M_b = \frac{0.050 \text{ mole} / \text{liter} \times 0.0250 \text{ liter}}{2 \times 0.0200 \text{ liter}}$$

$$M_b = 0.0312 \text{ mole} / \text{liter or } 0.0312 \text{ M}$$

BUFFER SOLUTIONS

Buffer solutions are equilibrium systems that resist changes in acidity and maintain constant pH when acids or bases are added to them. A typical laboratory buffer can be prepared by mixing equal molar quantities of a weak acid such as $HC_2H_3O_2$ and its salt, $NaC_2H_3O_2$. When a small amount of a strong base such as NaOH is added to the buffer, the acetic acid reacts (and consumes) most of the excess OH^- ion. The OH^- ion reacts with the H^+ ion from the acetic acid, thus reducing the H^+ ion concentration in this equilibrium:

$$HC_2H_3O_2 \rightleftharpoons H^+ + C_2H_3O_2^-$$

This reduction of H^+ causes a shift to the right, forming additional $C_2H_3O_2^-$ ions and H^+ ions. For practical purposes, each mole of OH^- added consumes 1 mole of $HC_2H_3O_2$ and produces 1 mole of $C_2H_3O_2^-$ ions.

When a strong acid such as HCl is added to the buffer, the H^+ ions react with the $C_2H_3O_2^-$ ions of the salt and form more undissociated $HC_2H_3O_2$. This does not alter the H^+ ion concentration. Proportional increases and decreases in the concentrations of $C_2H_3O_2^-$ and $HC_2H_3O_2$ do not significantly affect the acidity of the solution.

SALTS

A salt is an ionic compound containing positive ions other than hydrogen ions and negative ions other than hydroxide ions. The usual method of preparing a particular salt is by neutralizing the appropriate acid and base to form the salt and water.

The methods for preparing salts are:

1. *Neutralization reaction.* An acid and base neutralize each other to form the appropriate salt and water. For example:

$$2HCl + Ca(OH)_2 \rightarrow \quad CaCl_2 + 2H_2O$$
$$\text{acid} + \text{base} \quad \rightarrow \quad \text{salt} + \text{water}$$

2. *Single replacement reaction.* An active metal replaces hydrogen in an acid. For example:

$$Mg(s) + H_2SO_4 \text{ (aq)} \quad MgSO_4 \text{ (aq)} + H_2(g)$$

3. *Direct combination of elements.* An example of this method is the combination of iron and sulfur. In this reaction small pieces of iron are heated with powdered sulfur. For example:

$$Fe + S \rightarrow FeS$$
$$\text{iron (II) sulfide}$$

4. **Double replacement.** When solutions of two soluble salts are mixed, they form an insoluble salt compound. For example:

$$AgNO_3 \text{ (aq)} + NaCl \text{ (aq)} \qquad NaNO_3 \text{ (aq)} + AgCl \text{ (s)}$$

5. **Reaction of a metallic oxide with a nonmetallic oxide.** For example:

$$MgO + SiO_2 \rightarrow MgSiO_3$$

The naming of salts is discussed on page 81.

AMPHOTERIC SUBSTANCES

Some substances, such as the HCO_3^- ion, the HSO_4^- ion, the H_2O molecule, and the NH_3 molecule, can act as either a proton donor (acid) or a proton receiver (base), depending upon which substances they come in contact with. These substances are said to be *amphoteric*. Amphoteric substances donate protons in the presence of strong bases and accept protons in the presence of strong acids.

This can be illustrated by the reactions of the bisulfate ion, HSO_4^-:

With a strong acid, the HSO_4^- accepts a proton:

$$HSO_4^- + H^+ \rightarrow H_2SO_4$$

With a strong base, the HSO_4^- donates a proton:

$$HSO_4^- + OH^- \rightarrow H_2O + SO_4^{2-}$$

Chapter
11
Review

- **QUESTIONS**

1. The difference between HCl and $HC_2H_3O_2$ as acids is (a) the first has less hydrogen in solution   (b) the second has more ionized hydrogen   (c) the first is highly ionized   (d) the second is highly ionized.

2. The hydronium ion is represented as (a) H_2O^+ (b) H_3O^+ (c) HOH^+ (d) H^-.

3. H_2SO_4 is a strong acid because it is (a) slightly ionized   (b) unstable   (c) an organic compound   (d) highly ionized.

4. In the reaction of an acid with a base the common ionic reaction involves ions of (a) hydrogen and hydroxide   (b) sodium and chloride   (c) hydrogen and hydronium   (d) hydroxide and nitrate.

5. The pH of an acid solution is (a) 3   (b) 7   (c) 9   (d) 10.

6. The pH of a solution with a hydrogen ion concentration of 1×10^{-3} is (a) +3   (b) −3   (c) ±3   (d) 1 + 3.

7. According to the Brønsted Theory, an acid is (a) a proton donor   (b) a proton acceptor   (c) an electron donor   (d) an electron acceptor.

8. A buffer solution (a) changes pH rapidly with the addition of an acid (b) does not change pH at all (c) resists changes in pH (d) changes pH only with the addition of a strong base.

9. The point at which a titration is complete is called the (a) end point (b) equilibrium point (c) calibrated point (d) chemical point.

10. If 10 mL of 1 M HCl was required to titrate a 20-mL NaOH solution of unknown concentration to its end point, what was the concentration of the NaOH? (a) 0.5 M (b) 1.5 M (c) 2 M (d) 2.5 M

• ANSWERS:

1. (c)	3. (d)	5. (a)	7. (a)	9. (a)
2. (b)	4. (a)	6. (a)	8. (c)	10. (a)

10. The formula:

$$M_{acid} \times V_{acid} = M_{base} \times V_{base}$$
$$1\ M \times 10\ mL = x\ M \times 20\ mL$$
$$\underline{x = 0.5M}$$

Terms You Should Know

acid	conjugate base	pH
amphoteric	end point	salt
base	indicator	volumetric analysis
buffer solution	litmus	
conjugate acid	neutralization	

Arrhenius Theory
Brønsted Theory
Lewis Theory

12

Oxidation-Reduction and Electrochemistry

IONIZATION

In the early 1830s, Michael Faraday discovered that water solutions of certain substances conduct an electric current. He called these substances *electrolytes*. Our definition today of an electrolyte is much the same. It is a substance that dissolves in water to form a solution that will conduct an electric current.

The usual apparatus to test for this conductivity is a lightbulb placed in series with two prongs that are immersed in the solution tested.

FIGURE 35. *Conductivity of Electrolytes*

Using this type of apparatus, we can classify solutions as good, moderate, or poor electrolytes. If they do not conduct at all, they are called *nonelectrolytes*. The table below gives a classification of some common substances.

WATER SOLUTIONS TESTED FOR CONDUCTIVITY

Good	Poor	Nonelectrolyte
Sodium chloride	Acetic acid	Sugar
Hydrochloric acid	Ammonium hydroxide	Benzene

The reason that these substances conduct with varying degrees is related to the number of ions in solution.

The ionic lattice substances like sodium chloride are dissociated by the water molecules so that the individual positive and negative ions are dispersed throughout the solution. In the case of a covalent bonded substance the degree of polarity determines the extent to which it will be ionized. The water molecules, which are polar themselves, can help weaken and finally break the polar covalent bonds by clustering around the substance. When the ions are formed in this manner, the process is called *ionization;* when the ionic lattice comes apart, it is called *dissociation.* Substances that are nonelectrolytes are usually bonded so that the molecule is a nonpolar molecule. The polar water molecule cannot orient itself around the molecule and cause its ionization.

Let us see how the current was carried through the apparatus in Figure 35. The electricity causes one electrode to become positively charged and one negatively charged. These are called the *anode* and *cathode,* respectively. If the solution contains ions, they will be attracted to the electrode with the charge opposite to their own. This means that the positive ions migrate to the cathode, so they are referred to as *cations.* The negative ions migrate toward the positive anode, so they are called *anions.* When these ions arrive at the respective electrodes, the negative ions give up electrons and the positive ions accept electrons, and we have a completed path for the electric current. The more highly ionized the substance, the more current flows and the brighter the lightbulb glows. It was the Swedish chemist Arrhenius who proposed a theory to explain the behavior of electrolytes in aqueous solutions.

OXIDATION-REDUCTION

In the discussion of acids and metals reacting to produce hydrogen (page 92), the activities of metals compared with the activity of hydrogen are shown in the form of a chart. A more inclusive presentation of this activity of metals, called the *electromotive series,* is shown in Table 10.

From this chart you notice that zinc is above copper and, therefore, more active. This means zinc can displace copper ions in a solution of copper sulfate.

$$Zn^0 + Cu^{2+} + SO_4^{2-} \rightarrow Cu^0 + Zn^{2+} + SO_4^{2-}$$

The zero indi-   ⟶ The zinc ion
cates the zinc   carries a +2
atom is neutral.   charge.

The Zn atom must have lost 2 electrons to become Zn^{2+} ions.

$$Zn^0 \rightarrow Zn^{2+} + 2e^- \text{ (electrons)}$$

At the same time, the Cu^{2+} must have gained 2 electrons to become the Cu^0 atom.

$$Cu^{2+} + 2e^- \text{ (electrons)} \rightarrow Cu^0$$

These two equations are called *half-reactions.* The loss of electrons by the zinc is called oxidation; the gain of electrons by the copper ion, reduction. It is important to remember that the gain of electrons is *reduction* and the loss of electrons is *oxidation.*

The metal elements that lose electrons easily and become positive ions are placed high in the electromotive series. The metal elements that lose electrons with more difficulty are placed lower on the chart.

The energy required to remove electrons from metallic atoms can be assigned numerical values called *electrode potentials.* The energy required for the reduction of nonmetals is shown in Table 11.

STANDARD HALF-CELL VOLTAGES

Element			Oxidation Electron Reaction	Electrode Potential,* E_{ox}^0
Potassium			$K \rightarrow K^+ + e^-$	+2.93 V
Calcium			$Ca \rightarrow Ca^{2+} + 2e^-$	+2.87 V
Sodium	Increasing	Increasing	$Na \rightarrow Na^+ + e^-$	+2.71 V
Magnesium	Tendency	Tendency	$Mg \rightarrow Mg^{2+} + 2e^-$	+2.37 V
Aluminum	for	for	$Al \rightarrow Al^{3+} + 3e^-$	+1.67 V
Zinc	Atoms	Ions	$Zn \rightarrow Zn^{2+} + 2e^-$	+0.76 V
Iron	to	to Gain	$Fe \rightarrow Fe^{2+} + 2e^-$	+0.44 V
Tin	Lose	Electrons	$Sn \rightarrow Sn^{2+} + 2e^-$	+0.14 V
Lead	Electrons	and	$Pb \rightarrow Pb^{2+} + 2e^-$	+0.13 V
Hydrogen	and	Form	$H_2 \rightarrow 2H^+ + 2e^-$	0.00 V
Copper	Form	Atoms	$Cu \rightarrow Cu^{2+} + 2e^-$	−0.34 V
Mercury	Positive	of the	$2Hg \rightarrow Hg_2^{2+} + 2e^-$	−0.79 V
Silver	Ions	Metal	$Ag \rightarrow Ag^+ + e^-$	−0.80 V
Mercury			$Hg \rightarrow Hg^{2+} + 2e^-$	−0.85 V
Gold			$Au \rightarrow Au^{3+} + 3e^-$	−1.50 V

*A measure in volts of the tendency of atoms to gain or lose electrons.

TABLE 10. Electromotive Series

Element			Reduction Electron Reaction	E_{red}^0
Fluorine	Increasing Tendency	Increasing Tendency	$F_2 + 2e^- \rightarrow 2F^-$	+2.87 V
Chlorine	for Atoms to Gain	for Ions to Lose	$Cl_2 + 2e^- \rightarrow 2Cl^-$	+1.36 V
Bromine	Electrons and Form	Electrons and Form	$Br_2 + 2e^- \rightarrow 2Br^-$	+1.09 V
Iodine	Negative Ions	Atoms	$I_2 + 2e^- \rightarrow 2I^-$	+0.54 V

TABLE 11. Activity of Nonmetals

These voltages depend on the nature of the reaction, the concentrations of reactants and products, and the temperature. Throughout this discussion, we use standard concentrations, that is, all ions or molecules in aqueous solution are at a concentration of 1 molar. Furthermore, all gases taking part in the reactions are at 1 atmosphere pressure, and the temperature is 25°C. The voltage measured under these conditions is called *standard voltage*.

The currently accepted convention is to give the potentials of half-reactions as reduction processes. For example:

$$2H^+ + 2e^- \rightarrow H_2$$

The E^0 values corresponding to these half-reactions are called *standard reduction potentials*. The most common ones are given in Table Ⓛ in the Reference Section at the back of the book.

Some oxidation electrode potentials are shown in the last column of Table 10. Notice that hydrogen is used as the standard with an electrode potential of zero. These values help you to predict what reactions will occur and how readily they will occur. These reactions are called *voltaic cell reactions*.

The following examples will clarify the use of electrode potentials.

If magnesium reacts with chlorine, we can write the equation

$$Mg + Cl_2 \rightarrow MgCl_2$$

The two half-reactions with the electrode potentials would be:

Oxidation half-reaction:	$Mg \rightarrow Mg^{2+} + 2e^-$	$E^0 = +2.37$ V
Reduction half-reaction:	$Cl_2 + 2e^- \rightarrow 2Cl^-$	$E^0 = +1.36$ V
Net reaction:	$Mg + Cl_2 \rightarrow Mg^{2+} + 2Cl^-$	$E^0 = +3.73$ V

In the net reaction, E^0 is a positive number. This indicates that the reaction occurs spontaneously.

You should also note that the total number of electrons lost in oxidation is equal to the total number of electrons gained so that the net reaction (arrived at by adding the two reactions) does not contain any electrons.

Another example is sodium reacting with chlorine:

$$2Na + Cl_2 \rightarrow 2NaCl$$

The two half-reactions would be:

Oxidation half-reaction:	$2Na \rightarrow 2Na^+ + 2e^-$	$E^0 = +2.71$ V
Reduction half-reaction:	$Cl_2 + 2e^- \rightarrow 2Cl^-$	$E^0 = +1.36$ V
Net reaction:	$2Na + Cl_2 \rightarrow 2Na^+ + 2Cl^-$	$E^0 = +4.07$ V

Note: The standard potentials (E^0) are *not* multiplied by the coefficients in calculating the E^0 for the reaction.

Again, the E^0 for this reaction is positive and the reaction is spontaneous.

The next example shows a negative E^0.

Copper metal placed in an acid solution would be shown as follows:

Oxidation:	$Cu \rightarrow Cu^{2+} + 2e^-$	$E^0 = -0.34$ V
Reduction:	$2H^+ + 2e^- \rightarrow H_2$	$E^0 = 0.00$ V
Net reaction:	$Cu + 2H^+ \rightarrow Cu^{2+} + H_2$	$E^0 = -0.34$ V

Since E^0 is negative, we know the reaction will not take place.

You notice in Table $\textcircled{L}$ (page 340) that the reduction reactions with their E^0 values are shown for metals. If you must use the oxidation reactions of these metals, the equation must be reversed, and the sign of E^0 changed. An example of this is placing a piece of copper into a solution of silver ions.

Oxidation (reversed):	$Cu \rightarrow Cu^{2+} + 2e^-$	$E^0 = -0.34$ V
Reduction:	$2Ag^+ + 2e^- \rightarrow 2Ag^0$	$E^0 = +0.80$ V
	$Cu + 2Ag^+ \rightarrow Cu^{2+} + 2Ag^0$	$E^0 = +0.46$ V

This reaction would occur spontaneously since E^0 is positive.

NON-STANDARD-STATE CELL POTENTIALS

You are undoubtedly aware that the voltage of a battery does not remain constant. After the battery is in use, the voltage begins to decrease. The reason is related to the changes in concentration of the reactants and products that are causing the electron flow. For the voltaic standard mentioned earlier, the assumed concentration of 1 molar solutions was the basis of the standard voltages given. A German scientist, Walter Nernst, developed a mathematical relationship that enables us to calculate the cell voltages and direction of a spontaneous reaction at other than standard-state concentrations. For a general oxidation-reduction reaction

$$aA + bB \rightleftharpoons cC + dD$$

the *Nernst equation* has the form

$$E_{cell} = E^0_{cell} - \frac{0.059}{n} \log \frac{[C]^c[D]^d}{[A]^a[B]^b}$$

where E^0_{cell} is the standard-state cell voltage, n is the number of electrons exchanged in the equation for the reaction, and 0.059 is a constant at 298 K. The concentrations are all in molarities. Concentrations of solids and liquid solvents are considered constant and are not included in the expression.

Notice from this mathematical expression that, if the concentrations of the reactants are increased or if the concentrations of the products decreased, the voltage will increase ($E > E^0$). If the concentrations of the products are increased or the concentrations of reactants decreased, the voltage will decrease ($E < E^0$).

• SAMPLE PROBLEM:

Calculate the cell voltage of

$$Zn(s) \mid Zn^{2+}(0.001 \text{ M}) \parallel Ag^+(0.1 \text{ M}) \mid Ag(s)$$

(Each side of the $\parallel$ represents the half-reaction characters, with the concentration when appropriate.)

The reactions would be

Cathode reduction reaction:	$2Ag^+ + 2e^- \rightarrow 2Ag(s)$	$E^0 = +0.80$ V
Anode oxidation reaction:	$Zn(s) \rightarrow Zn^{2+} + 2e^-$	$E^0 = +0.76$ V
Cell reaction:	$2Ag^+ + Zn(s) \rightleftharpoons Zn^{2+} + 2Ag$	$E^0 = +1.56$ V

Substituting the known values in the Nernst equation, solve for the E_{cell}:

$$E_{cell} = E_{cell}^0 - \frac{0.059}{n} \log \frac{[Zn^{2+}]}{[Ag^+]^2}$$

$$E_{cell} = 1.56 - \frac{0.059}{2} \log \frac{[10^{-3}]}{[10^{-1}]^2}$$

$$E_{cell} = 1.56 - 0.03 \log 10^{-1}$$

$$E_{cell} = 1.56 - 0.03$$

$$E_{cell} = 1.59 \text{ V}$$

ELECTROLYTIC CELLS

Reactions that do not occur spontaneously can be forced to take place by supplying energy with an external current. These reactions are called *electrolytic reactions*. Some examples of this type of reaction are electroplating, electrolysis of salt solution, electrolysis of water, and electrolysis of molten salts. An example of the electrolytic setup is shown in Figure 36.

FIGURE 36. Electrochemical Reactions

If this solution contained Cu^{2+} ions and Cl^- ions, the half-reactions would be:

Anode half-reaction:

Oxidation: $\qquad\qquad\qquad 2Cl^- \rightarrow Cl_2(g) + 2e^- \qquad E^0 = -1.36 \text{ V}$

Cathode half-reaction:

Reduction: $\qquad\qquad\qquad Cu^{2+} + 2e^- \rightarrow Cu^0 \qquad\qquad E^0 = +0.34 \text{ V}$

$$\overline{Cu^2 + 2Cl^- \rightarrow Cu^0 + Cl_2(g) \qquad E^0 = -1.02 \text{ V}}$$

Notice that the E^0 for this reaction is negative so an outside source of energy must be used to make it occur.

In *electroplating*, where electrolysis is used to coat a material with a layer of metal, the object to be plated is made the cathode in the reaction. A bar of the plating metal is made the anode, and the solution contains ions of the plating metal.

For example, silver nitrate, $AgNO_3$, can be used to silver plate. When dissolved in H_2O, it forms a solution containing silver ions, Ag^+. Assume a metal fork is made the cathode and a bar of silver is made the anode. When the current is switched on, the positive silver ions in the solution are attracted to the fork. When the silver ions make contact, they are reduced and change from ions to atoms of silver. These atoms gradually form a metallic coating on the fork. At the anode, oxidation occurs, and the anode itself is oxidized. These two half reactions are:

Cathode reaction: $Ag^+ + e^- \rightarrow Ag$
Anode reaction: $Ag \rightarrow Ag^+ + e^-$

The silver ions formed at the anode replace those in the solution that are plated onto the fork during the cathode reaction.

Another example is the electrolysis of a water solution of sodium chloride. This water solution contains chloride ions, which are attracted to the anode and set free as chlorine molecules. The cathode reaction is somewhat more complicated. Although the sodium ions are attracted to the cathode, they are not set free as atoms. Remember that water can ionize to some extent, and the electromotive series shows that the hydrogen ion is reduced more easily than the sodium ion. Therefore, hydrogen, *not* sodium, is set free at the cathode. The reaction can be summarized in this way:

Cathode reaction: $2H_2O + 2e^- \rightarrow H_2 \ (g) + 2OH^-$
Reduction: $2Cl^- \rightarrow 2e^- + Cl_2$
_____
Net Reaction: $2H_2O + 2Cl^- \rightarrow H_2(g) + Cl_2(g) + 2OH^-$

Another example of electrolysis is the decomposition of water by using an apparatus like the one shown in Figure 37.

The solution in this apparatus contains distilled water and a small amount of H_2SO_4. The reason for adding H_2SO_4 is to make the solution an electrolyte since distilled water alone will not conduct an electric current. The solution, therefore, contains ions of H^+, HSO_4^-, and SO_4^{2-}.

FIGURE 37. Electrolysis of Water

The hydrogen ions (H^+) migrate to the cathode where they are reduced to hydrogen atoms and form hydrogen molecules (H_2) in the form of a gas. The SO_4^{2-} and HSO_4^- migrate to the anode but are not oxidized since the oxidation of water occurs more readily. These ions then are merely spectator ions. The oxidized water reacts as shown in the half-reaction below.

The half-reactions can be shown as follows:

Cathode reaction:

Reduction: $\qquad 4H_2O + 4e^- \rightarrow 2H_2(g) + 4OH^-$

Anode reaction:

Oxidation: $\qquad\qquad 2H_2O \rightarrow O_2(g) + 4H^+ + 4e^-$

___

Net Reaction: $\qquad\qquad 2HOH \rightarrow 2H_2(g) + O_2(g)$

Notice that the equation shows 2 volumes of hydrogen gas are released while only 1 volume of oxygen gas is liberated. This agrees with the discussion of the composition of water in Chapter 7.

APPLICATIONS OF ELECTROCHEMICAL CELLS (COMMERCIAL VOLTAIC CELLS)

One of the most common voltaic cells is the ordinary "dry cell" used in flashlights. Its makeup is shown in the drawing below, along with anode, cathode, and paste reactions.

Anode reaction:
$Zn(s) \rightarrow Zn^{2+}(aq) + 2e^-$

Cathode reaction:
$MnO_2(s) + 4NH_4^+(aq) + e^- \rightarrow$

$Mn^{3+}(aq) + 4NH_3(aq) + 2 H_2O$

The automobile lead storage battery is also a voltaic cell. When it discharges, the reactions are as shown in the drawing below.

Anode reaction:
$Pb(s) = HSO_4^- \rightarrow PbSO_4 + H^+ + 2e^-$

Cathode reaction:
$PbO_2(s) + HSO_4^- \rightarrow + 2 H^+ + 2e^- \rightarrow$

$PbSO_4(s) + 3 H_2O$

QUANTITATIVE ASPECTS OF ELECTROLYSIS

Relationship Between Quantity of Electricity and Amount of Products

The amounts of products liberated at the electrodes of an electrolytic cell are related to the quantity of electricity passed through the cell and the electrode reactions.

In electrolysis, 1 mole of electrons is called a *faraday* of electric charge. In the reaction of electrolysis of molten NaCl, 1 faraday will liberate 1 mole of sodium atoms.

Cathode reaction: $Na^+ + e^- \rightarrow Na(s)$

At the same time, 1 mole of Cl^- ions at the anode will form 1 mole of chlorine atoms and thus 0.5 mole of chlorine molecules.

If the metallic ion had been Ca^{2+}, 1 faraday would release only 0.5 mole of calcium atoms since each calcium ion requires 2 electrons, as shown here:

$$Ca^{2+} + 2e^- \rightarrow Ca(s)$$

• SAMPLE PROBLEM:

How many moles of electrons are required to reduce 2.93 g of nickel ions from melted $NiCl_2$?

Solution

The reaction is

$$Ni^{2+} + 2e^- \rightarrow Ni(s)$$

The 2.93 g represents

$$2.93\,\cancel{g} \times \frac{1 \text{ mole Ni}}{58.7\,\cancel{g}} = 0.05 \text{ mole Ni}$$

Since 2 moles of electrons are needed to produce 1 mole of Ni(s),

$$0.05 \;\cancel{\text{mole Ni}} \times \frac{2 \text{ moles of electrons}}{1 \;\cancel{\text{mole Ni}}} = 0.1 \text{ mole of electrons}$$

Balancing Redox Equations Using Oxidation Numbers

In general, many chemical reactions are so simple that, once the reactants and products are known, the equations can be readily written. Many redox reactions, however, are of such complexity that the process of writing the equations by trial and error is time consuming. In these cases the operation can be simplified by limiting the change involved to the actual electron shift, and balancing that—an operation that usually can be done without difficulty.

That part of the process being accomplished, the remainder of the equation can easily be adjusted to it. There are several methods by which this can be done, and a number of techniques that can be employed for each method. We will show two methods: the electron shift method and the ion-electron method.

• THE ELECTRON SHIFT METHOD:

First Example:

$$\begin{array}{ccccccccc} +1 & -1 & +4 & -2 & +1 & -2 & +2 & -1 & 0 \\ HCl & + & MnO_2 & \rightarrow & H_2O & + & MnCl_2 & + & Cl_2 \end{array}$$

1. The molecular formulas are written in a statement of the reaction, and the oxidation states are assigned to the elements. These appear in the above equation.

2. Inspection shows that the oxidation state of the manganese atoms has changed from +4 to +2, and that the oxidation state of the chlorine atoms which emerge from the action as a free element has changed from −1 to 0:

$$1\;(Mn^{4+} + 2e^+ \rightarrow Mn^{2+})$$
$$2\;(Cl^{1-} \qquad \rightarrow Cl^0 + 1e^-)$$

3. The reason for multiplying the two half-reactions by 1 and 2 is to balance the electrons gained with the electrons lost.

4. It appears that $1\ Mn^{4+} + 2Cl^{1-} \rightarrow 1Mn^{2+} + 2Cl^{0}$

5. Placing these coefficients in the equation gives

$$2HCl + 1MnO_2 \rightarrow *?\ H_2O + 1MnCl_2 + Cl_2$$

6. From the numbers thus established, the remaining coefficients can easily be deduced. Two *more* molecules of HCl are required to furnish the chlorine for $MnCl_2$, and the two atoms of oxygen in MnO_2 form $2H_2O$.

7. *The final equation is:* $4HCl + MnO_2 \rightarrow 2H_2O + MnCl_2 + Cl_2$.

Now consider a more complicated reaction. Only the elements that show a change in oxidation state are indicated.

Second Example:

$$\overset{-1}{HCl} + \overset{+7}{KMnO_4} \rightarrow H_2O + KCl + \overset{+2}{MnCl_2} + \overset{0}{Cl_2}$$

The electron change is: $5\ (Cl^{-1}\qquad = Cl^{0} + 1e^{-})$

$$1\ (Mn^{7+} + 5e^{-}\quad = Mn^{2+})$$

Therefore, $\quad \overline{5HCl + 1KMnO_4 = \tfrac{5}{2}Cl_2 + 1MnCl_2}$

$$\boxed{\text{5 atoms of chlorine}}$$

equal $\tfrac{5}{2}Cl_2$ molecules

Multiply by 2 to avoid fractions gives:

$$10HCl + 2KMnO_4 \rightarrow 5Cl_2 + 2MnCl_2$$

Substituting in the original equation:

$$10HCl + 2KMnO_4 \rightarrow *?\ H_2O + *?\ KCl + 2MnCl_2 + 5Cl_2$$

For the remainder of the coefficients, $2KMnO_4$ produces $2KCl$ and $8H_2O$, while $2KCl + 2MnCl_2$ call for six additional molecules of HCl. So finally the expression becomes

$$16HCl + 2KMnO_4 \rightarrow 8H_2O + 2KCl + 2MnCl_2 + 5Cl_2$$

• ION-ELECTRON METHOD:

The second method is more complex but seems to represent the true mechanism of such reactions more closely.

In this method, only units that actually have individual existence (atoms, molecules, or ions) in the particular reaction being studied are taken into consideration. The principal oxidizing agent and the principal reducing agent are chosen from these (by a method to be indicated later) and the electron loss or gain of these two principal actors is then determined by taking into consideration the fact that, since electrons can neither be created nor destroyed, electrons lost by one of these actors must be gained by the other. This is accomplished by using two separate partial equations representing the changes undergone by each of the two principal actors.

Probably the best way to show how the method operates will be to follow in detail the steps taken in balancing the example of an actual oxidation-reduction reaction.

---

*? indicates that this coefficient is not yet known.

First Example:

Assume that the equation $K_2CrO_4 + HCl \rightarrow KCl + CrCl_3 + H_2O + Cl_2$ is to be balanced.

1. Determine which of the substances present are involved in the oxidation-reduction.

    This is done by listing all substances present on each side of the arrow and the crossing out of those that appear on both sides of the arrow without any portion being changed in any way.

    $\cancel{K^+}, CrO_4^{2-}, H^+, Cl^- \rightarrow \cancel{K^+}, Cl^-, Cr^{3+}, H_2O, Cl_2$

    Note that Cl^- is not crossed out although it appears on both sides because some of the Cl^- from the left side appears in a changed form, namely, Cl_2, on the right.

    The two substances on the left side that are *not* crossed out are the ones involved in the oxidation-reduction.

    If, as *in this case* there are *more* than two, disregard H^+, OH^-, or H_2O; this will leave the two principal actors.

2. Indicate in two as yet unbalanced partial equations the fate of each of the two active agents thus:

    $CrO_4^{2-} \rightarrow Cr^{3+}$
    $Cl^- \rightarrow Cl_2$

3. Balance these equations chemically, inserting any substance necessary.

    $CrO_4^{2-} + 8H^+ \rightarrow Cr^{3+} + 4H_2O$
    $2Cl^- \rightarrow Cl_2$

    In the upper partial equation, H^+ had to be added in order to remove the oxygen from the CrO_4^{2-} ion.

    H+ always is used for this purpose in acid solutions. If the solution is basic, H_2O must be used for this purpose, thus:

    $CrO_4^{2-} + 4H_2O \rightarrow Cr^{3+} + 8OH^-$

4. Balance these equations electrically by adding electrons *on either side* so that the total electric charge is the same on the left and right sides, thus:

    $CrO_4^{2-} + 8H^+ + 3e^- \rightarrow Cr^{3+} + 4H_2O$
    $2Cl^- \rightarrow Cl_2 + 2e^-$

5. Add these partial equations.

    But before we add we must realize that electrons can neither be created nor destroyed. Electrons are gained in one of these equations and lost in the other. THOSE GAINED IN THE ONE MUST COME FROM THE OTHER. Therefore we must multiply each of these equations through by numbers so chosen that the number of electrons gained in one equation will be the same as the number lost in the other, thus:

    $2[CrO_4^{2-} + 8H^+ \; 3e^- \rightarrow Cr^{3+} + 4H_2O]$
    $3[2Cl^- \rightarrow Cl_2 + 2e^-]$

THIS IS THE KEY TO THE METHOD!

Adding the multiplied equations, we get:

$$2CrO_4^{2-} + 16H^+ + 6Cl^- \rightarrow 2Cr^{3+} + 8H_2O + 3Cl_2$$

This sum tells us all that really happens in the oxidation-reduction, but if a conventionally balanced equation is desired for problem purposes or for some other reason, it can be obtained by carrying the coefficient into the original skeleton equation, thus:

$$2K_2CrO_4 + 16HCl \rightarrow 4KCl + 2CrCl_3 + 8H_2O + 3Cl_2$$
$$\uparrow$$
(4 inserted by inspection)

Second Example:

$$2KMnO_4 + H_2O + KI \rightarrow 2MnO_2 + 2KOH + KIO_3$$

$$2[MnO_4^- + 2H_2O + 3e^- \rightarrow MnO_2 + 4OH^-]$$

$$I^- + 6OH^- \rightarrow IO_3^- + 3H_2O + 6e^-$$

Adding: $$\overline{2MnO_4^- + I^- + H_2O \rightarrow 2MnO_2 + IO_3^- + 2OH^-}$$

Note that, since this takes place in the basic solution, H_2O was used to remove oxygen in the upper partial.

Note also that in the second partial it was necessary to ADD oxygen and that this was done by means of OH^- ion.

If the solution had been *acid*, then H_2O would have been used for this purpose, thus:

$$I^- + 3H_2O \rightarrow IO_3^- + 6H^+ + 6e^-$$

In general, you will meet three types of partial equations:

1. Where electrons only are needed to balance, as in the second partial of the first example above.

2. Where oxygen must be removed from an ion.

   In acid solution, H^+ is used for this purpose as in the first partial of the first example above.

   In basic solution, H_2O is used for this as in the first partial of the second example above.

3. Where oxygen must be added to an ion.

   In basic solution, OH^- is used for this as in the second partial of the second example above.

   In acid solution, H_2O is used for this as in:

   $$SO_3^{2-} + H_2O \rightarrow SO_4^{2-} + 2H^+ + 2e^-$$

• TRY THESE PROBLEMS:

1. $Zn + HNO_3 \rightarrow Zn(NO_3)_2 + NH_4NO_3 + H_2O$ Ans. 4, 10, 4, 1, 3
2. $Cu + HNO_3 \rightarrow Cu(NO_3)_2 + H_2O + NO$ Ans. 3, 8, 3, 4, 2
3. $KMnO_4 + HCl \rightarrow KCl + MnCl_2 + H_2O + Cl_2$ Ans. 2, 16, 2, 2, 8, 5

• ANSWERS

1. The half-reactions balanced chemically:

$$Zn \rightarrow Zn^{2+}$$
$$NO_3^- + 10H^+ \rightarrow NH_4^+ + 3H_2O$$

Balancing electrically and adding:

$$4(Zn \rightarrow Zn^{2+} + 2e^-)$$
$$\frac{NO_3^- + 10H^+ + 8e^- \rightarrow NH_4^+ + 3H_2O}{4Zn + NO_3^- + 10H^+ \rightarrow 4Zn^{2+} + NH_4^+ + 3H_2O}$$

In the equation:

$$\underline{4}Zn + \underline{10}HNO_3 \rightarrow \underline{4}Zn(NO_3)_2 + NH_4NO_3 + \underline{3}H_2O$$

2. The half-reactions balanced chemically:

$$Cu^0 \rightarrow CU^{2+}$$
$$NO_3^{1-} + 4H^+ \rightarrow NO + 2H_2O$$

Balancing electrically and adding:

$$3(Cu^0 \rightarrow Cu^{2+} + 2e^-)$$
$$\frac{2(NO_3^{1-} + 4H^+ + 3e^- \rightarrow NO + 2H_2O)}{3Cu + 8H^+ + 2NO_3^{3-} \rightarrow 3Cu^{2+} + 2NO + 4H_2O}$$

In the equation:

$$\underline{3}Cu + \underline{8}HNO_3 \rightarrow \underline{3}Cu(NO_3)_2 + \underline{4}H_2O + \underline{2}NO$$

3. The half-reactions balanced chemically:

$$MnO_4^- + 8H^+ \rightarrow Mn^{2+} + 4H_2O$$
$$2Cl^- \rightarrow Cl_2$$

Balancing electrically and adding:

$$2(MnO_4^- + 8H^+ + 5e^- \rightarrow Mn^{2+} + 4H_2O)$$
$$\frac{5(2\,Cl^- \rightarrow Cl_2^0 + 2e^-)}{2MnO_4^- + 16H^+ + 10Cl^- \rightarrow 2Mn^{2+} + 8H_2O + 5Cl_2}$$

In the equation:

$$\underline{2}KMnO_4 + \underline{16}HCl \rightarrow \underline{2}KCl + \underline{2}MnCl_2 + \underline{8}H_2O + \underline{5}Cl_2$$

sets this as 2
by inspection

Chapter

12

Review

• QUESTIONS

1. Which of the following when dissolved in water and placed in the conductivity apparatus would cause the light to glow? (a) table salt   (b) ethyl alcohol   (c) sugar   (d) glycerine

2. The reason that a current could flow was because (a) ions combine to form molecules   (b) molecules migrate to the charge plates   (c) ions migrate to the charge plates   (d) sparks cross the gap.

3. The extent of ionization depends on the (a) nature of the solvent   (b) nature of the solute   (c) concentration of the solution   (d) temperature of the solution   (e) all the above.

4. Which of the following is TRUE? (a) The number of positive ions in solution equals the number of negative ions.   (b) The positive ions are called anions. (c) The positive ions are called cathodes.   (d) The total positive charge equals the total negative charge.   (e) None of the above.

5. The hydronium ion is represented as   (a) H_2O^+ (b) H_3O^+ (c) HOH^+ (d) H^-.

6. In the electrolysis of copper chloride, the substance liberated at the anode is (a) copper   (b) chlorine   (c) hydrogen   (d) copper chloride.

7. Ions are particles that exist (a) only in water solutions   (b) in some crystals (c) in polar covalent compounds   (d) in covalent compounds that are not polar.

8. Ionic compounds will conduct an electric current when they are (a) solidified (b) melted   (c) frozen   (d) dehydrated.

9. The cathode in a direct-current circuit is an electrode that is (a) always negative   (b) always positive   (c) always neutral   (d) sometimes negative and sometimes positive.

10. Electrolysis of a dilute solution of sodium chloride results in the cathode product (a) sodium   (b) chlorine   (c) hydrogen   (d) oxygen.

11. Electromotive series:

$$Zn^0 = Zn^{2+} + 2e^- \qquad E^0 = +0.76 \text{ V}$$
$$Au^0 = Au^{3+} + 3e^- \qquad E^0 = -1.42 \text{ V}$$

If a gold foil were placed in a solution containing Zn ions, the reaction potential would be (a) –1.34 V   (b) –2.18 V   (c) –0.66 V   (d) +2.18 V   (e) +1.34 V.

12. If the reaction potential is positive, this indicates that (a) the reaction will not occur   (b) the reaction will occur and give off energy   (c) the reaction will occur if heat or energy is added   (d) the reaction will power an outside alternating electric current.

• **ANSWERS:**

1. (a)	5. (b)	9. (a)
2. (c)	6. (b)	10. (c)
3. (e)	7. (b)	11. (b)
4. (d)	8. (b)	12. (b)

Terms You Should Know

anions	electrode potential	Nernst equation
anode	electrolyte	oxidation
cathode	electrolytic reaction	reduction
cations	electromotive series	standard voltage
dissociation	ionization	

13

Some Representative Groups and Families

In the following section, a brief description is given of some of the important and representative groups of elements usually referred to in most first-year chemistry courses.

SULFUR FAMILY

Since we discussed oxygen in Chapter 5, the most important element in this family left to discuss is sulfur.

Sulfur is found free in the volcanic regions of Japan, Mexico, and Sicily. It was removed from the rock mixtures by heating in retorts or furnaces.

In 1867 sulfur was discovered in underground deposits in Louisiana and Texas. The overlying layer of earth was "quicksand," which prohibited the usual mining operations. An American chemist, Herman Frasch, devised an ingenious plan for its removal, which bears his name, the Frasch process. It consists of drilling a hole and inserting three concentric pipes. Through the outer pipe, superheated water (at 160°C) is forced into the sulfur bed. This melts the sulfur. Compressed air, which is forced down the innermost pipe, causes the liquid sulfur to rise up through the middle pipe to the surface. At the surface the liquid sulfur is pumped into vast molds in which it solidifies.

Sulfur has three allotropic forms. Their differences are summarized in Table 12.

	ALLOTROPIC FORM		
Characteristic	Rhombic	Monoclinic	Amorphous
Shape	Rhombic or octahedral crystals	Needle-shaped monoclinic crystals	Noncrystalline
Color	Pale-yellow opaque, brittle	Yellow, waxy, translucent, brittle	Dark, tough, elastic
Preparation	Crystallizes from CS_2 solution of sulfur	Crystallizes from molten sulfur as it cools	Obtained by quick-cooling very hot liquid sulfur
Stability	Stable below 96°C	Stable between 96° and 119°C	Unstable; slowly changes to rhombic

TABLE 12. Allotropic Forms of Sulfur

Sulfur is used in making sprays to control plant diseases and harmful insects. It is used in rubber to give rubber added hardness or "vulcanize" it. Sulfur enters into the preparation of medicines, matches, gunpowder, and fireworks. Its most important use is making sulfuric acid, which is often referred to as the "king of chemicals" because of its widespread industrial uses.

Sulfuric Acid

Important properties of sulfuric acid. Sulfuric acid ionizes in two steps:

$$H_2SO_4 + H_2O \rightleftharpoons H_3O^+ + HSO_4^- \quad Ka_1 \text{ is very large.}$$
$$HSO_4^- + H_2O \rightleftharpoons H_3O^+ + SO_4^{2-} \quad Ka_2 \text{ is very small.}$$

to form a strong acid solution. The ionization is more extensive in a dilute solution. Most hydronium ions are formed in the first step as indicated by the longer forward reaction arrow. Salts formed with the HSO_4^- (bisulfate ion) are called *acids salts;* the SO_4^{2-} (sulfate ion) forms *normal salts.*

Sulfuric acid reacts like other acids, as shown below:

With active metals: $Zn + H_2SO_4 \rightarrow ZnSO_4 + H_2(g)$ (for dilute H_2SO_4)
With bases: $2NaOH + H_2SO_4 \rightarrow Na_2SO_4 + 2H_2O$
With metal oxides: $MgO + H_2SO_4 \rightarrow MgSO_4 + H_2O$
With carbonates: $CaCO_3 + H_2SO_4 \rightarrow CaSO_4 + H_2O + CO_2(g)$

Sulfuric acid has other particular characteristics as shown below:
As an oxidizing agent:

$$Cu + 2H_2SO_4 \text{ (concentrated)} \rightarrow CuSO_4 + SO_2(g) + 2H_2O$$

As a dehydrating agent with carbohydrates:

$$C_{12}H_{22}O_{11}(sugar) \xrightarrow[H_2SO_4]{Conc.} 12C + 11H_2O$$

Important uses of H₂SO₄

1. Making other acids (because of its high boiling point, 338°C):

$$NaCl + H_2SO_4 \xrightarrow{heat} NaHSO_4 + HCl(g)$$

(With more heat and an excess of salt, the second hydrogen ion can also be removed to form the salt Na_2SO_4.)

2. Washing objectionable colors from gasoline made by the "cracking" process.
3. Dehydrating agent in making explosives, dyes, and drugs.
4. Freeing metals of iron or steel of scale and rust. This is called "pickling."
5. Electrolyte in ordinary lead storage batteries.
6. Used in precipitating bath in making rayon.

Other Important Compounds of Sulfur

Hydrogen sulfide is a colorless gas having an odor of rotten eggs. It is fairly soluble in water and poisonous in rather small concentrations. It can be prepared by reacting ferrous sulfide with an acid, such as dilute HCl:

$$FeS + 2HCl \rightarrow FeCl_2 + H_2S(g)$$

It does burn in excess oxygen to form compounds of water and sulfur dioxide. If insufficient oxygen is available, some free sulfur will form. It is only a weak acid in a water solution. Hydrogen sulfide is used widely in qualitative laboratory tests since many metallic sulfides precipitate with recognizable colors. These sulfides are sometimes used as paint pigments. Some common sulfides and their colors are:

ZnS—White
CdS—Bright yellow
As₂S₃—Lemon yellow
Sb₂S₃—Orange
CuS—Black
HgS—Black
PbS—Brown-black

Another important compound of sulfur is *sulfur dioxide*. It is a colorless gas with a suffocating odor.

The structure of sulfur dioxide is a good example of a *resonance* structure. Its molecule is depicted in Figure 38.

FIGURE 38. Sulfur Dioxide Molecule

You will notice that the covalent bonds between sulfur and oxygen are shown in one drawing as single bonds and in the other as double bonds. This signifies that the bonds between the sulfur and oxygens have been shown by experimentation to be neither single nor double bonds, but "hybrids" of the two. Sulfur trioxide also is a resonance structure.

"——" indicates a covalent bond

HALOGEN FAMILY

The common members of the halogen family are shown in Table 13 with some important facts concerning them.

Item	Fluorine	Chlorine	Bromine	Iodine
Molecular formula	F_2	Cl_2	Br_2	I_2
Atomic number	9	17	35	53
Activity	Most active	←——— to ———→		Least active
Outer energy level structure	7 electrons	7 electrons	7 electrons	7 electrons
State and color at room temperature	Gas—pale yellow	Gas—green	Liquid—dark red	Solid—purplish black crystals

TABLE 13. Halogen Family

Because all the halogens lack one electron in their outer principal energy level, they usually are acceptors of electrons (oxidizing agents). Fluorine is the most active nonmetal in the periodic chart.

Some Important Halides and Their Uses

Hydrochloric acid —common acid prepared in the laboratory by reacting sodium chloride with concentrated sulfuric acid. It is used in many important industrial processes.

Silver bromide and silver iodide —halides used on photographic film. Light intensity is recorded by developing as black metallic silver those portions of the film upon which the light fell during exposure.

Hydrofluoric acid —acid used to etch glass by reacting with the SiO_2 to release silicon fluoride gas. Also used to frost lightbulbs.

NITROGEN FAMILY

The most common member of this family is nitrogen itself. It is a colorless, odorless, tasteless, and rather inactive gas that makes up about four-fifths of the air in our atmosphere. The inactivity of N_2 gas can be explained by the fact that the two atoms of nitrogen are bonded by three covalent bonds $(:N::N:)$ that require a great deal of energy to break. Since nitrogen must be "pushed" into combining with other elements, many of its compounds tend to decompose violently with a release of the energy that went into forming them.

Nature "fixes" nitrogen, or makes nitrogen combine, by a nitrogen-fixing bacteria found in the roots of beans, peas, clover, and other leguminous plants. Discharges of lightning also cause some nitrogen fixation with oxygen to form nitrogen oxides.

Nitric Acid

An important compound of nitrogen is nitric acid. This acid is useful in making dyes, celluloid film, and many of the lacquers on cars.

Its physical properties are: it is a colorless liquid (when pure), it is one and one-half times as dense as water, it has a boiling point of 86°C, the commercial form is about 68% pure, and it is miscible with water in all proportions.

Its outstanding chemical properties are: the dilute acid shows the usual properties of an acid except it rarely produces hydrogen when it reacts with metals, and it is quite unstable and decomposes as follows:

$$4HNO_3 \rightarrow 2H_2O + 4NO_2(g) + O_2(g)$$

Because of this ease of decomposition, nitric acid is a good oxidizing agent. When it reacts with metals, the nitrogen product formed will depend on the conditions of the reaction, especially the concentration of the acid, the activity of the metal, and the temperature. If the nitric acid is concentrated and the metal is copper, the principal reduction product will be nitrogen dioxide (NO_2), a heavy, red-brown gas with a pungent odor.

$$Cu + 4HNO_3 \rightarrow Cu(NO_3)_2 + 2NO_2(g) + 2H_2O$$

With dilute nitric acid, this reaction would be:

$$3Cu + 8HNO_3 \rightarrow 3Cu(NO_3)_2 + 4H_2O + 2NO(g)$$

The product, called nitric oxide (NO), is colorless and is immediately oxidized in air to NO_2 gas.

With still more dilute nitric acid, considerable quantities of nitrous oxide (N_2O) are formed; with an active metal like zinc, the product may be the ammonium ion (NH_4^+).

When nitric acid is mixed with hydrochloric acid, the mixture is called *aqua regia* because of its ability to dissolve gold. The great activity of aqua regia is due to the formation of the nitrosyl chloride (NOCl) and chlorine (Cl_2). The Cl_2 reacts with gold to form the soluble compound of gold.

Other Important Compounds of Nitrogen

Ammonia is one of the oldest known compounds of nitrogen. In times past, it was prepared by distilling leather scraps, hoofs, and horns. It can be prepared in the laboratory by heating an ammonium salt with a strong base. An example is:

$$Ca(OH)_2 + (NH_4)_2SO_4 \rightarrow CaSO_4 + 2H_2O + 2NH_3(g)$$

Industrial methods include the Haber process, which combines nitrogen and hydrogen (discussed further on page 161):

$$3H_2 + N_2 \rightarrow 2NH_3$$

and the destructive distillation of coal, which gives off NH_3 as a by-product.

The physical properties of ammonia are: it is a colorless, pungent gas that is extremely soluble in water; it is lighter than air and has a high critical temperature (the critical temperature of a gas is the temperature above which it cannot be liquefied by pressure alone).

Because ammonia can be liquefied easily and it has a high heat of vaporization, one of its uses is as a refrigerant. Also, a water solution of ammonia is used as a household cleaner to cut grease. Another use of ammonia is as a fertilizer.

Some other nitrogen compounds are summarized in Table 14.

Name	Formula	Notes of Importance
Nitrate ion	NO_3^-	1. Nitrates of metals can be decomposed with heat to release oxygen.
		2. Nitrate test—add freshly prepared ferrous sulfate solution; hold test tube at an angle and slowly add concentrated sulfuric acid down the side; if a brown ring forms just above the sulfuric acid level, it indicates a positive test for nitrate ion. (Could repeat using acetic instead of sulfuric acid; if brown ring again forms, it indicates a nitrite.)
Nitrous oxide	N_2O	1. Colorless; $1\frac{1}{2}$ times heavier than air; slight sweet smell and odor; supports combustion much like oxygen.
		2. Sometimes used as an anesthetic called "laughing gas."
Nitric oxide	NO	1. Colorless, slightly heavier than air, does not support combustion, is easily oxidized to NO_2 when exposed to air.
Nitrogen dioxide	NO_2	1. Reddish brown gas, 1.6 times heavier than air, very soluble in water, and extremely poisonous; has unpleasant odor.
Nitrogen tetroxide	N_2O_4	1. When NO_2 is cooled, its brown color fades to a pale yellow (N_2O_4). Two molecules of NO_2 combine to form this new compound.

TABLE 14. Some Nitrogen Compounds

Other Members of the Nitrogen Family

As you proceed down this family (see Table 15) the elements change from gases to volatile solids, and then to less volatile solids. At the same time acid-forming properties are decreasing while base-forming properties are increasing. These changes can be attributed to the increased atomic radii and the decreased ability to attract and hold the outer energy level electrons.

Name	Symbol	Atomic Number	Properties and Uses
Nitrogen	N	7	(Discusssed)
Phosphorus	P	15	Two allotropic forms: white (kindling temperature near room temperature, thus kept under water) and red. When burned in oxygen, forms P_2O_5, the acid anhydride of phosphoric acid. When P_2O_5 reacts with water, it first forms metaphosphoric acid, HPO_3, then more water reacts to form the orthophosphoric acid. Phosphorus is used in making matches.
Arsenic	As	33	Slightly volatile gray solid. Most abundant compound is As_2O_3, which is used as a preservative, in medicines, and in glass making (highly poisonous).
Antimony	Sb	51	Lustrous gray solid. Forms very weak acid, mostly basic solutions.
Bismuth	Bi	83	Almost wholly metallic properties. Shiny, dense metal with slight reddish tinge. Low melting point; therefore used in alloys for electric fuses and fire-sprinkler systems. "Woods metal" is the most outstanding of these (melting point 70°C).

TABLE 15. The Nitrogen Family

METALS

Properties of Metals

Some physical properties of metals are: they have metallic luster, they can conduct heat and electricity, they can be pounded into sheets (malleable), they can be drawn into wires (ductile), most have a silvery color, they have densities usually between 7 and 14 grams per cubic centimeter, and none is soluble in any ordinary solvent without a chemical change.

The general chemical properties of metals are: they are electropositive, and the more active metallic oxides form bases, although some metals form amphoteric hydroxides that can react as both acid and base.

Metals can be removed from their ores by various means depending on their degrees of activity (see Table 16).

Occurrence	Activity Decreasing	Metal	Ease of Reduction	Reduced by
In combined form only		Potassium Calcium Sodium Magnesium Aluminum	Very difficult	Electrolysis
Usually combined		Manganese Zinc Chromium	Difficult	Carbon with diminishing difficulty
		Iron Cadmium Cobalt Nickel		
Free and combined		Tin Lead HYDROGEN Antimony Bismuth Arsenic Copper	Easy	Carbon
		Mercury Silver	Very easy	By heating oxides
Free		Platinum Gold	Easiest	All compounds decomposed by heat

TABLE 16. *Means of Extracting Metals*

In some cases the ore is concentrated before it is reduced. This might be done by a merely physical means, such as flotation, in which the ore is crushed, mixed with a water solution containing a suitable oil (e.g., pine oil), and then vigorously agitated so that a frothy mass of oil rises to the top with the adhering metal-bearing portion of the ore. The froth is removed from the top of the machine and filtered to recover the metal ore.

A chemical process is used to convert sulfide or carbonate ores to oxides. It is called *roasting* and consists of heating the compounds in the air. For example:

$$2ZnS + 3O_2 \rightarrow 2ZnO + 2SO_2(g)$$

and

$$PbCO_3 \rightarrow PbO + CO_2(g)$$

These oxides can then be treated with appropriate reducing agents to free the metal.

The following table summarizes the properties of the first two families of metals found in Group IA or 1 and Group IIA or 2 of the periodic chart. The table shows the distinct similarities in the properties and bonding of these metals. These similarities are directly related to the similarities that exist in the outer orbital electron configuration.

ALKALI METALS

Element	Outer Orbital	Bonding and Charge	Physical Properties
Lithium Li	$2s^1$	Ionic, 1+	Silvery white
Sodium Na	$3s^1$	Ionic, 1+	Silvery white
Potassium K	$4s^1$	Ionic, 1+	Silvery white
Rubidium Rb	$5s^1$	Ionic, 1+	Silvery white
Cesium Cs	$6s^1$	Ionic, 1+	Silvery white
Francium Fr	$7s^1$	Ionic, 1+	Silvery white

ALKALINE-EARTH METALS

Element	Outer Orbital	Bonding and Charge	Physical Properties
Beryllium Be	$2s^2$	Ionic, 2+	Gray-white
Magnesium Mg	$3s^2$	Ionic, 2+	Silvery white
Calcium Ca	$4s^2$	Ionic, 2+	Silvery white
Strontium Sr	$5s^2$	Ionic, 2+	Silvery white
Barium Ba	$6s^2$	Ionic, 2+	Yellow white; lumpy
Radium Ra	$7s^2$	Ionic, 2+	White; radioactive

Some Important Reduction Methods

The Hall process is the electrolytic method of preparing *aluminum.* It uses molten cryolite (Na_3AlF_6) as the solvent for bauxite ore ($Al_2O_3 \cdot 2H_2O$). This mixture is electrolyzed in a rectangular iron tank lined with carbon that acts as the cathode. The anodes are heavy carbon rods suspended in the tank. The aluminum released at the cathode is tapped periodically and allowed to flow out of the tank so that the process can continue.

The outstanding properties of aluminum are its light weight with high strength, its ability to form an aluminum oxide coating that protects this metal from further oxidation, and its ability to conduct an electric current.

The Dow process is an electrolytic method of preparing *magnesium* from seawater. The magnesium salts are precipitated with $Ca(OH)_2$ solution to form $Mg(OH)_2$. The hydroxide is then converted to a chloride by adding hydrochloric acid to the hydroxide. This chloride is mixed with potassium chloride, melted, and electrolyzed to release the magnesium at the cathode and chlorine at the anode.

Magnesium has the important property of being a light, rigid, and inexpensive metal. It, too, can form a protective adherent oxide coating.

Another major metal prepared by electrolysis is *copper*. After the ore is concentrated by flotation and roasted to change sulfides to oxides, it is heated in a converter to get impure copper called "blister" copper. This copper is then used as anodes in the electrolysis apparatus, in which pure copper is used to make the cathode plates. As the process progresses, the copper in the anode becomes ions in the solution that are reduced at the cathode as pure copper.

Copper has wide usage in electrical appliances and wires because of its high conductivity. It is widely used in alloys of bronze (copper and tin) and brass (copper and zinc).

Iron ore is refined by reduction in a blast furnace, that is, a large, cylinder-shaped furnace charged with iron ore (usually hematite, Fe_2O_3), limestone, and coke. A hot air blast, often enriched with oxygen, is blown into the lower part of the furnace through a series of pipes called tuyeres. The chemical reactions that occur can be summarized as follows:

Burning coke:
$$2C + O_2 \rightarrow 2CO(g)$$
$$C + O_2 \rightarrow CO_2(g)$$

Reduction of CO_2:
$$CO_2 + C \rightarrow 2CO(g)$$

Reduction of ore:
$$Fe_2O_3 + 3CO \rightarrow 2Fe + 3CO_2(g)$$
$$Fe_2O_3 + 3C \rightarrow 2Fe + 3CO(g)$$

Formation of slag:
$$CaCO_3 \rightarrow CaO + CO_2(g)$$
$$CaO + SiO_2 \rightarrow CaSiO_3$$

The molten iron from the blast furnace is called "pig" iron.

From pig iron, the molten metal might undergo one of three steel-making processes that burn out impurities and set the contents of carbon, manganese, sulfur, phosphorus, and silicon. Often nickel and chromium are alloyed in steel to give particular properties of hardness needed for tool parts. The three most important means of making steel involve the basic oxygen, the open-hearth, and the electric furnace. The first two methods are the most common.

The basic oxygen furnace uses a lined "pot" into which the molten pig iron is poured. Then a high-speed jet of oxygen is blown from a water-cooled lance into the top of the pot. This "burns out" impurities to make a batch of steel rapidly and cheaply.

The open-hearth furnace is a large oven containing a dish-shaped area to hold the molten iron, scrap steel, and other additives with which it is charged. Alternating blasts of flame are directed across the surface of the melted metal until the proper proportions of additives are established for that "heat" so that the steel will have particular properties needed by the customer. The tapping of one of these furnaces holding 50 to 400 tons of steel is a truly beautiful sight.

The final method of making steel involves the electric furnace. This method uses enormous amounts of electricity through graphite cathodes that are lowered into the molten iron to purify it and produce a high grade of steel.

Alloys

An *alloy* is a mixture of two or more metals. In a mixture certain properties of the metals involved are affected. Some of these are:

1. Melting point   The melting point of an alloy is lower than that of its components.
2. Hardness   An alloy is usually harder than the metals that compose it.
3. Crystal structure   The size of the crystalline particles in the alloy determine many of the physical properties. The size of these particles can be controlled by heat treatment. If the alloy cools slowly, the crystalline particles tend to be larger. Thus, by heating and cooling an alloy, its properties can be altered considerably.

Metalloids

In the preceding sections, representative metals and nonmetals have been reviewed, along with the properties of each. Some elements, however, are difficult to classify as one or the other. One example is carbon. The diamond form of carbon is a poor conductor, yet the graphite form conducts fairly well. Neither form looks metallic, so carbon is classified as a nonmetal.

Silicon looks like a metal. However, its conductivity properties are closer to those of carbon.

Since some elements are neither distinctly metallic nor clearly nonmetallic, a third class, called the *metalloids*, is recognized.

The properties of metalloids are intermediate between those of metals and those of nonmetals. Although most metals form ionic compounds, metalloids as a group may form ionic or covalent bonds. Under certain conditions pure metalloids conduct electricity, but do so poorly, and are thus termed *semiconductors*. This property makes the metalloids important in microcircuitry.

The metalloids are located in the periodic table along the heavy dark line that starts alongside boron and drops down in steplike fashion between the elements found lower in the table (see the chart below).

	13 IIIA	14 IVA	15 VA	16 VIA	17 VIIA
	5 B	6 C	7 N	8 O	9 F
12 IIB	13 Al	14 Si	15 P	16 S	17 Cl
30 Zn	31 Ga	32 Ge	33 As	34 Se	35 Br
48 Cd	49 In	50 Sn	51 Sb	52 Te	53 I
80 Hg	81 Ti	82 Pb	83 Bi	84 Po	85 At

Location of Metalloids

• QUESTIONS

1. The most active nonmetallic element is (a) chlorine   (b) fluorine   (c) oxygen   (d) sulfur.

2. The order of decreasing activity of the halogens is (a) Fl, Cl, I, Br   (b) F, Cl, Br, I   (c) Cl, F, Br, I   (d) Cl, Br, I, F.

3. A light-sensitive substance used on photographic films has the formula (a) AgBr   (b) CaF_2 (c) CuCl   (d) $MgBr_2$.

4. Sulfur dioxide is the anhydride of (a) hydrosulfuric acid   (b) sulfurous acid   (c) sulfuric acid   (d) hyposulfurous acid.

5. The charring action of sulfuric acid is due to its being   (a) a strong acid   (b) an oxidizing agent   (c) a reducing agent   (d) a dehydrating agent.

6. Ammonia is prepared commercially by the (a) Ostwald process   (b) arc process   (c) Haber process   (d) contact process.

7. A nitrogen compound that has a color is (a) nitric oxide   (b) nitrous oxide   (c) nitrogen dioxide   (d) ammonia.

8. If a student heats a mixture of ammonium chloride and calcium hydroxide in a test tube, he will detect (a) no reaction   (b) the odor of ammonia   (c) the odor of rotten eggs   (d) nitric acid fumes.

9. The difference between ammonia and ammonium is (a) an electron   (b) a neutron   (c) a proton   (d) radioactivity.

10. An important ore of iron is (a) bauxite   (b) galena   (c) hematite   (d) smithsonite.

11. A reducing agent used in the blast furnace is (a) $CaCO_3$ (b) CO   (c) O_2 (d) SiO_2

12. Aluminum is extracted from its oxide by (a) roasting   (b) reduction with carbon   (c) smelting   (d) electrolysis.

• ANSWERS:

1. (b)	7. (c)
2. (b)	8. (b)
3. (a)	9. (c)
4. (b)	10. (c)
5. (d)	11. (b)
6. (c)	12. (d)

Terms You Should Know

acid salt	Haber process	resonance structure
allotropic forms	metalloids	rhombic, monoclinic,
alloys	normal salt	amorphous
electric furnace	pig iron	semiconductors

Carbon and Organic Chemistry

CARBON

Forms of Carbon

The element carbon occurs in three allotropic forms: diamond, graphite, and amorphous (although some evidence shows the amorphous forms have some crystalline structure).

The diamond form has a close-packed crystal structure that gives it its property of extreme hardness. In it each carbon is bonded to four other carbons in a tetrahedron arrangement like this:

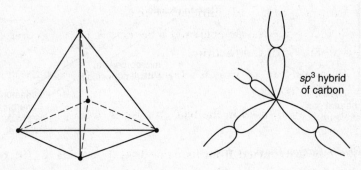

sp^3 hybrid of carbon

It has been possible to make synthetic diamonds in machines that subject carbon to extremely high pressures and temperatures. Most of these diamonds are used for industrial purposes, such as dies.

The graphite form is made up of planes of hexagonal structures that are weakly bonded to the planes above and below. This explains graphite's slippery feeling and makes it useful as a dry lubricant. Its structure can be seen below. Graphite also has the property of being an electrical conductor.

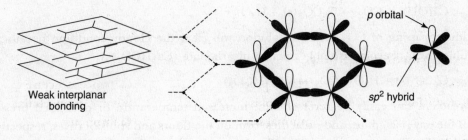

Weak interplanar bonding

p orbital

sp^2 hybrid

Some of the amorphous forms of carbon are charcoal, coke, bone black, and lampblack. Characteristically, none has a definite crystalline structure.

The process of destructive distillation is used in the preparation of both charcoal and coke. This process involves the heating of a substance in an enclosed container so that volatile materials, which would cause wood and coal to burn with a flame, can be forced out. After these substances are removed by this process, the charcoal and coke merely glow when burned. Many of these volatile substances have proved to be useful by-products of destructive distillation. From the destructive distillation of coal we get coal gas, coal tar, and ammonia. Coal gas can be used as a fuel; coal tars are the primary source of most synthetic dyes and many other useful organic compounds; and ammonia is also a useful chemical.

Carbon Dioxide

Carbon dioxide is a widely distributed gas that makes up 0.04 percent of the air. There is a cycle that keeps this figure relatively stable. It is shown in Figure 40.

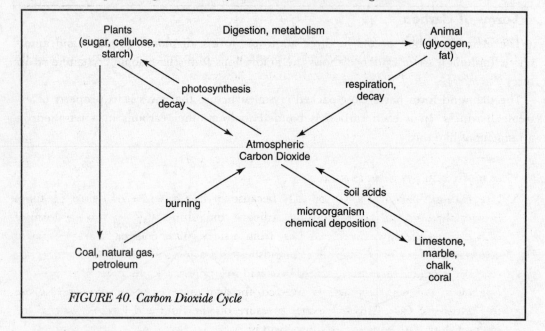

FIGURE 40. *Carbon Dioxide Cycle*

Laboratory preparation of CO$_2$. The usual laboratory preparation consists of reacting calcium carbonate (marble chips) with hydrochloric acid, although any carbonate or bicarbonate and any common acid could be used. The gas is collected by water displacement or air displacement. (See Figure 41.)

The test for carbon dioxide consists of passing it through limewater (Ca(OH)$_2$). If CO$_2$ is present the limewater turns cloudy because of the formation of a white precipitate of finely divided CaCO$_3$:

$$Ca(OH)_2 + CO_2 \rightarrow CaCO_3(s) + H_2O$$

Continued passing of CO$_2$ into the solution will clear the cloudy condition because the insoluble CaCO$_3$ becomes soluble calcium bicarbonate (Ca(HCO$_3$)$_2$).

$$CaCO_3(s) + H_2O + CO_2 \rightarrow Ca^{2+}(HCO_3)_2^-$$

This reaction can easily be reversed with increased temperature or decreased pressure. This is the way stalagmites and stalactites form on the floors and roofs of caves, respectively. The ground water containing calcium bicarbonate is deposited on the roof and floor of the cave and decomposes into solid calcium carbonate formations.

FIGURE 41. *Laboratory Preparation of Carbon Dioxide*

Important uses of CO₂

1. Since CO_2 is the acid anhydride of carbonic acid, it forms the acid when reacted with soft drinks, thus making them "carbonated" beverages.

$$CO_2 + H_2O \rightarrow H_2CO_3$$

2. Solid carbon dioxide ($-78°C$), or "dry ice," is used as a refrigerant because it has the advantage of not melting into a liquid but sublimes and in the process absorbs 3 times as much heat per gram as ice.

3. Fire extinguishers make use of CO_2 because of its properties of being $1\frac{1}{2}$ times heavier than air and not supporting ordinary combustion. It is used in the form of CO_2 extinguishers, which release CO_2 from a steel cylinder in the form of a gas to smother the fire. The soda-type extinguisher uses a solution of sodium bicarbonate ($NaHCO_3$) and a small bottle of sulfuric acid in a cylinder, which must be inverted to operate it. When the cylinder is inverted, the acid mixes with the solution to cause the release of CO_2. This gas exerts pressure on the fluid and forces it out of the extinguisher. The equation in ionic form is:

$$2Na^+ + 2HCO_3^- + 2H^+ + SO_4^{2-} \rightarrow 2Na^+ + SO_4^{2-} + 2\,H_2O + CO_2(g)$$

The foam-type extinguisher is similar to the above. It contains a charge of a water solution of sodium bicarbonate and licorice extract and a small bottle containing a water solution of alum (mainly $Al_2(SO_4)_3$), which acts as an acid. When these are mixed, CO_2 is liberated but is trapped in a sticky foam formed by the licorice and alum.

4. Leavening agents are usually made up of a dry acid-forming salt, dry baking soda ($NaHCO_3$), and a substance to keep these ingredients dry, such as starch or flour. Some of the acid-forming salts are alum [$NaAl(SO_4)_2$] in alum baking powders, potassium acid tartrate ($KHC_4H_4O_6$) in cream of tartar baking powders, and sodium hydrogen phosphate (NaH_2PO_4) or calcium hydrogen phosphate [$Ca(H_2PO_4)_2$] in the phosphate baking powders. In the moist batter, the acid reacts with the baking soda to release CO_2 as the leavening agent. This gas is trapped and raises the batter, which hardens in this raised condition.

Baking soda, when used independently, must have some source of acid, such as buttermilk, molasses, or sour milk, to cause the release of CO_2. The simple ionic equation is:

$$H^+ + HCO_3^- \rightarrow H_2O + CO_2(g)$$

5. Plants use up CO_2 in the photosynthesis process. In this process, chlorophyll (the catalyst) and sunlight (the energy source) must be present. The reactants and products of this reaction are:

$$6CO_2 + 6H_2O \rightarrow C_6H_{12}O_6 + 6O_2(g)$$
simple
sugar

OR

$$6CO_2 + 5H_2O \rightarrow C_6H_{10}O_5 + 6O_2(g)$$
cellulose

ORGANIC CHEMISTRY

Organic chemistry may be defined simply as the chemistry of the compounds of carbon. Since Friedrich Wöhler synthesized urea in 1828, chemists have synthesized thousands of carbon compounds in areas of dyes, plastic, textile fibers, medicines, and drugs. The number of organic compounds have been estimated to be in the neighborhood of a million and constantly increasing.

The carbon atom (atomic number 6) has four electrons in its outermost energy level, which show a tendency to be shared (electronegativity of 2.5) in covalent bonds. By this means, carbon bonds to other carbons, hydrogens, halogens, oxygen, and other elements to form the many compounds of organic chemistry.

Hydrocarbons

Hydrocarbons, as the name implies, are compounds containing only carbon and hydrogen in their structure. The simplest hydrocarbon is methane, CH_4. As previously mentioned, this type of formula, which shows the kinds of atoms and their respective numbers, is called an *empirical* formula. In organic chemistry this is not sufficient to identify the compound it is used to represent. For example, the empirical formula C_2H_6O could denote either an ether or an ethyl alcohol. For this reason, a *structural* formula is used to indicate how the atoms are arranged in the molecule. The ether of C_2H_6O looks like this:

```
      H       H
      |       |
  H—C—O—C—H
      |       |
      H       H
```

whereas the ethyl alcohol is represented by this structural formula:

```
      H   H
      |   |
  H—C—C—OH
      |   |
      H   H
```

To avoid ambiguity, structural formulas are more often used than empirical formulas in organic chemistry. The structural formula of methane is

$$
\begin{array}{c}
\text{H} \\
| \\
\text{H---C---H} \\
| \\
\text{H}
\end{array}
$$

Alkane Series (saturated)

Methane is the first member of a hydrocarbon series called the alkanes (or paraffin series). The general formula for this series is C_nH_{2n+2}, where n is the number of carbons in the molecule. Table 17 provides some essential information about this series. Since many other organic structures use the stem of the alkane names, you should learn these names and structures well. Notice that, as the number of carbons in the chain increases, the boiling point also increases. The first four alkanes are gases at room temperature; the subsequent compounds are liquid, then become more viscous with increasing length of the chain.

Since the chain is increased by a carbon and two hydrogens in each subsequent molecule, the alkanes are referred to as a *homologous* series.

The alkanes are found in petroleum and natural gas. They are usually extracted by fractional distillation, which separates the compounds by varying the temperature so that each vaporizes at its respective boiling point. Methane, which forms 90% of natural gas, can also be prepared in the laboratory by heating soda lime (containing NaOH) with sodium acetate:

$$NaC_2H_3O_2 + NaOH \rightarrow CH_4(g) + Na_2CO_3$$

When the alkanes are burned with sufficient air, the compounds formed are CO_2 and H_2O. An example would be:

$$2C_2H_6 + 7O_2 \rightarrow 4CO_2(g) + 6H_2O(g)$$

The alkanes can be reacted with halogens so that hydrogens are replaced by a halogen atom:

Some common substitution compounds of methane are:

methyl chloride   chloroform   carbon tertrachloride

monochloromethane   trichloromethane   tetrachloromethane

Naming alkane substitutions. When an alkane hydrocarbon has an end hydrogen removed, it is referred to as an alkyl substituent or group. The respective name of each is the alkane name with *-ane* replaced by *-yl.*

Name	Formula	Number of Structural Isomers	Structure	Boiling Point (°C)
Methane	CH_4	1		-162
Ethane	C_2H_6	1		-89
Propane	C_3H_8	1		-42
n-Butane	C_4H_{10}	2		0
n-Pentane	C_5H_{12}	3		36
n-Hexane	C_6H_{14}	5	$CH_3-CH_2-CH_2-CH_2-CH_2-CH_3$	69
n-Heptane	C_7H_{16}	7	$CH_3-CH_2-CH_2-CH_2-CH_2-CH_2-CH_3$	98
n-Octane	C_8H_{18}	18	$CH_3-CH_2-CH_2-CH_2-CH_2-CH_2-CH_2-CH_3$	126
n-Nonane	C_9H_{20}	35	$CH_3-CH_2-CH_2-CH_2-CH_2-CH_2-CH_2-CH_2-CH_3$	151
n-Decane	$C_{10}H_{22}$	75	$CH_3-CH_2-CH_2-CH_2-CH_2-CH_2-CH_2-CH_2-CH_2-CH_3$	174

TABLE 17. The Alkanes

Alkane	*Alkyl Group*	*Compound*
methane	methyl	methyl bromide
butane	butyl	butyl chloride

One method of naming a substitution product is to use the alkyl substituent or group name for the respective chain and the halide as shown above.

The IUPAC system uses the name of the longest carbon chain as the parent chain. The carbon atoms are numbered in the parent chain to indicate where branching or substitution takes place. The direction of numbering is chosen so that the lowest numbers possible are given to the side chains. The complete name of the compound is arrived at by first naming the attached group, each one being prefixed by the number of the carbon to which it is attached, and then the parent alkane. If a particular group appears more than once, the appropriate prefix (di, tri, and so on) is used to indicate how many times the group appears. A carbon atom number must be used to indicate the position of each such group. If two or more of the same group are attached to the same carbon atom, the number of the carbon atom is repeated. If two or more different substituted groups are in a name, they are arranged alphabetically.

- **EXAMPLE 1**   2, 2-dimethylbutane

- **EXAMPLE 2**   1,1-dichloro-3-ethyl-2, 4-dimethylpentane

• **EXAMPLE 3**   2-iodo-2-methylpropane

Cycloalkanes. Starting with propane in the alkane series, it is possible to get a ring form by attaching the two chain ends. This reduces the number of hydrogens by two.

This hydrocarbon is called cyclopropane.

Since all the members of the alkane series have single covalent bonds, this series and all such structures are said to be *saturated.*

If the hydrocarbon molecule contains double or triple covalent bonds, it is referred to as *unsaturated.*

Alkene Series (unsaturated)

The alkene series has a double covalent bond between two adjacent carbon atoms. The general formula of this series is C_nH_{2n}. In naming these compounds, the suffix of the alkane is replaced by *-ene.* Two examples:

ethene (common name: ethylene)

propene (common name: propylene)

If the double bond occurs on an interior carbon, the chain is numbered so that the position of the double bond is designated by the lowest possible number assigned to the first doubly bonded carbon. For example:

H H   H
| | |
H—C—C—C=C—C—H   2-pentene
| | | | |
H H H H H

The bonding is more complex in the double covalent bond than in the single bonds in the molecule. Using the orbital pictures of the atom, we can show this as follows:

sp^2 orbital

p orbital

The two *p* lobes attached above and below constitute *one* bond called a pi (π) bond.

The *sp*² orbital bonds between the carbons and with each hydrogen are referred to as sigma (σ) bonds.

Alkyne Series (unsaturated)

The alkyne series has a triple covalent bond between two adjacent carbons. The general formula of this series is C_nH_{2n-2}. In naming these compounds, the alkane suffix is replaced by *-yne*. A few examples:

H—C≡C—H   ethyne (common name: acetylene)

H—C≡C—C—H   propyne

The orbital structure can be shown as follows:

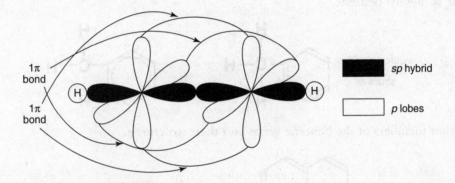

1π bond

1π bond

sp hybrid

p lobes

The bonds formed by the *p* orbitals and the one bond between the *sp* orbitals make up the triple bond.

Aromatics

The aromatic compounds are unsaturated ring structures. The basic formula of this series is C_nH_{2n-6}, and the simplest compound is benzene (C_6H_6). The benzene structure is a resonance structure that is represented like this:

Note: The carbon-to-carbon bonds are neither single nor double bonds but hybrid bonds. This is called *resonance.*

The orbital structure can be represented like this:

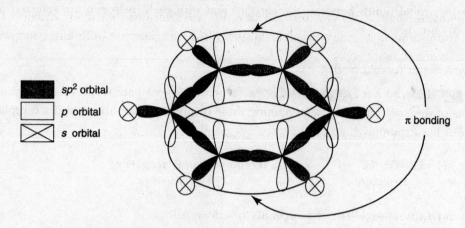

sp² orbital
p orbital
s orbital

π bonding

Most of the aromatics have an aroma, thus the name "aromatic."

The C_6H_5 group is a substituent called phenyl. This is the benzene structure with one hydrogen missing. If the phenyl substituent adds a methyl group, the compound is called toluene or methyl benzene.

Abbreviation for the benzene structure

or

Two other members of the benzene series and their structures:

$C_{10}H_8$ naphthalene

$C_{14}H_{10}$ anthracene

The IUPAC system of naming benzene derivatives, as with chain compounds, involves numbering the carbon atoms in the ring in order to pinpoint the locations of the side chains. However, if only two groups are substituted in the benzene ring, the compound formed will be a benzene derivative having three possible isomeric forms. In such cases, the prefixes *ortho-*, *meta-*, and *para-*, abbreviated as *o-*, *m-*, and *p-*, may be used to name the isomers. In the ortho- structure, the two substituted groups are located on adjacent carbon atoms. In the meta- structure, they are separated by one carbon atom. In the para- structure, they are separated by two carbon atoms.

1,2-Dimethylbenzene
o-xylene

1,3-Dimethylbenzene
m-xylene

1,4-Dimethylbenzene
p-xylene

Isomers

Many of the chain hydrocarbons can have the same formula, but their structures may differ. For example, butane is the first compound that can have two different structures or *isomers* for the same formula.

n-butane   isobutane

This isomerization can be shown by the following equation:

$$CH_3—CH_2—CH_2—CH_3 \xrightarrow[70–100°C]{AlCl_3} CH_3—\overset{\displaystyle CH_3}{\overset{|}{CH}}—CH_3$$

butane   isobutane

The isomers have different properties, both physical and chemical, from the normal structure.

Hydrocarbon Derivatives

Alcohols

The simplest alcohols are alkanes that have one or more hydrogen atoms replaced by the hydroxyl group,—OH. This is called its functional group.

• METHANOL

Methanol is the simplest alcohol. Its structure is

$$H—\overset{\displaystyle H}{\overset{|}{\underset{|}{\underset{\displaystyle H}{C}}}}—OH \quad \text{Methanol or wood alcohol}$$

Properties and uses. Methanol is a colorless, flammable liquid with a boiling point of 65°C. It is miscible with water, is exceedingly poisonous, and can cause blindness if taken internally. It can be used as a fuel, as a solvent, and as a denaturant to make ethyl alcohol, unsuitable for drinking.

• ETHANOL

Ethanol is the best known and most used alcohol. Its structure is

   ethanol

[Notice that the alcohol names are derived from the alkanes by replacing the *e* ending with *-ol.*]

Its common names are ethyl alcohol or grain alcohol.

Properties and uses. Ethanol is a colorless, flammable liquid with a boiling point of 78°C. It is miscible with water and is a good solvent for a wide variety of substances (these solutions are often referred to as "tinctures"). It is used as an antifreeze because of its low freezing point, –115°C, and for making acetaldehyde and ether.

• OTHER ALCOHOLS

Isomeric alcohols have similar formulas but different properties because of their differences in structure. If the —OH is attached to an end carbon, the alcohol is called a primary alcohol. If attached to a "middle" carbon, it is called a secondary alcohol. Some examples:

1-propanol
propyl alcohol

2-propanol
isopropyl alcohol

The number in front of the name indicates to which carbon the —OH ion is attached.

Other alcohols with more than one —OH group:

$$\begin{array}{c} H \quad H \\ | \quad | \\ H-C-C-H \\ | \quad | \\ OH \ OH \end{array}$$

ethylene glycol
1,2-ethanediol

A colorless liquid, high boiling point, low freezing point. Used as permanent antifreeze in automobiles.

$$\begin{array}{c} H \quad H \quad H \\ | \quad | \quad | \\ H-C-C-C-H \\ | \quad | \quad | \\ OH \ OH \ OH \end{array}$$

glycerine
glycerol
1,2,3-propanetriol

Colorless liquid, odorless, viscous, sweet taste. Used to make nitroglycerine, resins for paint, and cellophane.

Aldehydes

The functional group of an aldehyde is the $-C\overset{\displaystyle O}{\underset{\displaystyle H}{\diagup}}$, formyl group. The general formula is RCHO, where R represents a hydrocarbon radical.

Preparation from an alcohol. Aldehydes can be prepared by the oxidation of an alcohol. This can be done by inserting a hot copper wire into the alcohol. A typical reaction is:

methanol
(methyl alcohol)

methanal
(formaldehyde)

The middle structure is an intermediate structure, but since two hydroxyl groups do not stay attached to the same carbon, it changes to the aldehyde by a water molecule "breaking away."

The aldehyde name is derived from the alcohol name by dropping the -*ol* and adding -*al*. Ethanol forms ethanal (acetaldehyde) in the same manner.

Organic Acids or Carboxylic Acids

The functional group of an organic acid is the , carboxyl group. The general formula is R—COOH.

Preparation from an aldehyde. Organic acids can be prepared by the mild oxidation of an aldehyde. The simplest acid is methanoic acid, which is present in ants, bees, and other insects. A typical reaction is:

methanal   methanoic acid
(formaldehyde)   (formic acid)

Notice that in the IUPAC system the name is derived from the alkane stem by adding -*oic*. Ethanal can be oxidized to ethanoic acid:

ethanal   ethanoic acid
                (acetic acid)

Acetic acid, as ethanoic acid is commonly called, is a mild acid that, in the concentrated form, is called glacial acetic acid. Glacial acetic acid is used in many industrial processes, such as making cellulose acetate. Vinegar is a 4% to 8% solution of acetic acid that can be made by fermenting alcohol.

Summary of Oxygen Derivatives

Functional Group

R—H → R—Cl → R—OH → R¹CHO → R¹—COOH

hydrocarbon   chlorine   alcohol   aldehyde   acid
             substitution
             product   (ending -*ol*) (ending -*al*) (ending -*oic*)

Note: R^1 indicates a hydrocarbon chain different from R by one less carbon in the chain.
An actual example using ethane:

$$C_2H_6 \xrightarrow{Cl_2} C_2H_5Cl \xrightarrow{NaOH} C_2H_5OH \xrightarrow{[O]} CH_3CHO \xrightarrow{[O]} CH_3COOH$$

ethane   monochloro-   ethanol   ethanal   ethanoic
         ethane   acid

Ketones

When a secondary alcohol is slightly oxidized, it forms a compound having the functional group , called a ketone. The R^1 indicates that this group need not be the same as R. An example is:

2-propanol   2-propanone
(isopropyl alcohol)   (acetone)

In the IUPAC method the name of the ketone has the ending *-one* with a digit indicating the carbon that has the double-bonded oxygen. Another method of designating a ketone is to name the radicals on either side of the ketone structure and use the word *ketone*. In the preceding reaction, the product would be dimethyl ketone.

Ethers

When a primary alcohol, such as ethanol, is dehydrated with sulfuric acid, an ether forms. The functional group is $R—O—R^1$, in which R^1 may be the same hydrocarbon group as shown below in the first example, or a different hydrocarbon group as shown in example 2.

ethanol   ethanol   ethoxyethane
(ethyl ether or diethyl ether)

2. Another ether with unlike groups, $R—O—R^1$:

ethoxypropane
(ethyl propyl ether)

In the IUPAC method the ether name is made up of the first substituent's stem, then *-oxy*, and then the alkane name for the second substituents.

Amines and Amino Acids

The group NH_2^- is found in the amide ion and the amino group. Under the proper conditions, this ion can replace a hydrogen in a hydrocarbon compound. The resulting compound is called an *amine*. Some examples:

methylamine

aniline

In *amides*, the NH_2^- group replaces a hydrogen in the carboxyl group. When naming amides, the *-ic* of the common name or the *-oic* of the IUPAC name of the parent acid is replaced by *-amide*. See below.

Amino acids are organic acids that contain one or more amino groups. The simplest uncombined amino acid is glycine, or amino acetic acid, NH_2CH_2—COOH. More than 20 amino acids are known, about half of which are necessary in the human diet because they are needed to make up the body proteins.

          acetamide
          ethanamide

Esters

Esters are often compared to inorganic salts because their preparations are similar. To make a salt, you react the appropriate acid and base. To make an ester, you react the appropriate organic acid and alcohol. For example:

The name is made up of the alkyl substituent of the alcohol and the acid name, in which *-ic* is replaced with *-ate*.

The general equation is:

$$\overset{*}{R}O—H + R^1CO—OH \rightarrow R^1CO\overset{*}{O}—R + HOH$$

    alcohol   acid   ester

Esters usually have a sweet smell and are used in perfumes and flavor extracts.

Chapter

14

Review

• QUESTIONS

1. Carbon atoms usually   (a) lose 4 electrons   (b) gain 4 electrons   (c) form 4 covalent bonds   (d) share the 2 electrons in the first principal energy level.

2. Coke is produced from bituminous coal by   (a) cracking   (b) synthesis   (c) substitution   (d) destructive distillation.

3. The usual method for preparing carbon dioxide in the laboratory is (a) heating a carbonate   (b) fermentation   (c) reacting an acid and a carbonate   (d) burning carbonaceous materials.

4. The precipitate formed when carbon dioxide is bubbled into limewater is (a) $CaCl_2$ (b) H_2CO_3 (c) CaO   (d) $CaCO_3$.

5. The "lead" in a lead pencil is (a) bone black   (b) graphite and clay   (c) lead oxide   (d) lead peroxide.

6. The first and simplest alkane is (a) ethane   (b) methane   (c) C_2H_2 (d) methene   (e) CCl_4.

7. Slight oxidation of a primary alcohol gives (a) a ketone   (b) an organic acid   (c) an ether   (d) an aldehyde   (e) an ester.

8. The characteristic group of the organic ester is   (a) —CO—   (b) —COOH   (c) —CHO   (d) —O—   (e) —COO—.

9. The organic acid that can be made from ethanol is (a) acetic acid   (b) formic acid   (c) C_3H_7OH (d) found in bees and ants   (e) butanoic acid.

10. An ester can be prepared by the reaction of (a) two alcohols   (b) an alcohol and an aldehyde   (c) an alcohol and an organic acid   (d) an organic acid and an aldehyde   (e) an acid and a ketone.

11. Compounds that have the same composition but differ in structural formulas (a) are used for substitution products   (b) are called isomers   (c) are called polymers   (d) have the same properties   (e) are usually alkanes.

12. Ethene is the first member of the (a) alkane series   (b) saturated hydrocarbons   (c) alkyne series   (d) unsaturated hydrocarbons   (e) aromatic hydrocarbons.

• ANSWERS

1. (c)	5. (b)	9. (a)
2. (d)	6. (b)	10. (c)
3. (c)	7. (d)	11. (b)
4. (d)	8. (e)	12. (d)

Terms You Should Know

alcohol	amine	ether
aldehyde	amino acid	hydrocarbon
alkane	aromatics	isomer
alkene	ester	ketone
alkyne		

Nucleonics

RADIOACTIVITY

The discovery of radioactivity came in 1896 (less than two months after Wilhelm Röntgen had announced the discovery of X rays). Röntgen had pointed out that the X rays came from a spot on a glass tube where a beam of electrons, in his experiments, was hitting, and that this spot simultaneously showed strong fluorescence. It occurred to Antoine-Henri Becquerel and others that X rays might in some way be related to fluorescence (emission of light when exposed to some exciting agency) and to phosphorescence (emission of light *after* the exciting agency is removed).

Becquerel accordingly tested a number of phosphorescent substances to determine whether they emitted X rays while phosphorescing. He had no success until he tried a compound of uranium. Then he found that the uranium compound, whether or not it was allowed to phosphoresce by exposure to light, continuously emitted something that could penetrate lightproof paper and even thicker materials.

Becquerel determined that the compounds of uranium and the element itself produced *ionization* in the surrounding air. Thus either the ionizing effect, as indicated by the rate of discharge of a charged electroscope, or the degree of darkening of a photographic plate, could be used to measure the intensity of the invisible emission. Moreover, the emission from the uranium was continuous and perhaps even permanent, and required no energy from any external source. Yet, probably because of the current interest and excitement over X rays, Becquerel's work received little attention until early in 1898, when Marie and Pierre Curie entered the picture (Pierre Curie was one of Becquerel's colleagues in Paris).

Searching for the source of the intense radiation in uranium ore, Marie and Pierre Curie used tons of it to isolate very small quantities of two new elements, radium and polonium, both radioactive. Along with Becquerel, the Curies shared the Nobel Prize in physics in 1903. Unfortunately, Pierre Curie died soon afterward from a tragic accident with a horse-drawn street cart. Marie Curie carried on their work and was chosen to succeed him at the Sorbonne as the first female professor at that university. In 1911, she received a second Nobel Prize for her work in chemistry with the newly discovered elements radium and polonium. She founded the Radium Institute in Paris and devoted much of her time to the application of radioactivity to mediums.

The Nature of Radioactive Emissions

While the early separation experiments were in progress, an understanding was slowly being gained of the nature of the spontaneous emission from the various radioactive elements. Becquerel thought at first that there were simply X rays, but THREE different kinds of radioactive emission, now called *alpha particles, beta particles,* and *gamma rays,* were soon found. We now know that alpha particles are positively charged particles of helium nuclei, beta particles are streams of high-speed electrons, and gamma rays are high-energy radiations similar to X rays. The emission of these three types of radiation are depicted below.

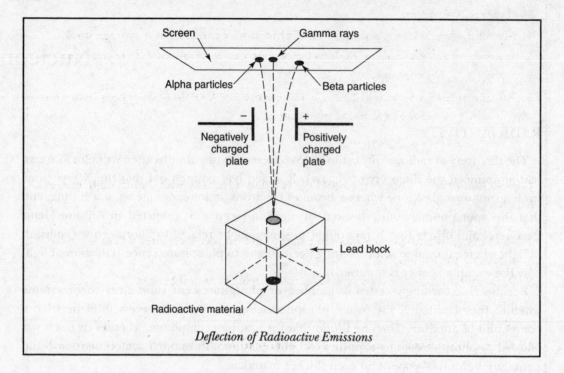

Deflection of Radioactive Emissions

The important characteristics of each type of radiation can be summarized as follows:

Alpha Particle   (helium nucleus $^4_2$He) Positively charged, 2+

1. Ejection reduces the atomic number by 2, the atomic weight by 4 amu.
2. High energy, relatively low velocity.
3. Range: about 5 cm in air.
4. Shielding needed: stopped by the thickness of a sheet of paper, skin.
5. Interactions: produces about 100,000 ionizations per centimeter; repelled by the positively charged nucleus; attracts electrons, but does not capture them until its speed is much reduced.
6. An example: Thorium-230 has an unstable nucleus and undergoes radioactive decay through alpha emission. The nuclear equation that describes this reaction is:

$$^{230}_{90}\text{Th} \rightarrow ^4_2\text{He} + ^{226}_{88}\text{Ra}$$

In a decay reaction like this, the initial element (thorium-230) is called the parent nuclide and the resulting element (radium-226) is called the daughter nuclide.

Beta Particle　　(fast electron) Negatively charged, 1–

1. Ejected when a *neutron* decays into a proton and an electron.
2. High velocity, low energy.
3. Range: about 12 m.
4. Shielding needed: stopped by 1 cm of aluminum or thickness of average book.
5. Interactions: weak due to high velocity, but produces about 100 ionizations per centimeter.
6. An example: Protactinium-234 is a radioactive nuclide that undergoes beta emission. The nuclear equation is:

$$^{234}_{91}\text{Pa} \rightarrow\ ^{234}_{92}\text{U} +\ ^{0}_{-1}\text{e}$$

Gamma Radiation　　(electromagnetic radiation identical with light; high energy) No charge

1. Beta particles and gamma rays are usually emitted together; after a beta is emitted, a gamma ray follows.
2. Arrangement in nucleus is unknown. Same velocity as visible light.
3. Range: no specific range.
4. Shielding needed: about 13 cm of lead.
5. Interactions: weak of itself . . . gives energy to electrons, which then perform the ionization.

Methods of Detection of Alpha, Beta, and Gamma Rays

All methods of detection of these radiations rely on their ability to ionize. There are a number of methods in common use.

1. Photographic plate. The fogging of a photographic emulsion led to the discovery of radioactivity. If this emulsion is viewed under a high-power microscope, it is seen that beta and gamma rays cause the silver bromide grains to develop in a scattered fashion. However, alpha particles, owing to the dense ionization they produce, leave a definite track of exposed grains in the emulsion. Hence not only is the alpha particle detected, but also its range (in the emulsion) can be measured. Special emulsions are capable of showing the beta-particle tracks.

2. Scintillation counter. A fluorescent screen (e.g., ZnS) will show the presence of electrons and X rays, as we have already seen. If the screen is viewed with a magnifying eyepiece, small flashes of light, called scintillations, will be observed. By observing the scintillations, one not only can detect the presence of alpha particles, but also can actually count them.

3. The cloud chamber. One of the most useful instruments for detecting and measuring radiation is the Wilson cloud chamber. Its operation depends on the well-known fact that moisture tends to condense around ions (the probable explanation for some formation of clouds in the sky). If an enclosed region of air is saturated with water vapor (this is always the case if water is present) and the air is cooled suddenly, it becomes supersaturated, that is, it contains for the instant more water vapor than it can hold permanently, and a fog of water droplets develops around the ions in the chamber.

Cloud Chamber

For example, an alpha particle traveling through such a super-saturated atmosphere supplies a trail of ions on which water droplets will condense. This trail is thus made visible. These trails are usually photographed by a camera that operates whenever a piston moves downward causing the air to expand and cool.

4. *The bubble chamber.* This device utilizes a liquid such as ether, ethyl alcohol, pentane, or propane, which is superheated to well above the boiling point. When the pressure is released quickly, the liquid is in a highly unstable condition, ready to boil violently. An ionizing particle passing through the liquid at this instant leaves a trail of tiny bubbles that may be photographed.

5. *The electroscope.* One of the simplest and one of the first used in work on radioactivity consists of two thin gold leaves that are charged by a battery with the same type of charge. Radioactive materials produce ions in proportion to their radioactivity. A negatively charged electroscope becomes discharged when ions in the air take electrons from the electroscope, causing the leaves to collapse gradually.

Electroscope

6. *The Geiger counter.* This instrument is perhaps the most widely used at the present time for counting individual radiation. It consists of a fine wire of tungsten mounted along the axis of a tube that contains a gas at reduced pressure. A difference of potential of about 1,000 volts is applied in such a way as to make the metal tube negative with respect to the wire. The voltage is high enough so that the electrons produced are accelerated by the electric field. Near the wire, where the field is strongest, the accelerated particles produce more ions, positive ones going to the walls, negative ones being collected by the wire. Any particle that will produce an ion gives rise to the same avalanche of ions, so the type of particle cannot be identified. However, each individual particle can be detected.

Source of Radioactivity

The nuclei of uranium, radium, and other radioactive elements are continually disintegrating. It should be emphasized that spontaneous disintegration produces the gas known as radon. The time it takes for half of the material to disintegrate is called its *half-life*.

For example, for radium, we know that, on the average, half of all the radium nuclei present will have disintegrated to radon in 1,590 years. In another 1,590 years, half of this remainder will decay, and so on. When a radium atom disintegrates, it loses an alpha particle, which eventually, upon gaining two electrons, becomes a neutral helium atom. The remainder of the atom becomes radon.

Such a conversion of an element to a new element (because of a change in the number of protons) is called a *transmutation*. This transmutation can be produced artificially by bombarding the nuclei of a substance with various particles from a particle accelerator, such as the cyclotron.

The following uranium-radium disintegration series shows how a radioactive atom may change when it loses each kind of particle. Note that an atomic number is shown by a subscript ($_{92}$U), and the isotopic mass by a superscript (^{238}U). The alpha particle is shown by the Greek symbol "α" and the beta particle by "β".

$$^{238}_{92}\text{U} \xrightarrow{-\alpha} \ ^{234}_{90}\text{Th} \xrightarrow{-\beta} \ ^{234}_{91}\text{Pa} \xrightarrow{-\beta} \ ^{234}_{92}\text{U} \xrightarrow{-\alpha} \ ^{230}_{90}\text{Th} \xrightarrow{-\alpha}$$

$$^{226}_{88}\text{Ra} \xrightarrow{-\alpha} \ ^{222}_{86}\text{Rn} \xrightarrow{-\alpha} \ ^{218}_{84}\text{Po} \xrightarrow{-\alpha} \ ^{214}_{82}\text{Pb} \xrightarrow{-\beta} \ ^{214}_{83}\text{Bi} \xrightarrow{-\beta}$$

$$^{214}_{84}\text{Po} \xrightarrow{-\alpha} \ ^{210}_{82}\text{Pb} \xrightarrow{-\beta} \ ^{210}_{83}\text{Bi} \xrightarrow{-\beta} \ ^{210}_{84}\text{Po} \xrightarrow{-\alpha} \ ^{206}_{82}\text{Pb (stable)}$$

Particle Lost	*Weight Change*	*Atomic Number Change*
Alpha (α)	Loses 4 amu	Decreases by 2
Beta (β)	None	Increases by 1

This can be shown graphically in the radioactive decay series below.

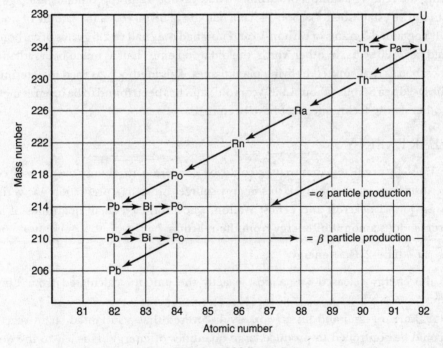

Uranium-238 Decay Series

The stability of an atom seems to be related to binding energy. The binding energy is the amount of energy released when a nucleus is formed from its component particles. If you add the mass of the components and compare this sum with the actual mass of the nucleus formed, there will be a small difference in these figures. This difference in mass can be converted to its energy equivalent using Einstein's equation, $E = mc^2$, where E is the energy, m is the mass, and c is the velocity of light. It is this energy that is called the binding energy.

It has been found that the lightest and heaviest elements have the smallest binding energy per nuclear particle and thus are less stable than the elements of intermediate atomic masses, which have the greatest binding energy.

The relationship of even-odd number of protons to the number of neutrons affects the stability of a nucleus. Many of the stable nuclei have even numbers of protons and neutrons, while stability is less frequent in nuclei that have an even number of protons and an odd number of neutrons, or vice versa. Only a few stable nuclei are known that have odd numbers of protons and neutrons.

Radioactive Dating

A helpful adaptation of radioactive decay is in the determination of the age of a substance such as rocks and relics that have bits of organic material trapped in them. Because carbon-14 has a half-life of about 5,700 years and occurs in the remains of organic materials, it has been useful in dating these materials. A small percentage of CO_2 in the atmosphere contains carbon-14. The stable isotope of carbon is carbon-12. Carbon-14 is a beta emitter and decays to form nitrogen-14:

$$^{14}_{6}C \rightarrow ^{14}_{7}N + ^{0}_{-1}e$$

In any living organism, the ratio of carbon-14 to carbon-12 is the same as in the atmosphere because of the constant interchange of materials between organism and surroundings. When an organism dies, this interaction stops, and the carbon-14 gradually decays to nitrogen. By comparing the relative amounts of carbon-14 and carbon-12 in the remains, the age of the organism can be established. Carbon-14 has a half-life of 5,700 years. If a sample of wood had originally contained 5 grams of carbon-14 and now had only half or 2.5 grams of carbon-14, its age would be 5,700 years. In other words, the old wood emits half as much beta radiation per gram of carbon as that emitted by living plant tissues. This method was used to determine the age of the Dead Sea Scrolls (about 1,900 years old) and has been found to be in agreement with several other dating techniques that have been used.

NUCLEAR ENERGY

Since Einstein predicted that matter could be converted to energy over a half century ago, scientists have tried to unlock this energy source. Einstein's prediction was verified in 1932 by Sir John Cockcroft and Ernest Walton, who produced small quantities of helium and a tremendous amount of energy from the reaction

$$^{7}_{3}Li + ^{1}_{1}H \rightarrow 2^{4}_{2}He + energy$$

In fact, the energy released was almost exactly the amount calculated from Einstein's equation.

In 1942, Enrico Fermi and his co-workers discovered that a sustained chain reaction of fission could be controlled to produce large quantities of energy. This led to the development of the nuclear fission bomb, which brought World War II to an end.

Conditions for Fission

When fissionable material, like ^{235}U, is bombarded with a "slow" neutron, fission occurs, giving off different fission products. An example of such a reaction is shown in Figure 42. As long as one of the released neutrons produces another reaction, the chain reaction will continue. If each fission starts more than one reaction, the process becomes tremendously powerful in a very short time. This occurs in the atomic bomb.

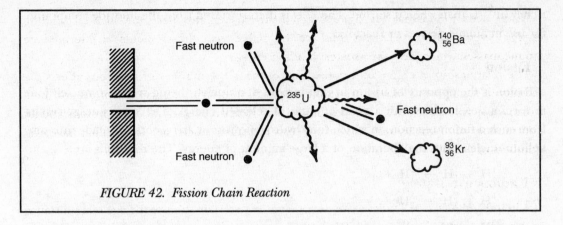

FIGURE 42. *Fission Chain Reaction*

To obtain the neutron "trigger" for this reaction, neutrons must be slowed down so that they do not pass through the nucleus without effect. This "slowing down" or moderating is best done by letting fast neutrons collide with many relatively light atoms, such as hydrogen, deuterium, and carbon. Graphite, paraffin, ordinary water, and "heavy water" (containing deuterium instead of ordinary hydrogen) are all suitable moderators.

Nuclear fission can be made to occur in an uncontrolled explosion or in a controlled nuclear reactor. In both cases enough fissionable material must be present so that, once the reaction starts, it can at least sustain itself. This amount of fissionable material is called the *critical mass*. In the bomb, the number of reactions increases tremendously, whereas in a reactor the rate of fissions is controlled.

The typical nuclear reactor or "pile" is made up of the following kinds of material:

1. Fissionable material—sustains the chain reaction.

2. Moderator—slows down fission neutrons.

3. Control rods (cadmium or boron steel rods)—absorb excess neutrons and control rate of reaction.

4. Concrete encasement—provides shielding from radiation.

There are many variations of this basic reactor as reactors become more efficient. Many United States' cities now receive some of their electrical power from nuclear power stations.

Methods of Obtaining Fissionable Material

The natural abundance of ^{235}U is extremely small: about 0.0005% of the earth's crust, with 0.7% natural uranium. Therefore the problem is to separate this isotope from the rest of the more abundant isotopes. Two methods have been used to isolate ^{235}U: the electromagnetic process and gaseous diffusion.

The electromagnetic process is based on the principle of the mass spectrograph. Gaseous ions with similar charges but different masses are passed between the poles of a magnet. The lighter ones are deflected more than the heavier ones and are thus collected in separate compartments. These large mass separators are called "calutrons" and were used extensively to collect the first uranium-235 isotopes.

The diffusion process was found to be faster and less expensive than the electromagnetic process. This process makes use of the gas diffusion principle that light molecules diffuse faster than heavier ones. When uranium is converted into gaseous uranium hexafluoride, UF_6, and allowed to diffuse through miles of porous partitions under reduced pressure, the lighter $^{235}UF_6$ will diffuse faster than the $^{238}UF_6$. Thus the gas coming through first will be richer in ^{235}U than when it started. The ^{235}U is then retrieved from the fluoride compound for use in atomic devices or reactors.

Fusion

Fusion is the opposite of fission in that instead of a nucleus being split, two nuclei join to form a new nucleus with a great amount of released energy. The sun's energy results from such a fusion reaction, in which four hydrogen protons are eventually made into one helium nucleus with the liberation of a large amount of energy. The reactions are:

$$^1_1H + ^1_1H \rightarrow ^2_1H + ^0_1e$$

$$^1_1H + ^2_1H \rightarrow ^3_2He$$

$$^3_2He + ^3_2He \rightarrow ^4_2He + 2^1_1H$$

$$^3_2He + ^1_1H \rightarrow ^4_2He + ^0_1e$$

So far, nuclear fusion has been produced only in an uncontrolled explosion, the hydrogen bomb. Scientists in every major country of the world are attempting to find a means of controlling a sustained fusion reaction.

In the meantime, nuclear fission is being adapted to produce energy for electrical power stations, merchant ships, aircraft carriers, and submarines. Experimentation is proceeding on the nuclear aircraft engine and a nuclear rocket engine.

NEW SUBATOMIC PARTICLES

As scientists continue the search for more information about the atom and its makeup, they have been especially interested in the subatomic particles. In the 1960s, with more powerful accelerators to investigate the atom, they found more than 100 different subatomic particles. More recently, research has suggested a new order that brings a simplicity to the particles that compose matter. By studying the tracks made on photographic plates by the particles that result from nuclear collisions, Gell-Mann at the California Institute of Technology and Zweig at CERN in Switzerland arrived at a new way of accounting for these subatomic particles. Their theory is that all small particles are combinations of still smaller particles called quarks. A quark is a particle with a fractional charge of either 1/3 or 2/3 the charge of an electron. This theory simplifies the multitude of particles.

At first three quarks were postulated, but the number has grown to six. The usual subatomic particles presented in a first-year chemistry book account for only two quarks. Each quark is described or named by its particular characteristic, which is referred to as its "flavor." The following table lists the six quarks and their fractional charges.

SIX KINDS OF QUARKS

Name	Electric Charge
Up	2/3
Down	−1/3
Charmed	2/3
Strangeness	−1/3
Top (truth)	2/3
Bottom (beauty)	−1/3

The two quarks described as "up" and "down" are thought to constitute most of the ordinary matter in the universe. Scientists have never isolated individual quarks; they attribute their failure to the magnitude of the binding force that holds the quarks together.

The electron is now believed to be composed of smaller, elementary particles, called *leptons*. This is also true of the neutrino, which is similar in mass to the electron but has a neutral charge.

Chapter
15
Review

• QUESTIONS

1. Radioactive changes differ from ordinary chemical changes because radioactive changes   (a) involve changes in the nucleus   (b) are explosive   (c) absorb energy   (d) release energy.

2. Isotopes of uranium have different   (a) atomic numbers   (b) atomic masses   (c) numbers of planetary electrons   (d) numbers of protons.

3. Pierre and Marie Curie discovered   (a) oxygen   (b) hydrogen   (c) chlorine   (d) radium.

4. The number of protons in the nucleus of an atom of atomic number 32 is   (a) 4   (b) 32   (c) 42   (d) 73.

5. Atoms of ^{235}U and ^{238}U differ in structure by three   (a) electrons   (b) isotopes   (c) neutrons   (d) protons.

6. In a hydrogen bomb, hydrogen is converted into   (a) uranium   (b) helium   (c) barium   (d) plutonium.

7. The use of radioactive isotopes has already produced promising results in the treatment of certain types of   (a) cancer   (b) heart disease   (c) pneumonia   (d) diabetes.

8. The emission of a beta particle results in a new element with the atomic number   (a) increased by 1   (b) increased by 2   (c) decreased by 1   (d) decreased by 2.

9. A substance used as a moderator in a nuclear reactor is (a) marble   (b) hydrogen   (c) tritium   (d) graphite.

10. A self-sustaining nuclear fission chain reaction depends upon the release of   (a) protons   (b) neutrons   (c) electrons   (d) alpha particles.

11. Einstein's formula for the Law of Conservation of Mass and Energy states that $E =$ _____ .

12. The energy of the sun is thought to be produced by the fusion of hydrogen atoms into _____ .

13. Could a mass spectrograph (or calutron) be used to separate ^{54}Mn from ^{54}Cr? _____ .

14. The release of energy from the combination of two nuclei into one is called _____ .

15. Write the equation for the first observed artificial transmutation by Rutherford: _____ .

Choose an answer from this list that best matches the numbered statements below.

A. Crooke   H. gamma ray
B. alpha particle   I. gaseous diffusion
C. beta particle   J. neptunium
D. Becquerel   K. plutonium
E. Einstein   L. ^{235}U
F. chain reaction   M. ^{238}U
G. deuteron

16. Helium nucleus:
17. Stable isotope of uranium:
18. One method of separating isotopes:
19. Most penetrating ray from radioactive decay:
20. Mass-energy conversion:

• **ANSWERS:**

1. (a)	8. (a)	15. $^{14}_{7}N + ^{4}_{2}He \rightarrow ^{17}_{8}O + ^{1}_{1}H$
2. (b)	9. (d)	16. B
3. (d)	10. (b)	17. M
4. (b)	11. mc^2	18. I
5. (c)	12. He	19. H
6. (b)	13. No	20. E
7. (a)	14. fusion	

Terms You Should Know

alpha particle	fission	lepton
beta particle	fluorescence	phosphorescence
bubble chamber	fusion	quark
cloud chamber	gamma ray	transmutation
critical mass	Geiger counter	
electroscope	half-life	

Representative Laboratory Setups

SOME BASIC SETUPS

The figures in this section show some of the basic laboratory setups used in beginning chemistry. The purpose of these diagrams is to review the basic techniques of assembling equipment with regard to some knowledge about the reactants and products involved.

In the equations, the letters in parentheses have the following meanings:

(s) = solid   (ℓ) = liquid   (g) = gas

1. *Preparation of a gaseous product, nonsoluble in water, by water displacement from solid reactants.*

FIGURE 43. Preparation of Oxygen

- **EXAMPLE:**

    Preparation of oxygen (O_2).

    $$2KClO_3(s) + MnO_2(s) \rightarrow 2KCl + 3O_2(g) + MnO_2$$

2. *Preparation of a gaseous product, nonsoluble in water, by water displacement from at least one reactant in solution.*

(*Note:* Purpose of thistle tube shown in Figure 43:

1. Introduction of more liquid without "opening" the reacting vessel.
2. Safety valve to indicate blocked delivery tube by the rise of liquid in the thistle tube.)

FIGURE 44. Preparation of Carbon Dioxide, Nitric Oxide, and Hydrogen

• EXAMPLES:

Preparation of carbon dioxide (CO_2).

$$CaCO_3(s) + 2HCl(\ell) \rightarrow CaCl_2 + H_2O + CO_2(g)$$

Preparation of nitric oxide (NO).

$$3Cu(s) + 8 \text{ dilute } HNO_3(\ell) \rightarrow 3Cu(NO_3)_2 + 4H_2O + 2NO(g)$$

Preparation of hydrogen (H_2).

$$Zn(s) + 2HCl(\ell) \rightarrow ZnCl_2 + H_2(g)$$

3. *Preparation of a gaseous product heavier than air that can best be collected by the upward displacement of air.*

FIGURE 45. Preparation of Chlorine and Hydrogen Chloride

Basis of determining when the bottles are filled could be the color (such as chlorine—yellowish green); some indicator in the last bottle (such as hypo, sodium thiosulfate, which turns milky yellow when Cl_2 bubbles into it); a litmus solution turning pink when HCl bubbles into it; or wafting to check odor (such as SO_2).

- **EXAMPLES:**

    Preparation of chlorine (Cl_2).

    $$MnO_2(s) + 4HCl(\ell) \rightarrow 2H_2O + MnCl_2 + Cl_2(g)$$

    Preparation of hydrogen chloride (HCl).

    $$NaCl(s) + H_2SO_4(\ell) \rightarrow NaHSO_4 + HCl(g)$$

Using thistle tube with stopcock to control flow of liquid reactant:

Glass stopcock to control flow of liquid reactant

Card cover

FIGURE 46. *Preparation of Sulfur Dioxide and Hydrogen Sulfide*

- **EXAMPLES:**

    Preparation of sulfur dioxide (SO_2).

    $$Na_2SO_3(s) + H_2SO_4(\ell) \rightarrow Na_2SO_4 + H_2O + SO_2(g)$$
    $$\text{strong, irritating odor}$$

    Preparation of hydrogen sulfide (H_2S).

    $$2HCl(\ell) + FeS(s) \rightarrow FeCl_2 + H_2S(g)$$
    $$\text{rotten egg odor}$$

4. *Preparation of a gaseous product that is soluble in water and lighter than air by the downward displacement of air.*

- **EXAMPLE:**

    Preparation of ammonia (NH_3).

    $$2NH_4Cl(s) + Ca(OH)_2(s) \rightarrow CaCl_2 + 2H_2O + 2NH_3(g)$$

FIGURE 47. Preparation of Ammonia

5. *Distillation of a liquid.*

Thermometer

Liebig
condenser

Water out

Water in

Distillate

FIGURE 48. Distillation of a Liquid

Removes dissolved impurities, which remain in the flask. Does not remove volatile materials, which vaporize and pass over into the distillate.

6. *Separation of a mixture by chromatography.*

FIGURE 49. *Chromatography Setup*

• EXAMPLE:

Chromatography is a process used to separate parts of a mixture. The component parts separate as the solvent carrier moves past the spot by capillary action. Because of variations in solubility, attraction to the filter paper, and density, each fraction moves at a different rate. Once separation occurs, the fractions are either identified by their color or removed for other tests. A usual example is the use of Shaeffer Skrip Ink No. 32, which separates into yellow, red, and blue streaks of dyes.

7. *Measuring potentials in electrochemical cells.*

FIGURE 50. *Potentiometer Setup for Measuring Potential*

• **EXAMPLE:**

The voltmeter in this zinc-silver electrochemical cell would read approximately 1.56 volts. This means that the Ag to Ag^+ half-cell has 1.56V more electron-attracting ability than the Zn to Zn^{2+} half-cell. If the potential of the zinc half-cell were known, the potential of the silver half-cell could be determined by adding 1.56V to the potential of the zinc half-cell. In a setup like this, only the difference in potential between two half-cells can be measured.

8. *Titration setup.*

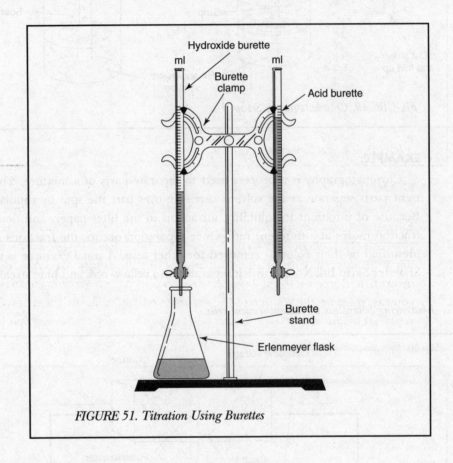

FIGURE 51. Titration Using Burettes

• **EXAMPLE:**

The titration of NaOH solution of unknown concentration in one burette with 0.1 M HCl in the other.

Introduce approximately 15 mL of NaOH into the flask, and add an indicator such as litmus or phenolphthalein. Add the HCl slowly with constant swirling. When color change occurs and is retained, record the amount used. To find the *M* of the NaOH use this formula:

$$M_{acid} \times V_{acid} = M_{base} \times V_{base}$$

9. *Replacement of hydrogen by a metal.*

FIGURE 52. Eudiometer Apparatus

- **EXAMPLE:**

    Measure the mass of a strip of magnesium with an analytical balance to the nearest 0.001 g. Using a strip with a mass of about 0.040 g produces about 40 mL of H_2. Pour 5 mL of concentrated HCl into the eudiometer, and slowly fill the remainder with water. Try to minimize mixing. Lower the coil of Mg strip into the tube, invert it, and lower it to the bottom of the battery jar. After the reaction is complete, you can measure the volume of the gas released and calculate the mass of hydrogen replaced by the magnesium. (Refer to Chapter 5 for a discussion of gas laws.)

10. *Drying gases to remove water vapor.*

FIGURE 53. CaCl₂ Drying Tube

Calcium chloride is hydroscopic and absorbs water vapor as the gas passes through this tube.

11. *Using displaced water to measure the volume of a gas that is produced.*

FIGURE 54. *Displacing Water with an Insoluble Gas*

- **EXAMPLE:**

In this setup, before the gas begins to be generated, you should release the pinch clamp and gently blow in the tube at *A* to get water up and into the remainder of the system. Close the pinch clamp once water starts to flow at *B*; then connect the system to the generator. You are ready to read the quantity of gas evolved into the flask by reading the volume of water displaced into the graduated cylinder.

SUMMARY OF QUALITATIVE TESTS

I. Identification of Gases

Gas	Test	Result
Ammonia NH_3	1. Smell cautiously. 2. Test with litmus. 3. Expose to HCl fumes.	1. Sharp odor. 2. Red litmus turns blue. 3. White fumes form (NH_4Cl).
Carbon dioxide CO_2	1. Pass through limewater, $Ca(OH)_2$.	1. White precipitate forms, $CaCO_3$.
Carbon monoxide CO	1. Burn it and pass product through limewater, $Ca(OH)_2$.	1. White precipitate forms, $CaCO_3$.
Chlorine Cl_2	1. Smell cautiously 2. Observe color.	1. Irritating odor. 2. Yellowish green.
Hydrogen H_2	1. Allow it to mix with some air, then ignite. 2. Burn it—trap product.	1. Gas explodes. 2. Burns with blue flame—product H_2O turns cobalt chloride paper from blue to pink.

Gas	Test	Result
Hydrogen chloride HCl	1. Smell cautiously. 2. Exhale over the gas. 3. Dissolve in water and test with litmus. 4. Add $AgNO_3$ to the solution.	1. Choking odor. 2. Vapor fumes form. 3. Blue litmus turns red. 4. Forms white precipitate.
Hydrogen sulfide H_2S	1. Smell cautiously. 2. Test with moist lead acetate paper.	1. Rotten egg odor. 2. Turns brown-black (PbS).
Nitric oxide NO	1. Expose to the air.	1. Colorless gas turns reddish brown.
Nitrous oxide N_2O	1. Insert glowing splint. 2. Add nitric oxide gas.	1. Bursts into flame. 2. Remains colorless.
Oxygen O_2	1. Insert glowing splint. 2. Add nitric oxide gas.	1. Bursts into flame. 2. Turns reddish brown.
Sulfur dioxide SO_2	1. Smell cautiously. 2. Allow it to bubble into purple potassium permanganate solution.	1. Choking odor. 2. Solution becomes colorless.

II. Identification of Negative Ions

Ion	Test	Result
Acetate $C_2H_3O_2^-$	Add conc. H_2SO_4 and warm gently.	Odor of vinegar released.
Bromide Br^-	Add chlorine water and some CCl_4; shake.	Reddish brown color concentrated in CCl_4 layer.
Carbonate CO_3^-	Add HCl acid; pass released gas through limewater.	White cloudy ppt. forms (ppt. = precipitate).
Chloride Cl^-	1. Add silver nitrate solution. 2. Then add nitric acid, later followed by ammonium hydroxide.	1. White ppt. forms. 2. Precipitate insol. in HNO_3 but dissolves in NH_4OH.
Hydroxide OH^-	Test with red litmus paper.	Turns blue.
Iodide I^-	Add chlorine water and some CCl_4; shake.	Purple color concentrated in CCl_4 layer.
Nitrate NO_3^-	Add freshly made ferrous sulfate sol., then conc. H_2SO_4 carefully down the side of the tilted tube.	Brown ring forms at junction of layers.
Nitrite NO_2^-	Add dilute H_2SO_4.	Brown fumes (NO_2) released.
Sulfate SO_4^-	Add sol. of $BaCl_2$, then HCl.	White ppt. forms; insoluble in HCl.

Ion	Test	Result
Sulfide S^{2-}	Add HCl and test gas released with lead acetate paper.	Gas, with rotten egg odor, turns paper brown-black.
Sulfite SO_3^{2-}	Add HCl and pass gas into purple $KMnO_4$ sol.	Solution turns colorless.

III. Identification of Positive Ions

Ion	Test	Result
Ammonium NH_4^+	Add strong base (NaOH); heat gently.	Odor of ammonia.
Ferrous Fe^{2+}	Add sol. of potassium ferricyanide, $K_3Fe(CN)_6$.	Dark blue ppt. forms (Turnball's blue).
Ferric Fe^{3+}	Add sol. of potassium ferrocyanide, $K_4Fe(CN)_6$.	Dark blue ppt. forms (Prussian blue).
Hydrogen H^+	Test with blue litmus paper.	Turns red.

IV. Qualitative Tests for Metals

Flame tests. Carefully clean a platinum wire by dipping it into dil. HNO_3 and heating in the Bunsen flame. Repeat until the flame is colorless. Dip heated wire into the substance being tested (either solid or solution), and then hold it in the hot outer part of the Bunsen flame.

Compound of	Color of Flame
Na	Yellow
K	Violet (use cobalt-blue glass to screen out Na impurities)
Li	Crimson
Ca	Orange-red
Ba	Green
Sr	Bright red

Hydrogen sulfide tests. Bubble hydrogen sulfide gas through the solution of a salt of the metal being tested. Check color of the precipitate formed.

Compound of	Color of Sulfide Precipitate
Lead (Pb)	Brown-black (PbS)
Copper (Cu)	Black (CuS)
Silver (Ag)	Black (Ag_2S)
Mercury (Hg)	Black (HgS)
Nickel (Ni)	Black (NiS)
Iron (Fe)	Black (FeS)
Cadmium (Cd)	Yellow (CdS)
Arsenic (As)	Light yellow (As_2S_3)
Antimony (Sb)	Orange (Sb_2S_3)
Zinc (Zn)	White (ZnS)
Bismuth (Bi)	Brown (Bi_2S_3)

*Practice
SAT II:
Chemistry
Tests*

GENERAL INFORMATION

SAT II: Chemistry is planned to test the principles and concepts drawn from the factual material found largely in inorganic chemistry and, to a much lesser extent, in organic chemistry. Only a few questions are asked concerning industrial or analytical chemistry.

According to a description of the test by the College Entrance Examination Board, it includes kinetic molecular theory and the three states of matter; atomic theory and structure and the periodic table; nuclear reactions; quantitative relations as applied to chemical formulas and equations; chemical bonding and molecular structure, and their relations to properties; the nature of chemical reactions, including acid-base reactions, oxidation-reduction reactions, ionic reactions, and other chemical changes occurring in solution; energy changes accompanying chemical reactions; interpretation of chemical equilibria and reaction rates; solution phenomena; electrochemistry, nuclear chemistry, and radiochemistry; physical and chemical properties of the more familiar metals, transition elements, and nonmetals and of the more familiar compounds; understanding and interpretation of laboratory procedures and observations.

You will be provided with a periodic table to use during the test. All necessary information regarding atomic numbers and atomic masses is given on the chart. Plan to use the chart provided with the diagnostic test in the front of this book when taking each practice test.

SAT II: Subject Test in Chemistry has 85 multiple-choice questions that are distributed in the following manner among the topics covered:

Topics Covered*	Approximate Percentage of Test	Approximate Number of Questions
I. ATOMIC AND MOLECULAR STRUCTURE: • Atomic Theory and Structure, including periodic relationships • Nuclear Reactions • Chemical Bonding and Molecular Structure	23	20
II. STATES OF MATTER: • Kinetic Molecular Theory of Gases • Solutions, including concentration units, solubility, conductivity, and colligative properties	14	12
III. REACTION TYPES: • Acids and Bases • Oxidation-reduction, including electrochemical cells • Precipitation	14	12
IV. STOICHIOMETRY: Including the mole concept, Avogadro's number, empirical and molecular formulas, percentage composition, stoichiometric calculations, and limiting reagents	14	12
V. EQUILIBRIUM AND REACTION RATES: Equilibrium, including mass action expressions, ionic equilibria, and Le Châtelier's principle; factors affecting rates of reaction	7	6
VI. THERMODYNAMICS: Including energy changes in chemical reactions and physical processes, Hess's Law, and randomness	7	6
VII. DESCRIPTIVE CHEMISTRY: Physical and chemical properties of elements and their more familiar compounds, including simple examples from organic chemistry, and environmental chemistry	14	12
VIII. LABORATORY: Equipment, measurement, procedures, observations, safety, calculations, and interpretation of results	7	6

Note: Every edition contains approximately five questions on equation balancing and/or predicting products of chemical reactions. These are distributed among the various content categories.

*Reprinted by permission of Educational Testing Service.

BASIC ABILITIES TESTED FOR

1. The ability to demonstrate an understanding of basic scientific principles and concepts.

2. The ability to apply scientific concepts and principles.

3. The ability to handle quantitative relationships (to use mathematical relationships and to solve problems involving chemical principles and concepts).

4. The ability to organize and interpret information obtained by observation and experimentation so as to discern cause-and-effect relationships.

5. The ability to interpret so as to draw conclusions or make inferences from experimental data, including data presented in graphic or tabular form.

6. The ability to apply laboratory procedures.

Each of the following tests measures these abilities with various types of questions. Before attempting the tests, read the Introduction in the front of this book, especially the section entitled "Final Preparation—The Day before the Test."

ANSWER SHEET FOR PRACTICE TEST 1

Determine the correct answer for each question. Then, using a No. 2 pencil, blacken completely the oval containing the letter of your choice.

1. (A) (B) (C) (D) (E)
2. (A) (B) (C) (D) (E)
3. (A) (B) (C) (D) (E)
4. (A) (B) (C) (D) (E)
5. (A) (B) (C) (D) (E)
6. (A) (B) (C) (D) (E)
7. (A) (B) (C) (D) (E)
8. (A) (B) (C) (D) (E)
9. (A) (B) (C) (D) (E)
10. (A) (B) (C) (D) (E)
11. (A) (B) (C) (D) (E)
12. (A) (B) (C) (D) (E)
13. (A) (B) (C) (D) (E)
14. (A) (B) (C) (D) (E)
15. (A) (B) (C) (D) (E)
16. (A) (B) (C) (D) (E)
17. (A) (B) (C) (D) (E)
18. (A) (B) (C) (D) (E)
19. (A) (B) (C) (D) (E)
20. (A) (B) (C) (D) (E)
21. (A) (B) (C) (D) (E)
22. (A) (B) (C) (D) (E)
23. (A) (B) (C) (D) (E)

ON THE ACTUAL CHEMISTRY TEST, THE FOLLOWING TYPE OF QUESTION MUST BE ANSWERED ON A SPECIAL SECTION (LABELED "CHEMISTRY") AT THE LOWER LEFT-HAND CORNER OF PAGE 2 OF YOUR ANSWER SHEET. THESE QUESTIONS WILL BE NUMBERED BEGINNING WITH 101 AND MUST BE ANSWERED ACCORDING TO THE FOLLOWING DIRECTIONS.

CHEMISTRY* Fill in oval CE only if II is a correct explanation of I.

	I	II	CE*
101.	(T) (F)	(T) (F)	◯
102.	(T) (F)	(T) (F)	◯
103.	(T) (F)	(T) (F)	◯
104.	(T) (F)	(T) (F)	◯
105.	(T) (F)	(T) (F)	◯
106.	(T) (F)	(T) (F)	◯
107.	(T) (F)	(T) (F)	◯
108.	(T) (F)	(T) (F)	◯
109.	(T) (F)	(T) (F)	◯
110.	(T) (F)	(T) (F)	◯
111.	(T) (F)	(T) (F)	◯
112.	(T) (F)	(T) (F)	◯
113.	(T) (F)	(T) (F)	◯
114.	(T) (F)	(T) (F)	◯
115.	(T) (F)	(T) (F)	◯
116.	(T) (F)	(T) (F)	◯

ON THE ACTUAL CHEMISTRY TEST, THE RE-MAINING QUESTIONS MUST BE ANSWERED BY RETURNING TO THE SECTION OF YOUR AN-SWER SHEET YOU STARTED FOR CHEMISTRY.

40. (A) (B) (C) (D) (E)   56. (A) (B) (C) (D) (E)   71. (A) (B) (C) (D) (E)
41. (A) (B) (C) (D) (E)   57. (A) (B) (C) (D) (E)   72. (A) (B) (C) (D) (E)
42. (A) (B) (C) (D) (E)   58. (A) (B) (C) (D) (E)   73. (A) (B) (C) (D) (E)
43. (A) (B) (C) (D) (E)   59. (A) (B) (C) (D) (E)   74. (A) (B) (C) (D) (E)
44. (A) (B) (C) (D) (E)   60. (A) (B) (C) (D) (E)   75. (A) (B) (C) (D) (E)
45. (A) (B) (C) (D) (E)   61. (A) (B) (C) (D) (E)   76. (A) (B) (C) (D) (E)
46. (A) (B) (C) (D) (E)   62. (A) (B) (C) (D) (E)   77. (A) (B) (C) (D) (E)
47. (A) (B) (C) (D) (E)   63. (A) (B) (C) (D) (E)   78. (A) (B) (C) (D) (E)
48. (A) (B) (C) (D) (E)   64. (A) (B) (C) (D) (E)   79. (A) (B) (C) (D) (E)
49. (A) (B) (C) (D) (E)   65. (A) (B) (C) (D) (E)   80. (A) (B) (C) (D) (E)
50. (A) (B) (C) (D) (E)   66. (A) (B) (C) (D) (E)   81. (A) (B) (C) (D) (E)
51. (A) (B) (C) (D) (E)   67. (A) (B) (C) (D) (E)   82. (A) (B) (C) (D) (E)
52. (A) (B) (C) (D) (E)   68. (A) (B) (C) (D) (E)   83. (A) (B) (C) (D) (E)
53. (A) (B) (C) (D) (E)   69. (A) (B) (C) (D) (E)   84. (A) (B) (C) (D) (E)
54. (A) (B) (C) (D) (E)   70. (A) (B) (C) (D) (E)   85. (A) (B) (C) (D) (E)
55. (A) (B) (C) (D) (E)

Note: For all questions involving solutions and/or chemical equations, assume that the system is in water unless otherwise stated.

Part A

Directions: Each set of lettered choices below refers to the numbered statements or formulas immediately following it. Select the one lettered choice that best fits each statement or formula and then fill in the corresponding oval on the answer sheet. A choice may be used once, more than once, or not at all in each set.

Questions 1–9

Periodic Table (abbreviated)

^{3}Li					(D)	^{10}Ne
	(A)	$\cdots$		(C)		
(B)	^{20}Ca		$\cdots$			(E)

1. The most electronegative element

2. The element with a possible oxidation number of –2

3. The element that would react in a one-to-one ratio with (D)

4. The element with the smallest ionic radius

5. The element with the smallest first ionization potential

6. The element with a complete p orbital as its outermost energy level

7. A member of the alkali-metals family

8. An inert gas

9. The element that would react most actively when placed in water to form a strong base

Questions 10–12 refer to the following heating curve for water:

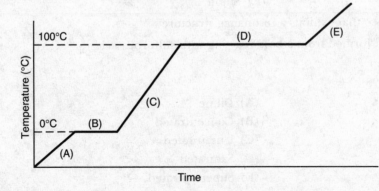

10. In which part of the curve is the state of H_2O only a solid?

11. When is the heat to change the state of H_2O greater?

12. Where is the temperature of H_2O changing at 1°C/cal/g?

Questions 13–15 refer to the following diagram:

Course of Reaction

13. Indicates the activation energy of this reaction

14. Indicates the portion of the curve that would be directly affected by the addition of a catalyst

15. Indicates the net energy released from the reaction

Questions 16–18

        (A) Iron
        (B) Gold
        (C) Sodium
        (D) Helium
        (E) Uranium

16. A metal element that resists reaction with acids

17. A monoatomic element with filled p orbitals

18. A transition element that occurs when the inner $3d$ orbital is partially filled

Questions 19–20

        (A) Radioactive isotope
        (B) Monoclinic crystal
        (C) Sulfur trioxide
        (D) Sulfate
        (E) Sulfite

19. A substance that exhibits a resonance structure

20. A product formed from a base reacting with sulfurous acid

Questions 21–23

        (A) Dilute
        (B) Concentrated
        (C) Unsaturated
        (D) Saturated
        (E) Supersaturated

21. The condition, unrelated to quantities, that indicates that the rate going into solution is equal to the rate coming out of solution

22. The condition that exists when a water solution that has been at equilibrium is heated to a higher temperature with a higher solubility, but no additional solute is added

23. The descriptive term that indicates there is a large quantity of solute, compared with the amount of solvent, in a solution

Part B

ON THE ACTUAL CHEMISTRY TEST, THE FOLLOWING TYPE OF QUESTION MUST BE ANSWERED ON A SPECIAL SECTION (LABELED "CHEMISTRY") AT THE LOWER LEFT-HAND CORNER OF PAGE 2 OF YOUR ANSWER SHEET. THESE QUESTIONS WILL BE NUMBERED BEGINNING WITH 101 AND MUST BE ANSWERED ACCORDING TO THE FOLLOWING DIRECTIONS.

<u>Directions</u>: Each question below consists of two statements, I in the left-hand column and II in the right-hand column. For each question, determine whether statement I is true or false <u>and</u> if statement II is true or false and fill in the corresponding T or F ovals on your answer sheet. <u>Fill in oval CE only if statement II is a correct explanation of statement I.</u>

Sample Answer Grid:

CHEMISTRY * Fill in oval CE only if II is a correct explanation of I.

	I	II	CE*
101.	Ⓣ Ⓕ	Ⓣ Ⓕ	◯

101. Nonmetallic oxides are usually acid anhydrides   BECAUSE   nonmetallic oxides form acids when placed in water.

102. When HCl gas and NH₃ gas come in contact, a white smoke forms   BECAUSE   NH₃ and HCl react to form a white solid, ammonium chlorate.

103. The reaction of barium chloride and sodium sulfate does not go to completion   BECAUSE   the compound barium sulfate is formed.

104. When two elements react exothermically to form a compound, the compound should be relatively stable   BECAUSE   the release of energy from a reaction indicates the compound formed is at a lower energy level than the reactants and thus relatively stable.

105. The ion of a nonmetallic atom is larger in radius than the atom   BECAUSE   each principal energy level above the first in the atom may have several subshells.

106. Oxidation and reduction occur together   BECAUSE   in redox reactions, electrons must be gained and lost.

107. Decreasing the atmospheric pressure on a pot of boiling water causes it to stop boiling   BECAUSE   changes in pressure are directly related to the boiling point of water.

108. Catalysts speed up or slow down a reaction   BECAUSE   catalysts change the energy released from a reaction.

109. Atoms of different elements can have the same mass number   BECAUSE   the atoms of each element have a characteristic number of protons in the nucleus.

110. Isotopes have different atomic numbers

    BECAUSE

    each isotope has a different number of neutrons.

111. $^{13}_{12}C$ and $^{12}_{12}C$ are isotopes of the element carbon

    BECAUSE

    isotopes of an element have the same number of protons in the nucleus.

112. The Cu^{2+} ion needs to be oxidized to form Cu metal

    BECAUSE

    oxidation is a gain of electrons.

113. The volume of a gas at 100°C and 600 mm Hg pressure will be decreased at STP

    BECAUSE

    decreasing temperature and increasing pressure will cause the volume to decrease.

114. The pH of a 0.01 molar solution of HCl will be 2

    BECAUSE

    HCl is made up of two essentially ionic particles.

115. Nuclear fusion on the sun converts hydrogen to helium with a release of energy

    BECAUSE

    some mass is converted to energy.

116. The "bullet" usually used to initiate the fusion of ^{235}U is a neutron

    BECAUSE

    capture of the neutron by the nucleus causes an unstable condition that leads to its disintegration.

Part C

<u>Directions</u>: Each of the questions or incomplete statements below is followed by five suggested answers or completions. Select the one that is best in each case and then fill in the corresponding oval on the answer sheet.

40. What is the approximate formula mass of $Ca(NO_3)_2$?
    (A) 70
    (B) 82
    (C) 102
    (C) 150
    (E) 164

41. In this reaction: $XClO_3 + A \rightarrow XCl + O_2(g) + A$, which substance is the catalyst?
    (A) X
    (B) $XClO_3$
    (C) A
    (D) XCl
    (E) O_2

42. The normal electron configuration for ethyne (acetylene) is
    (A) H:C::C:H
    (B) H:C:C:H
    (C) H·C:::C·H
    (D) H:C::C:H
    (E) H:C:C:H

43. According to the kinetic molecular theory, molecules increase in kinetic energy when they
    (A) are mixed with other molecules at lower temperature
    (B) are frozen into a solid
    (C) are condensed into a liquid
    (D) are melted from a solid to a liquid state
    (E) collide with each other in a container at a lower temperature

44. How many atoms are represented in the formula $Ca_3(PO_4)_2$?
    (A) 5
    (B) 8
    (C) 9
    (D) 12
    (E) 13

45. All of the following have covalent bonds EXCEPT
    (A) HCl
    (B) CCl_4
    (C) H_2O
    (D) CsF
    (E) CO_2

46. Which of the following is (are) the weakest attractive force?
    (A) van der Waals forces
    (B) coordinate covalent bonding
    (C) covalent bonding
    (D) polar covalent bonding
    (E) ionic bonding

47. Which of these resembles the molecular structure of the water molecule?

48. The two most important considerations in deciding whether a reaction will occur spontaneously are
    (A) the stability and state of the reactants
    (B) the energy gained and the heat evolved
    (C) the exothermic energy and the randomness of the products
    (D) the endothermic energy and the randomness of the products
    (E) the endothermic energy and the structure of the products

49. The reaction of an acid like HCl and a base like NaOH always
    (A) forms a precipitate
    (B) forms a volatile product
    (C) forms a soluble salt and water
    (D) forms a sulfate salt and water
    (E) forms a salt and water

50. The oxidation number of sulfur in H_2SO_4 is
    (A) +2
    (B) +3
    (C) +4
    (D) +6
    (E) +8

51. Balance this reaction by using the oxidation-reduction method of electron exchange:

    $$KMnO_4 + H_2SO_3 \rightarrow K_2SO_4 + MnSO_4 + H_2SO_4 + H_2O$$

    Which of the following partial equations is the correct reduction half-reaction for the *balanced* equation?
    (A) $5SO_3^{2-} + 5H_2O \rightarrow 5SO_4^{2-} + 2H^+ + 10e^-$
    (B) $2MnO_4^- + 16H^+ + 10e^- \rightarrow 2Mn^{2+} + 8H_2O$
    (C) $SO_3^{2-} \rightarrow SO_4^{2-} + 2e^-$
    (D) $SO_3^{2-} + 2H^+ \rightarrow SO_4^{2-} + H_2O + 2e^-$
    (E) $Mn^{7+} \rightarrow Mn^{2+} + 5e^-$

52. Acid solutions are illustrated by which of the following when placed into water?
    I. $HCl(g)$
    II. Excess H_3O^+
    III. $CuSO_4(s)$
    (A) I only
    (B) III only
    (C) I and II only
    (D) II and III only
    (E) I, II, and III

53. The property of matter that is independent of its surrounding conditions and position is
    (A) volume
    (B) density
    (C) mass
    (D) weight
    (E) velocity

54. Where are the highest ionization energies found in the periodic chart?
    (A) upper left corner
    (B) lower left corner
    (C) upper right corner
    (D) lower right corner
    (E) middle of transition elements

55. Which of the following pairs of compounds can be used to illustrate the Law of Multiple Proportions?
    (A) NO and NO_2
    (B) CH_4 and CO_2
    (C) ZnO_2 and $ZnCl_2$
    (D) NH_4 and NH_4Cl
    (E) H_2O and HCl

56. In this equilibrium reaction: $A + B \rightleftharpoons AB +$ heat (in a closed container), how could the forward reaction rate be increased?
    I. By increasing the concentration of AB
    II. By increasing the concentration of A
    III. By removing some of product AB
    (A) I only
    (B) III only
    (C) I and III only
    (D) II and III only
    (E) I, II, and III

57. In the reaction of sodium with water, the balanced equation has which of the following coefficients?
    I. 1
    II. 2
    III. 3
    (A) I only
    (B) III only
    (C) I and II only
    (D) II and III only
    (E) I, II, and III

58. If 10 liters of CO gas react with sufficient oxygen to react completely, how many liters of CO_2 gas are formed?
    (A) 5
    (B) 10
    (C) 15
    (D) 20
    (E) 40

59. If 49 grams of H_2SO_4 react with 80 grams of NaOH, how much reactant will be left over after the reaction is complete?
    (A) 24.5 g H_2SO_4
    (B) none of either compound
    (C) 20 g NaOH
    (D) 40 g NaOH
    (E) 60 g NaOH

60. The K_{sp} of AgCl is 1.9×10^{-10}. What is the molar concentration of Ag^+ in 1 liter of a saturated water solution of silver chloride?
    (A) 0.44×10^{-5}
    (B) 0.95×10^{-5}
    (C) 1.38×10^{-5}
    (D) 0.95×10^{-10}
    (E) 3.81×10^{-10}

61. If the density of a diatomic gas is 1.43 grams/liter, what is its gram-molecular mass?
    (A) 16 g
    (B) 32 g
    (C) 48 g
    (D) 64 g
    (E) 14.3 g

62. From 2 moles of $KClO_3$ how many liters of O_2 can be produced by decomposition of all the $KClO_3$?
    (A) 11.2
    (B) 22.4
    (C) 33.6
    (D) 44.8
    (E) 67.2

63. What value determines whether a reaction is spontaneous?
    (A) Gibbs free energy
    (B) Entropy
    (C) Kinetic energy
    (D) Enthalpy change
    (E) Heat of formation

Questions 64–68 refer to the following experimental setup and data:

Recorded data:

Before: CuO + porcelain boat = 62.869 g
         $CaCl_2$ + U tube   = 80.483 g

After: Porcelain boat + contents = 54.869 g
        $CaCl_2$ + U tube   = 89.483 g

Reactions: $Zn + 2HCl \rightarrow ZnCl_2 + H_2$
           then $H_2 + CuO \rightarrow H_2O + Cu$

64. What type of reaction occurred in the porcelain boat?
    (A) electrolysis
    (B) double displacement
    (C) reduction and oxidation
    (D) decomposition or analysis
    (E) combination or synthesis

65. Why was the $CaCl_2$ tube placed between the generator and the tube containing the porcelain boat?
    (A) to absorb evaporated HCl
    (B) to absorb evaporated H_2O
    (C) to slow down the gases released
    (D) to absorb the evaporated Zn particles
    (E) to remove the initial air that passes through the tube

66. How many grams of hydrogen were used in the formation of the water that was a product?
    (A) 1
    (B) 2
    (C) 4
    (D) 8
    (E) 9

67. What conclusion can you draw from this experiment?
    (A) Hydrogen diffuses faster than oxygen.
    (B) Hydrogen is lighter than oxygen.
    (C) The gram-molecular mass of oxygen is 32 g.
    (D) Water is a triatomic molecule with polar characteristics.
    (E) Water is formed in a ratio of 1 : 8 from hydrogen and oxygen.

68. Which of the following is an observation rather than a conclusion?
    (A) A substance is an acid if it changes litmus paper from blue to red.
    (B) A gas is lighter than air if it escapes from a bottle left mouth upward.
    (C) The gas H_2 forms an explosive mixture with air.
    (D) Air is mixed with hydrogen gas and ignited; it explodes.
    (E) An oil liquid is immiscible with water because it separates into a layer above the water.

69. When HCl fumes and NH_3 fumes are introduced into opposite ends of a long, dry glass tube, a white ring forms in the tube. Which answer explains this phenomenon?
    (A) NH_4Cl forms.
    (B) HCl diffuses faster.
    (C) NH_3 diffuses faster.
    (D) The ring occurs closer to the end into which HCl was introduced.
    (E) The ring occurs in the very middle of the tube.

70. The correct formula for calcium hydrogen sulfate is
    (A) CaH_2SO_4
    (B) $CaHSO_4$
    (C) $Ca(HSO_4)_2$
    (D) Ca_2HSO_4
    (E) $Ca_2H_2SO_4$

71. Ten grams of sodium hydroxide dissolved in 1 liter of water makes a solution that is
    (A) 0.25 M
    (B) 0.5 M
    (C) 1 M
    (D) 1.5 M
    (E) 4 M

72. For a saturated solution, which statement is true?

    (A) All dissolving has stopped.

    (B) Crystals begin to grow.

    (C) An equilibrium has been established.

    (D) Crystals of the solute will continue to dissolve.

    (E) The solute is exceeding its solubility.

73. In which of the following series is the pi bond present in the bonding structure?

    I. Alkane

    II. Alkene

    III. Alkyne

        (A) I only

        (B) III only

        (C) I and III only

        (D) II and III only

        (E) I, II, and III

Questions 74–76 refer to the following setup:

74. Why could you NOT use this setup for preparing H_2 if the generator contained Zn + vinegar?

    (A) Hydrogen would not be produced.

    (B) The setup of the generator is improper.

    (C) The generator must be heated with a burner.

    (D) The delivery tube setup is wrong.

    (E) The gas cannot be collected over water.

75. If the setup were corrected, which of these gases could NOT be collected over H_2O because of its solubility?

    (A) CO_2

    (B) NO

    (C) O_2

    (D) NH_3

    (E) CH_4

76. What is the percentage of oxygen in the formula mass of $Ca(NO_3)_2$?
    (A) 28
    (B) 42
    (C) 58
    (D) 96
    (E) 164

77. For the following reaction: $N_2O_4(g) \rightleftharpoons 2NO_2(g)$, the K_e expression is

    (A) $K_e = \dfrac{[N_2O_4]}{[NO_2]}$

    (B) $K_e = \dfrac{[N_2O_4]}{[NO_2]^2}$

    (C) $K_e = \dfrac{[NO_2]}{[N_2O_4]}$

    (D) $K_e = \dfrac{[NO_2]^2}{[N_2O_4]}$

    (E) $K_e = \dfrac{[N_2O_4]^2}{[NO_2]}$

78. What is the K_e for the above reaction if at equilibrium the concentration of N_2O_4 is 4×10^{-2} mole/liter and that of NO_2 is 2×10^{-2} mole/liter?
    (A) 1×10^{-2}
    (B) 2×10^{-2}
    (C) 4×10^{-2}
    (D) 4×10^{-4}
    (E) 8×10^{-2}

79. How much water, in liters, must be added to 0.5 liter of 6 M HCl to make it 2 M?
    (A) 0.33
    (B) 0.5
    (C) 1
    (D) 1.5
    (E) 2

80. Four grams of hydrogen are ignited with 4 grams of oxygen. How many grams of water can be formed?
    (A) 0.5
    (B) 2.5
    (C) 4.5
    (D) 8
    (E) 36

81. Which structure is an ester?

(A)

(B)

(C)

(D)

(E)

82. What piece of apparatus can be used to introduce more liquid into a reaction and also serve as a pressure valve?
    (A) stopcock
    (B) pinchcock
    (C) thistle tube
    (D) flask
    (E) condenser

83. Which formulas could represent the empirical formula and the molecular formula of a given compound?
    (A) CH_2O and $C_4H_6O_4$
    (B) CHO and $C_6H_{12}O_6$
    (C) CH_4 and C_5H_{12}
    (D) CH_2 and C_3H_6
    (E) CO and CO_2

84. The reaction $Fe \rightarrow Fe^{2+} + 2e^-$ (+0.44 volt) would occur spontaneously with which of the following?

    I. $Pb \rightarrow Pb^{2+} + 2e^-$ (+0.13 volt)

    II. $Cu \rightarrow Cu^{2+} + 2e^-$ (−0.34 volt)

    III. $2Ag + 2e^- \rightarrow 2Ag°$ (+0.80 volt)

    (A) I only

    (B) III only

    (C) I and III only

    (D) II and III only

    (E) I, II, and III

85. $2Na(s) + Cl_2(g) \rightarrow 2NaCl(s) + 822$ kJ

How much heat is released by the above reaction if 0.5 mole of sodium reacts completely with chlorine?

    (A) 205 kJ

    (B) 411 kJ

    (C) 822 kJ

    (D) 1,644 kJ

    (E) 3,288 kJ

STOP

IF YOU FINISH BEFORE ONE HOUR IS UP, YOU MAY GO BACK TO CHECK YOUR WORK OR COMPLETE UNANSWERED QUESTIONS.

1. (D) The most electronegative element, F, would be found in the upper right corner; the noble gases are exceptions at the far right.

2. (C) Elements in the Group with (C) have a possible oxidation number of –2.

3. (B) Elements in the Group with (B) react in a 1:1 ratio with elements in the Group with (D), since one has an electron to lose and the other one needs an electron to complete its outer energy level.

4. (A) (A) loses 2 electrons to form an ion whose remaining electrons, being close to the nucleus, are pulled in closer because of the unbalanced +2 charge.

5. (B) Since (B) has only one electron in the outer 4s orbital, it can more easily be removed than can an electron from the 3s orbital of (A), which is closer to the positive nucleus.

6. (E) All Group VIII elements have a comple *p* orbital as the outer energy level. This explains why these elements are "inert."

7. (B) The Group IA elements comprise the alkali-metals family.

8. (E) The Group VIII elements are the inert or noble gases.

9. (B) The alkali metals react with water to form a strong base. Activity increases down this family.

10. (A) H_2O is ice in Part A.

11. (D) H_2O changes state at Parts B and D. The heat of vaporization at D (540 cal/g) is greater than the heat of fusion (80 cal/g) at B.

12. (C) Water is heating at 1°C/cal/g in Part C.

13. (B) This is the energy needed to start the reaction.

14. (B) The addition of a catalyst affects only the part indicated by B, by either raising or lowering the energy needed to commence the reaction.

15. (C) The net energy released is the exothermic quantity indicated by C.

16. (B) Gold is known as a noble metal because of its resistance to acids. Aqua regia, a mixture of HNO_3 and HCl, will react with gold.

17. (D) Helium is the only element that is monoatomic in the molecular form.

18. (A) Iron has 5 electrons in the *d* orbitals, which are partially filled.

19. (C) Only sulfur trioxide has a resonance structure (as shown here):

20. (E) Sulfurous acid reacts with a base to form a sulfite salt.

21. (D) The condition described is the equilibrium that exists at saturation.

22. (C) With the increased temperature more solute may go into solution; therefore the solution is now unsaturated.

23. (B) The term "concentrated" means that there is a large amount of solute in the solvent.

101. (T, T, CE) Nonmetallic oxides are usually acid anhydrides, and they form acids in water.

102. (T, F) The white smoke formed is ammonium chloride, not ammonium chlorate.

103. (F, T) The reaction does go to completion since barium sulfate is a precipitate.

104. (T, T, CE) The product of an exothermic reaction is relatively stable because it is at a lower energy level than the reactants.

105. (T, T) The statements are both true but are not related.

106. (T, T, CE) Both the statements are true, and the reason explains the assertion.

107. (F, T) The statement is false while the reason is true. Decreasing the pressure on a boiling pot will only cause it to boil more vigorously.

108. (T, F) The statement is true, but the reason is false. Catalysts change the activation energy.

109. (T, T) The statements are true, but the reason doesn't explain the assertion.

110. (F, T) Isotopes do not have different atomic numbers; they have the same one. The number of neutrons in the nucleus, however, varies.

111. (T, T, CE) The two configurations of carbon are isotopes because they have the same atomic number.

112. (F, F) Both the statement and the reason are false.

113. (T, T, CE) To go from 100°C to 0°C decreases the volume as the gas gets colder; therefore, the temperature fraction expressed in kelvins must be $\frac{273}{373}$ to decrease the volume. To go from 600 mm to 760 mm of pressure increases the pressure, thus causing the gas to contract. The fraction must then be $\frac{600}{760}$ to cause the volume to decrease. You could use the formula

$$V_2 = V_1 \times \frac{T_2}{T_1} \times \frac{P_1}{P_2}$$

114. (T, T) Since HCl is a strong acid and ionizes completely in dilute solution of water, the [H$^+$] (H$_3$O$^+$ is the same thing) or molar concentration of a 0.01 molar concentration is 1×10^{-2} mole/liter.
pH = –log [H$^+$]
pH = –log [1×10^{-2}]
pH = –(–2) = 2
The pH is 2, but the reason, although true, does not explain the statement.

115. (T, T, CE) Both statements are true, and the reason explains the assertion.

116. (F, T) The statement should refer to fission, not fusion. The reason is true.

40. (E) The total formula mass is:
Ca = 40
2N = 28
6O = 96
Total 164

41. (C) The catalyst, by definition, is not consumed in the reaction and ends up in its original form as one of the products.

42. (D) The ethyne molecule is the first member of the acetylene series with a general formula of C$_n$H$_{2n-2}$. It contains a triple bond between the two C atoms: H : C⦂⦂C : H.

43. (D) Heating molecules increases their kinetic energy.

44. (E) 3Ca + 2P + 8O = 13 atoms.

45. (D) Cesium and fluorine are from the most electropositive and electronegative portions, respectively, of the periodic chart, and thus form an ionic bond by cesium giving an electron to fluorine to form the respective ions.

46. (A) Van der Waals forces are the weak attraction of the nuclear positive charge of one atom to the negative electron field of an adjacent atom. They are much weaker than the others named.

47. (D) The molecular structure of water is a polar covalent compound with the hydrogens 105° apart.

48. (C) The most important considerations for a spontaneous reaction are (1) that the reaction is exothermic so that once started it tends to continue on its own because of the energy released and (2) that reactions tend to go to the highest state of randomness.

49. (E) Normal H$^+$ acids and OH$^-$ bases form water and a salt.

50. (D) All compounds have a charge of 0. H usually has +1, and O usually has –2, so
H$_2$ = +2
O$_4$ = –8
S = x
Total = 0
$+2 - 8 + x = 0$
$x = -2 + 8 = +6$

51. (B) In the balanced equation the two half reactions are:

$5SO_3^{2-} + 5H_2O \rightarrow 5SO_4^{2-} + 10H^+ + 10e^-$
$2MnO_4^- + 16H^+ + 10e^- \rightarrow 2Mn^{2+} + 8H_2O$

$2KMnO_4 + 5H_2SO_3 \rightarrow$
$K_2SO_4 + 2MnSO_4 + 2H_2SO_4 + 3H_2O$

52. (E) HCl and H$_3$O$^+$ give acid solutions, as does CuSO$_4$, when it hydrolyzes in water. I, II, and III are correct.

53. (C) Mass is a constant and is not dependent on position or surrounding conditions.

54. (C) The complete outer energy level of electrons of the smallest inert gases have the highest ionization potential.

55. (A) Only NO and NO_2 fit the definition of the Law of Multiple Proportions, in which one substance stays the same and the other varies in units of whole integers.

56. (D) Increasing the concentration of one or both of the reactants and removing some of the product formed would cause the forward reaction to increase in rate to try to regain the equilibrium condition. II and III are correct.

57. (C) I and II are correct. The equation is $2Na + 2H_2O \rightarrow 2NaOH + H_2(g)$. The coefficients include a 1 and a 2.

58. (B) $2CO + O_2 \rightarrow 2CO_2$ indicates 2 volumes of CO react with 1 volume of O_2 to form 2 volumes of CO_2. Therefore 10 liters of CO form 10 liters of CO_2.

59. (D) $H_2SO_4 + 2NaOH \rightarrow 2H_2O + Na_2SO_4$ is the equation for this reaction. 1 mole H_2SO_4 = 98 g. 1 mole NaOH = 40 g. Then 49 g of H_2SO_4 = $\frac{1}{2}$ mole H_2SO_4. The equation shows that 1 mole of sulfuric reacts with 2 moles of sodium hydroxide or a ratio of 1:2. Therefore, $\frac{1}{2}$ mole of sulfuric reacts with 1 mole of sodium hydroxide in this reaction. 1 mole of NaOH we found to equal 40 g. Since 80 g of NaOH is given, 40 g of it will remain after the reaction has gone to completion.

60. (C) The K_{sp} = $[Ag^+][Cl^-]$ = 1.9×10^{-10}. Let x = the molar concentration of Ag^+ and Cl^-.

Then $(x)(x) = x^2 = 1.9 \times 10^{-10}$
$$x = \sqrt{1.9 \times 10^{-10}}$$
$$x = 1.38 \times 10^{-5}$$

61. (B) If 1.43 g is the weight of 1 liter, then the mass of 22.4 liters, which is the gram-molecular volume of a gas at STP, will give the gram-molecular mass. So 22.4 L $\times$ 1.43 g/L = 32 g, the gram-molecular mass.

62. (E) The equation is:

$2KClO_3 \rightarrow 2KCl + 3O_2$

This shows that 2 moles of $KClO_3$ yields 3 moles of O_2. Three moles of O_2 = 3×22.4 L = 67.2 L of O_2.

63. (A) Gibbs free energy combines the overall energy changes and the entropy change. The formula is $\Delta G = \Delta H - T\Delta S$. Only if ΔG is negative will the reaction be spontaneous in the forward direction.

64. (C) The CuO was reduced while the H_2 was oxidized, forming H_2O + Cu. The reaction is:

$CuO + H_2 \rightarrow H_2O + Cu$

65. (B) The purpose of this $CaCl_2$ tube is to absorb any water evaporated from the generator. If it were not present, some water vapor would pass through to the final drying tube and cause the weight of water gained there to be larger than it should be from the reaction only.

66. (A) Since 8 g of O_2 was lost by the CuO, the weight of water gained by the U-tube came from 8 g of O_2 and 1 g of H_2, to make 9 g of water that it absorbed.

67. (E) The weight ratio of water is 1 g of H_2 to 8 g of O_2 or 1:8. The other statements are true but are not conclusions from this experiment.

68. (D) This is the only observation of the group. All the others are conclusions. Remember that an observation is only what you see, smell, taste, or measure with a piece of equipment.

69. (A) The phenomenon is the formation of the white ring, which is NH_4Cl. Although measuring the distance traveled by each gas could be used to verify Graham's Law of Gaseous Diffusion, this was not asked. The relationship is that the diffusion rate is inversely proportional to the square root of the gas's molecular weight.

70. (C) Since Ca^{2+} and HSO_4^- combine, the formula is $Ca(HSO_4)_2$.

71. (A) NaOH is 40 g/mole. 10 g is 10/40 or 0.25 mole in 1 liter or .25 M.

72. (C) A saturated solution represents a condition where the solute is going into solution as rapidly as some solute is coming out of the solution.

73. (D) II and III are correct since the double-bonded carbons in the alkene and the triple-bonded carbons in the alkyne series have pi bonds.

74. (B) The thistle tube is not below the level of the liquid in the generator and

the gas would escape into the air. (Vinegar is an acid and would produce hydrogen.)

75. (D) NH_3 is very soluble and could not be collected in this manner. All others are not sufficiently soluble to hamper this method of collection.

76. (D) The formula weight (mass) is the total of (Ca =) 40 + (2N =) 28 + (6O =) 96 or a total of 164. Since oxygen is 96 amu of 164 total, the percentage is 96/164 × 100 = 58.5%.

77. (D) The K_e expression is made up of the concentration(s) of the products over those of the reactants, with the coefficients becoming exponents. So $K_e = \dfrac{[NO_2]^2}{[N_2O_4]}$.

78. (A) $K_e = \dfrac{[2\times10^{-2}]^2}{[4\times10^{-2}]} = \dfrac{4\times10^{-4}}{4\times10^{-2}}$
$= 1 \times 10^{-2}$.

79. (C) In dilution expressions $M_1 \times V_1 = M_2 \times V_2$ can be used. Substituting (6 M) (0.5 L) = (2 M) (x L) gives x = 1.5 L total volume. Since there was 0.5 L to begin with, an additional 1 L must be added.

80. (C) The reaction equation and information given can be set up like this:

Given   Given
4 g   4 g x g
$2H_2$ + O_2 → $2H_2O$
4 g   32 g   36 g

Studying this shows that the limiting element will be the 4 g of oxygen since 4 g of H_2 would require 32 g of O_2. The solution setup is

$\dfrac{4\,g\,O_2}{32\,g\,O_2} = \dfrac{x\,g\,H_2O}{36\,g\,H_2O}$
∴ = 4.5 g H_2O

81. (E) The functional group is $R-\overset{\overset{\displaystyle O}{\|}}{C}-R_1$. This appears only in (E).

82. (C) The thistle tube serves both these purposes.

83. (D) The empirical formula is a representation of the elements in their simplest ratio. Therefore, CH_2 is the simplest ratio of the molecular formula C_3H_6.

84. (E) I, II, and III would occur since their reduction reactions would be −.13, +.34, +.80 volt, respectively. These numbers added to +.44 volt give a positive E^0 for the reaction with Fe.

85. (A) In the equation, 2 moles of Na will release 822 kJ of heat. If only 0.5 mole of Na is consumed, only one-fourth as much heat will be released or $\frac{1}{4}$ × 822 = 205 kJ.

DIAGNOSING YOUR NEEDS

After taking Practice Test 1, check your answers against the correct ones. Then fill in the chart below.

In the space under each question number, place a check if you answered that question correctly.

- **EXAMPLE:**

If your answer to question 5 was correct, place a check in the appropriate box.

Next, total the check marks for each section and insert the number in the designated block. Now do the arithmetic indicated, and insert your percent for each area.

SUBJECT AREA	(✔) QUESTIONS ANSWERED CORRECTLY							
I. Atomic Theory and Structure, including periodic relationships	6	8	105	109	110	111	47	54
☐ No. of checks ÷ 8 × 100 = _____ %								
II. Nuclear Reactions							115	116
☐ No. of checks ÷ 2 × 100 = _____ %								
III. Chemical Bonding and Molecular Structure	3	17	18	19	45	46	55	70	73
☐ No. of checks ÷ 9 × 100 = _____ %								
IV. States of Matter and Kinetic Molecular Theory of Gases	10	11	12	107	108	113	43
☐ No. of checks ÷ 7 × 100 = _____ %								
V. Solutions, including concentration units, solubility, and colligative properties	21	103	60	71	79
☐ No. of checks ÷ 5 × 100 = _____ %								
VI. Acids and Bases	20	101	102	114	49	52	69
☐ No. of checks ÷ 7 × 100 = _____ %								
VII. Oxidation-Reduction and Electrochemistry	106	112	50	51	64	84
☐ No. of checks ÷ 6 × 100 = _____ %								

SUBJECT AREA

(✔) QUESTIONS ANSWERED CORRECTLY

VIII. Stoichiometry	40	44	57	58	59	61	62	66	76	80

☐ No. of checks ÷ 10 × 100 = _____%

IX. Reaction Rates	41	48

☐ No. of checks ÷ 2 × 100 = _____%

X. Equilibrium	22	56	72	77	78

☐ No. of checks ÷ 5 × 100 = _____%

XI. Thermodynamics: energy changes in chemical reactions, randomness, and criteria for spontaneity	13	14	15	104	63	85

☐ No. of checks ÷ 6 × 100 = _____%

XII. Descriptive Chemistry: physical and chemical properties of elements and their familiar compounds; organic chemistry; periodic properties	1	2	4	5	7	9
	16	23	42	53	81	83

☐ No. of checks ÷ 12 × 100 = _____%

XIII. Laboratory: equipment, procedures, observations, safety, calculations, and interpretation of results	65	67	68	74	75	82

☐ No. of checks ÷ 6 × 100 = _____%

PLANNING YOUR STUDY

The percentages give you an idea of how you have done on the various major areas of the test. Because of the limited number of questions on some parts, these percentages may not be as reliable as the percentages for parts with larger numbers of questions. However, you should now have at least a rough idea of the areas in which you have done well and those in which you need more study.

Start your study with the areas in which you are weakest. The corresponding chapters are indicated below:

Subject Area	Chapters to Review
I. Atomic Theory and Structure, including periodic relationships	2
II. Nuclear Reactions	15
III. Chemical Bonding and Molecular Structure	3, 4
IV. States of Matter and Kinetic Molecular Theory of Gases	1
V. Solutions, including concentration units, solubility, and colligative properties	7
VI. Acids and Bases	11
VII. Oxidation-reduction and Electrochemistry	12
VIII. Stoichiometry	5, 6
IX. Reaction Rates	9
X. Equilibrium	10
XI. Thermodynamics, including energy changes in chemical reactions, randomness, and criteria for spontaneity	8
XII. Descriptive Chemistry: physical and chemical properties of elements and their familiar compounds; organic chemistry; periodic properties	1, 2, 13, 14
XIII. Laboratory: equipment, procedures, observations, safety, calculations, and interpretation of results.	All lab diagrams, 16

ANSWER SHEET FOR PRACTICE TEST 2

Determine the correct answer for each question. Then, using a No. 2 pencil, blacken completely the oval containing the letter of your choice.

ON THE ACTUAL CHEMISTRY TEST, THE FOLLOWING TYPE OF QUESTION MUST BE ANSWERED ON A SPECIAL SECTION (LABELED "CHEMISTRY") AT THE LOWER LEFT-HAND CORNER OF PAGE 2 OF YOUR ANSWER SHEET. THESE QUESTIONS WILL BE NUMBERED BEGINNING WITH 101 AND MUST BE ANSWERED ACCORDING TO THE FOLLOWING DIRECTIONS.

CHEMISTRY* Fill in oval CE only if II is a correct explanation of I.

	I	II	CE*
101.	T F	T F	◯
102.	T F	T F	◯
103.	T F	T F	◯
104.	T F	T F	◯
105.	T F	T F	◯
106.	T F	T F	◯
107.	T F	T F	◯
108.	T F	T F	◯
109.	T F	T F	◯
110.	T F	T F	◯
111.	T F	T F	◯
112.	T F	T F	◯
113.	T F	T F	◯
114.	T F	T F	◯
115.	T F	T F	◯
116.	T F	T F	◯

ON THE ACTUAL CHEMISTRY TEST, THE REMAINING QUESTIONS MUST BE ANSWERED BY RETURNING TO THE SECTION OF YOUR ANSWER SHEET YOU STARTED FOR CHEMISTRY.

40. A B C D E   56. A B C D E   71. A B C D E
41. A B C D E   57. A B C D E   72. A B C D E
42. A B C D E   58. A B C D E   73. A B C D E
43. A B C D E   59. A B C D E   74. A B C D E
44. A B C D E   60. A B C D E   75. A B C D E
45. A B C D E   61. A B C D E   76. A B C D E
46. A B C D E   62. A B C D E   77. A B C D E
47. A B C D E   63. A B C D E   78. A B C D E
48. A B C D E   64. A B C D E   79. A B C D E
49. A B C D E   65. A B C D E   80. A B C D E
50. A B C D E   66. A B C D E   81. A B C D E
51. A B C D E   67. A B C D E   82. A B C D E
52. A B C D E   68. A B C D E   83. A B C D E
53. A B C D E   69. A B C D E   84. A B C D E
54. A B C D E   70. A B C D E   85. A B C D E
55. A B C D E

Practice Test

2

<u>Note</u>: For all questions involving solutions and/or chemical equations, assume that the system is in water unless otherwise stated.

Part A

<u>Directions</u>: Each set of lettered choices below refers to the numbered statements or formulas immediately following it. Select the one lettered choice that best fits each statement or formula and then fill in the corresponding oval on the answer sheet. A choice may be used once, more than once, or not at all in each set.

<u>Questions 1–4</u>

(A) Law of Definite Composition
(B) Nuclear fusion
(C) van der Waals forces
(D) Graham's Law of Diffusion (Effusion)
(E) Triple point

1. At a particular temperature and pressure, three states of a compound may coexist.
2. The combining of nuclei to release energy.
3. The rate of movement of hydrogen gas compared with oxygen gas is 4:1.
4. The molecules of nitrous oxide and nitrogen dioxide differ by a multiple of the mass of one oxygen.

<u>Questions 5–7</u> refer to the following diagram:

Initial state $C(s) + \frac{1}{2}O_2(g)$ (D)

-26.4 kcal (B)
$CO(g) + \frac{1}{2}O_2(g)$

-94.0 kcal (A)

-67.6 kcal (C)

$CO_2(g)$ (E)

5. The ΔH of the reaction to form CO from $C + O_2$
6. The ΔH of the reaction to form CO_2 from $CO + O_2$
7. The ΔH of the reaction to form CO_2 from $C + O_2$

Questions 8–11

(A) Hydrogen bond
(B) Ionic bond
(C) Polar covalent bond
(D) Pure covalent bond
(E) Metallic bond

8. The type of bond between atoms of potassium and chloride in a crystal of potassium chloride

9. The type of bond between the atoms in a nitrogen molecule

10. The type of bond between the atoms in a molecule of CO_2 (electronegativity difference = 1)

11. The type of bond between the atoms of calcium in a crystal of calcium

Questions 12–14 refer to the following diagram, which represents data obtained when a solid was heated at a constant rate below its melting point:

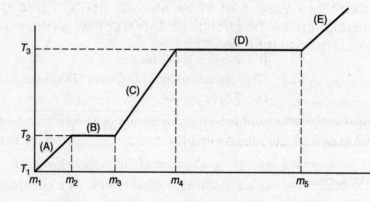

12. The part that represents the temperature at which the solid melts

13. The part that represents the section of the curve where the molecules are at the lowest kinetic energy

14. The part that represents the temperature at which boiling begins

Questions 15–23 refer to the following graph, which shows the variation of the first ionization potential with respect to increasing atomic numbers:

15. The atoms likely to react with water to release hydrogen

16. Nonmetals that are all found in the gaseous state at STP

17. The noble gases

18. The alkali metals

19. The half-filled condition of the p orbitals

20. The filled s orbitals with the exception of He

21. The beginning of pairing in the p orbitals

22. The most active metals

23. The filled p orbitals

Part B

ON THE ACTUAL CHEMISTRY TEST, THE FOLLOWING TYPE OF QUESTION MUST BE ANSWERED ON A SPECIAL SECTION (LABELED "CHEMISTRY") AT THE LOWER LEFT-HAND CORNER OF PAGE 2 OF YOUR ANSWER SHEET. THESE QUESTIONS WILL BE NUMBERED BEGINNING WITH 101 AND MUST BE ANSWERED ACCORDING TO THE FOLLOWING DIRECTIONS.

Directions: Each question below consists of two statements, I in the left-hand column and II in the right-hand column. For each question, determine whether statement I is true or false <u>and</u> if statement II is true or false and fill in the corresponding T or F ovals on your answer sheet. <u>Fill in oval CE only if statement II is a correct explanation of statement I.</u>

Sample Answer Grid:

CHEMISTRY * Fill in oval CE only if II is a correct explanation of I.

	I	II	CE*
101.	T F	T F	◯

101.	The structure of SO_3 is shown by using more than one structural formula	BECAUSE	SO_3 is very unstable and resonates between these possible structures.
102.	When the ΔG of a reaction at a given temperature is negative, the reaction occurs spontaneously	BECAUSE	when ΔG is negative, the ΔH is also negative.
103.	One mole of CO_2 has a greater mass than one mole of H_2O	BECAUSE	the molecular mass of CO_2 is greater than the molecular mass of H_2O.
104.	Hydrosulfuric acid is often used in qualitative tests	BECAUSE	it is a weak acid.
105.	Molecules of sodium chloride go into solution in water as ions	BECAUSE	the sodium has a +1 charge and the chloride has a −1 charge and they are hydrated by the water molecules.

106. In an equilibrium reaction, if the concentration of the reactants is increased, the reaction will increase its forward rate

BECAUSE

when a stress is applied to a reaction in equilibrium, the equilibrium shifts in the direction that opposes the stress.

107. The $\Delta H_{\text{reaction}}$ of a particular reaction can be arrived at by the summation of the $\Delta H_{\text{reaction}}$ values of two or more reactions that, added together, give the $\Delta H_{\text{reaction}}$ of the particular reaction

BECAUSE

the First Law of Thermodynamics states that the total energy of the universe is a constant.

108. In a reaction that has both a forward and a reverse reaction, A + B ⇌ AB, when only A and B are introduced into a reacting vessel, the forward reaction rate is the highest at the beginning and begins to decrease from that point until equilibrium is reached

BECAUSE

the reverse reaction does not begin until equilibrium is reached.

109. At equilibrium, the forward reaction and reverse reaction stop

BECAUSE

at equilibrium, the reactants and products have reached the equilibrium concentrations.

110. The hydrid orbital form of carbon in acetylene is believed to be the *sp* form

BECAUSE

it is a linear compound with a triple bond between the carbons.

111. The weakest of the bonds between molecules are coordinate covalent bonds

BECAUSE

they represent the weak attractive force of the electrons of one molecule for the positively charged nucleus of another.

112. A saturated solution is not necessarily concentrated

BECAUSE

dilute and *concentrated* are terms that relate only to the relative amount of solute dissolved in the solvent.

113. Lithium is the most active metal in the first group of the periodic chart

BECAUSE

it has only one electron in the outer energy level.

114. The anions migrate to the cathode in an electrochemical reaction

BECAUSE

positively charged ions are attracted to the negatively charged cathode.

115. The atomic number of a neutral atom that has a mass of 39 and has 19 electrons is 19

BECAUSE

the number of protons in a neutral atom is equal to the number of electrons.

116. For an element with an atomic number of 17, the most probable oxidation number is +1

BECAUSE

the outer energy level has a tendency to add one electron to itself.

Part C

Directions: Each of the questions or incomplete statements below is followed by five suggested answers or completions. Select the one that is best in each case and then fill in the corresponding oval on the answer sheet.

40. All of the following involve a chemical change EXCEPT
    (A) the formation of HCl from H_2 and Cl_2
    (B) the color change when NO is exposed to air
    (C) the formation of steam from burning H_2 and O_2
    (D) the solidification of "Crisco" at low temperatures
    (E) the odor of NH_3 when NH_4Cl is rubbed together with $Ca(OH)_2$ powder

41. When most fuels burn, the products include carbon dioxide and
    (A) hydrocarbons
    (B) hydrogen
    (C) water
    (D) hydroxide
    (E) hydrogen peroxide

42. In the metric system, the prefix *kilo-* means
    (A) 10^0
    (B) 10^{-1}
    (C) 10^{-2}
    (D) 10^2
    (E) 10^3

43. How many atoms are in 1 mole of water?
    (A) 3
    (B) 54
    (C) 6.02×10^{23}
    (D) $2(6.02 \times 10^{23})$
    (E) $3(6.02 \times 10^{23})$

44. Which of the following atoms normally forms monoatomic molecules?
    (A) Cl
    (B) H
    (C) O
    (D) N
    (E) He

45. The shape of a PCl_3 molecule is described as
    (A) bent
    (B) trigonal planar
    (C) linear
    (D) trigonal pyramidal
    (E) tetrahedral

46. The complete loss of an electron of one atom to another atom with the consequent formation of electrostatic charges is said to be
    (A) a covalent bond
    (B) a polar covalent bond
    (C) an ionic bond
    (D) a coordinate covalent bond
    (E) a pi bond between p orbitals

47. In the electrolysis of water, the cathode reaction is
    (A) $2H_2O(\ell) + 2e^- \rightarrow H_2(g) + 2OH^- + O_2(g)$
    (B) $2H_2O(\ell) \rightarrow \dfrac{1}{2}O_2(g) + 2H^+ + 2e^-$
    (C) $2OH^- + 2e^- \rightarrow O_2(g) + H_2(g)$
    (D) $2H^+ + 2e^- \rightarrow H_2(g)$
    (E) $2H_2O(\ell) + 4e^- \rightarrow O_2(g) + 2H_2(g)$

48. Which of the following particles has the least mass?
    (A) alpha particle
    (B) beta particle
    (C) proton
    (D) neutron
    (E) gamma ray

49. Which emission from a radioactive source is an electromagnetic wave and is not affected by an electric field?
    (A) alpha particle
    (B) beta particle
    (C) proton
    (D) neutron
    (E) gamma ray

50. What is the pH of an acetic acid solution if the $[H_3O^+] = 1 \times 10^{-4}$ mole/liter?
    (A) 1
    (B) 2
    (C) 3
    (D) 4
    (E) 5

51. The polarity of water is useful in explaining which of the following?
    I. The solution process
    II. The ionization process
    III. The high conductivity of distilled water
    (A) I only
    (B) II only
    (C) I and II only
    (D) II and III only
    (E) I, II, and III

52. When sulfur dioxide is bubbled through water, the solution will contain
    (A) sulfurous acid
    (B) sulfuric acid
    (C) hyposulfuric acid
    (D) persulfuric acid
    (E) anhydrous sulfuric acid

53. Four grams of hydrogen gas at STP contain
    (A) 6.02×10^{23} atoms
    (B) 12.04×10^{23} atoms
    (C) 12.04×10^{46} atoms
    (D) 1.2×10^{22} molecules
    (E) 12.04×10^{23} molecules

54. Analysis of a gas gave: C = 85.7% and H = 14.3%. If the formula mass of this gas is 42 amu, what are the empirical formula and the true formula?
    (A) CH ; C_4H_4
    (B) CH_2 ; C_3H_6
    (C) CH_3 ; C_3H_9
    (D) C_2H_2 ; C_3H_6
    (E) C_2H_4 ; C_3H_6

55. Which fraction would be used to correct a given volume of gas at 30°C to its new volume when it is heated to 60°C and the pressure is kept constant?

    (A) $\dfrac{30}{60}$

    (B) $\dfrac{60}{30}$

    (C) $\dfrac{273}{333}$

    (D) $\dfrac{303}{333}$

    (E) $\dfrac{333}{303}$

56. What would be the predicted freezing point of a solution that has 684 grams of sugar (1 mole = 342 grams) dissolved in 1,500 grams of water?
    (A) 1.86°C
    (B) –0.93°C
    (C) –1.39°C
    (D) –2.48°C
    (E) –2.79°C

57. What is the approximate pH of a 0.003 M solution of H_2SO_4?
    (A) 1
    (B) 2
    (C) 5
    (D) 9
    (E) 13

58. How many grams of NaOH are needed to make 100 grams of a 5% solution?
    (A) 2
    (B) 5
    (C) 20
    (D) 40
    (E) 95

59. For the Haber process: $N_2 + 3H_2 \rightleftharpoons 2NH_3$ + heat (at equilibrium), which statement concerning the reaction rate is true?
    (A) The reaction to the right will decrease when more N_2 is added.
    (B) The reaction to the right will decrease when the pressure is increased.
    (C) The reaction to the right will decrease when the temperature is increased.
    (D) The reaction to the right will decrease when more H_2 is added.
    (E) The reaction to the right will decrease when NH_3 is removed.

60. If you titrate 1 M H_2SO_4 solution against 50 milliliters of 1 M NaOH solution, what volume of H_2SO_4, in milliliters, will be needed for neutralization?
    (A) 10
    (B) 25
    (C) 40
    (D) 50
    (E) 100

61. How many grams of CO_2 can be prepared from 150 grams of calcium carbonate reacting with an excess of hydrochloric acid solution?
    (A) 11
    (B) 22
    (C) 33
    (D) 44
    (E) 66

Questions 62–63 refer to the following diagram:

62. The diagram represents a setup that may be used to prepare and collect
    (A) NH_3
    (B) NO
    (C) H_2
    (D) SO_3
    (E) CO_2

63. The apparatus would have to be heated with a Bunsen burner to be used for the preparation and collection of
    (A) HCl
    (B) H_2S
    (C) Br_2
    (D) H_2
    (E) NH_3

Questions 64–66 refer to the following experimental setup and data:

Recorded data:

Weight of U-tube .... 20.36 g
Weight of U-tube and calcium chloride   before .... 39.32 g
Weight of U-tube and calcium chloride   after...................................... 57.32 g
Weight of boat and copper oxide   before.. 30.23 g
Weight of boat and copper oxide   after.. 14.23 g
Weight of boat .... 5.00 g

64. What is the reason for the $CaCl_2$ drying tube?
    (A) generate water
    (B) absorb hydrogen
    (C) absorb water that evaporates from the flask
    (D) decompose the water from the flask
    (E) act as a catalyst for the combination of hydrogen and oxygen

65. What conclusion can be derived from the data collected?
    (A) Oxygen was lost from the $CaCl_2$.
    (B) Oxygen was generated in the U-tube.
    (C) Water was formed from the reaction.
    (D) Hydrogen was absorbed by the $CaCl_2$.
    (E) CuO was formed in the decomposition.

66. What is the ratio of the mass of water formed to the mass of hydrogen used in the formation of water?
    (A) 1:8
    (B) 1:9
    (C) 8:1
    (D) 9:1
    (E) 8:9

67. What is the mass, in grams, of 1 mole of $KAl(SO_4)_2 \cdot 12H_2O$?
    (A) 132
    (B) 180
    (C) 394
    (D) 474
    (E) 516

68. What mass of aluminum will be completely oxidized by 44.8 liters of oxygen?
    (A) 18 g
    (B) 37.8 g
    (C) 50.4 g
    (D) 72.0 g
    (E) 100.8 g

69. In general, when metal oxides react with water, they form solutions that are
    (A) acidic
    (B) basic
    (C) neutral
    (D) unstable
    (E) colored

Questions 70–72 refer to the following diagram:

70. The oxidation reaction will occur at
    (A) A
    (B) B
    (C) C
    (D) D
    (E) E

71. The apparatus at *C* is called the
    (A) anode
    (B) cathode
    (C) salt bridge
    (D) ion bridge
    (E) osmotic bridge

72. Starch solution added to beaker *D* after the reaction has occurred for some time will
    (A) remain clear
    (B) turn red
    (C) turn blue-black
    (D) cause the current to reverse
    (E) cause an increase in activity at the pole

73. How many liters of oxygen (STP) can you prepare from the decomposition of 213 grams of sodium chlorate?
    (A) 11.2
    (B) 22.4
    (C) 44.8
    (D) 67.2
    (E) 78.4

74. In this equation: $Al(OH)_3 + H_2SO_4 \rightarrow Al_2(SO_4)_3 + H_2O$, the coefficients of the balanced equation are
    (A) 1, 3, 1, 2
    (B) 2, 3, 2, 6
    (C) 2, 3, 1, 6
    (D) 2, 6, 1, 3
    (E) 1, 3, 1, 6

75. What is $\Delta H_{\text{reaction}}$ for the decomposition of 1 mole of sodium chlorate? (H_f° values are as follows: $NaClO_3(s) = -85.7$ kcal/mole, $NaCl(s) = -98.2$ kcal/mole, $O_2(g) = 0$ kcal/mole)
    (A) −183.9 kcal
    (B) −91.9 kcal
    (C) +45.3 kcal
    (D) +22.5 kcal
    (E) −12.5 kcal

76. Isotopes of an element are related because which of the following is (are) the same in these isotopes?
    I. Atomic mass
    II. Atomic number
    III. Arrangement of orbital electrons
    (A) I only
    (B) II only
    (C) I and II only
    (D) II and III only
    (E) I, II, and III

77. In the reaction of Zn with dilute HCl to form H_2, which of the following will increase the reaction rate?

I. Increasing the temperature

II. Increasing the exposed surface of Zn

III. Using a more active metal instead of Zn

(A) I only

(B) II only

(C) I and III only

(D) II and III only

(E) I, II, and III

78. The laboratory setup shown above can be used to prepare

(A) gas lighter than air

(B) gas heavier than air

(C) gas soluble in water

(D) gas insoluble in water

(E) gas that reacts with water

79. In this reaction: $CaCO_2 + 2HCl \rightarrow CaCl_2 + H_2O + CO_2$, if 3 moles of HCl are available to the reaction with an unlimited supply of $CaCO_3$, how many moles of CO_2 can be produced at STP?

(A) 1

(B) 1.5

(C) 2

(D) 2.5

(E) 3

80. A saturated solution of $BaSO_4$ at 25°C contains 3.9×10^{-5} mole/liter of Ba^{2+} ions. What is the K_{sp} of this salt?

(A) 3.9×10^{-5}

(B) 3.9×10^{-6}

(C) 2.1×10^{-7}

(D) 1.5×10^{-8}

(E) 1.5×10^{-9}

81. Which of the following will definitely cause the volume of a gas to increase?
    I. Decrease the pressure with the temperature held constant.
    II. Increase the pressure with a temperature decrease.
    III. Increase the temperature with a pressure increase.
    (A) I only
    (B) II only
    (C) I and III only
    (D) II and III only
    (E) I, II, and III

82. If 0.1 mole of K_2SO_4 were added to the solution in question 80, what would happen to the Ba^{2+} concentration?
    (A) It would increase.
    (B) It would decrease.
    (C) It would remain the same.
    (D) It would first increase, then decrease.
    (E) It would first decrease, then increase.

83. The number of oxygen atoms in 0.5 mole of $Al_2(CO_3)_3$ is
    (A) 4.5×10^{23}
    (B) 9.0×10^{23}
    (C) 3.6×10^{24}
    (D) 2.7×10^{24}
    (E) 5.4×10^{24}

Questions 84–85 refer to a solution of 0.100 M acetic acid.

84. What is the H_3O^+ concentration? ($K_a = 1.8 \times 10^{-5}$)
    (A) 1.8×10^{-5}
    (B) 1.8×10^{-4}
    (C) 1.3×10^{-2}
    (D) 1.3×10^{-3}
    (E) 0.9×10^{-3}

85. What is the percent dissociation of the acetic acid if the $[H_3O^+]$ is 1×10^{-3} mole/liter?
    (A) 0.01%
    (B) 0.1%
    (C) 1.0%
    (D) 1.5%
    (E) 2.0%

STOP

IF YOU HAVE FINISHED BEFORE ONE HOUR IS UP, YOU MAY GO BACK TO CHECK YOUR WORK OR COMPLETE UNANSWERED QUESTIONS.

1. (E) The triple point of a phase diagram shows that all three states can exist at this point.

2. (B) The combining of nuclei is called nuclear fusion.

3. (D) According to Graham's Law of Gaseous Diffusion (or Effusion), the rate of diffusion is inversely proportional to the square root of the molecular weight.
So.

$$\frac{\text{rate}_H}{\text{rate}_O} = \frac{\sqrt{\text{mol. mass O}_2}}{\sqrt{\text{mol. mass H}_2}} = \frac{\sqrt{32}}{\sqrt{2}} = \sqrt{\frac{16}{1}} = \frac{4}{1}$$

4. (A) The Law of Definite Composition states that, when compounds form, they always form in the same ratio by mass. Water, for instance, always forms in a ratio of 1:8 of hydrogen to oxygen by mass. For nitrous oxide (NO) and nitrogen dioxide (NO_2) the difference in molecular mass is one atomic mass of oxygen.

5. (B) The first step is the ΔH for $C + \frac{1}{2}O_2 \rightarrow CO$. It releases 26.4 kcal of heat. This is written as −26.46.

6. (C) This is the second step on the diagram.

7. (A) To arrive at the ΔH of this, you can take the total drop (−94.0 kcal) or algebraically add these reactions:

$C + \frac{1}{2}O_2$	$\rightarrow CO$	−26.4 kcal
$CO + \frac{1}{2}O_2$	$\rightarrow CO_2$	−67.6 kcal
$C + O_2 \rightarrow CO_2$		−94.0 kcal

8. (B) Potassium and chlorine have a large enough difference in their electronegativities to form ionic bonds. The respective positions of these two elements in the periodic chart also are indicative of the large difference in their electronegativity values.

9. (D) Two atoms of an element that forms a diatomic molecule always have a pure covalent bond between them since the electron attraction or electronegativity of each atom is the same.

10. (C) Electronegativity differences between 0.5 and 1.7 are usually indicative of polar covalent bonding. CO_2 is an interesting example of a *nonpolar molecule* with polar covalent bonds since the bonds are symmetrical in the molecule.

11. (E) Calcium is a metal and forms a metallic bond between atoms.

12. (B) In typical heating curves, the first of two plateaus indicates the change of state from a solid to a liquid.

13. (A) Since Part A is at a lower temperature, it represents an average lower kinetic energy.

14. (D) Boiling occurs at the very beginning of the second plateau.

15. (A) These are the alkali metals.

16. (B) Because of complete outer energy levels these elements have very low boiling points.

17. (B) The complete outer energy level makes it difficult to remove an electron; thus the high peaks would be the noble gases.

18. (A) With only one electron in the outer energy level, the alkali metals lose an electron relatively easily.

19. (E) There is a slight increase in the stability of the *p* orbitals when each has one electron—thus the little peaks at E.

20. (C) The filled *s* orbital shows a slight increase in stability—thus the little peaks at C.

21. (D) The dip at the D areas on the chart shows the first pairing of electrons in the *p* orbitals and indicates this first paired electron has a lower ionization potential than in the atom, which has only one electron in each *p* orbital.

22. (A) The most active metals, the alkalis, have the lowest ionization potentials.

23. (B) The filled *p* orbitals occur in the noble gases.

101. (T, F) Sulfur trioxide is shown by three structural formulas because each

bond is "hybrid" of a single and double bond. Resonance in chemistry does not mean that the bonds resonate between the structures shown in the structural drawing.

102. (T, F) When ΔG is negative in the Gibbs equation, the reaction is spontaneous. However, the total equation determines this, not just the ΔH. The Gibbs equation is:

$$\Delta G = \Delta H - T \Delta S.$$

103. (T, T, CE) One mole of each of the gases contains 6.02×10^{23} molecules, but their molecular masses are different. CO_2 is found by adding one C = 44 and two O = 32, or a total of 76 amu. The H_2O, however, adds up to two H = 2 plus one O = 16, or a total of 18 amu. So it is true that 1 mole of CO_2 is heavier than 1 mole of H_2O.

104. (T, T) Hydrosulfuric acid is a weak acid but is used in qualitative tests because of the distinctly colored precipitates of sulfides that it forms with many metallic ions.

105. (F, T) Sodium chloride is an ionic crystal, not a molecule, and its ions are hydrated by the polar water molecules.

106. (T, T, CE) This statement and its reason fit the rules of Le Châtelier's Principle.

107. (T, T, CE) The statement is true, and the reason is also true and explains the statement.

108. (T, F) The statement is true, but not the reason. In an equilibrium reaction, concentrations can be shown to progress like this:

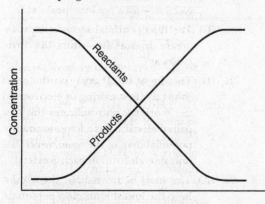

until equilibrium is reached. Then the concentrations stabilize.

109. (F, T) The forward and reverse reactions are occurring at equal rates when equilibrium is reached. The reactions do not stop. The concentrations remain the same at this point.

110. (T, T, CE) Since acetylene is known to be a linear molecule with a triple bond between the two carbons, the sp orbitals along the central axis with the hydrogens bonded on either end fit the experimental evidence.

111. (F, F) The weakest bonds between molecules are called van der Waals forces.

112. (T, T, CE) The terms *dilute* and *concentrated* merely indicate a relatively large amount of solvent and a small amount of solvent, respectively. You can have a dilute saturated solution if the solute is only slightly soluble.

113. (F, T) Cs is the most active Group I metal because it has (a) the largest atomic radius, thus making it easier to lose its outer energy level electron, and (b) the intermediate electrons help screen the positive attraction of the nucleus, also increasing the ease with which the outer electron is lost.

114. (F, T) The cations are positively charged ions and migrate to the cathode, while the anions are negatively charged and migrate to the anode.

115. (T, T, CE) There are as many electrons as there are protons in a neutral atom, and the atomic number represents the number of each.

116. (F, T) The first two principal energy levels fill up at 2 and 8 electrons, respectively. That leaves 7 electrons to fill the 3s and 3p orbitals like this: $3s^2$, $3p^5$. With only one electron missing in the 3p orbitals, the most likely oxidation number is –1.

40. (D) The solidification of "Crisco" is merely a physical change just like the formation of ice from liquid water at lower temperatures. All the others involve actual recombinations of the atoms and thus are chemical changes.

41. (C) Water is formed because most common fuels contain hydrogen in their structure.

42. (E) The other choices, in order, would be 1, $\frac{1}{10}$ or deci-, $\frac{1}{100}$ or centi-, and 100.

43. (E) One mole of any substance contains 6.02×10^{23} molecules. Since each water molecule is triatomic, there would be $3(6.02 \times 10^{23})$ atoms present.

44. (E) The "noble" gases are all monoatomic because of their complete outer energy level. A rule to help you remember diatomic gases is: Gases ending in *-gen* or *-ine* usually form diatomic molecules.

45. (D) By both the VSEPR (valence shell electron pair repulsion) method and the orbital structure method, the PCl_3 molecule is trigonal pyramidal like this:

46. (C) The complete loss and gain of electrons is an ionic bond. All other bonds indicated are "sharing of electrons" type bonds or some form of covalent bonding.

47. (D) The cathode reaction releases only H_2 gas. The half-reaction is as given in reaction D.

48. (B) The beta particle is a high-speed electron and has the smallest mass of the first four choices. Gamma rays are electromagnetic waves.

49. (E) A gamma ray is not affected by an electric field because it is an electromagnetic wave. The neutron is not an electromagnetic wave; it is a particle with no charge but the approximate mass of a proton.

50. (D) $pH = -\log[H_3O^+] = -\log[1 \times 10^{-4}] = -(-4) = 4$.

51. (C) Only I and II are true. Distilled water does not significantly conduct an electric current. The polarity of the water molecule is helpful in ionization and in causing substances to go into solution.

52. (A) SO_2 is the acid anhydride of H_2SO_3 or sulfurous acid. $H_2O + SO_2 \rightarrow H_2SO_3$.

53. (E) Four grams of hydrogen gas at STP represent 2 moles of hydrogen since 2 grams is the gram-molecular mass of hydrogen. Each mole of a gas contains 6.02×10^{23} molecules, and 2 moles would contain $2 \times 6.02 \times 10^{23}$ or 12.04×10^{23} molecules.

54. (B) To solve % composition problems, first divide the % given by the atomic mass:

$$12 \overline{)85.7} \qquad 1 \overline{)14.3}$$
$$7.14 \qquad 14.3$$

Then divide by the smallest quotient to get small whole numbers:

$$7.14 \overline{)7.14} \qquad 7.14 \overline{)14.3}$$
$$1.0 \qquad 2.0$$

So the empirical formula is CH_2. Since the molecular mass is 42 and the empirical formula has a molecular mass of 14, the true formula must be 3 times the empirical formula, or C_3H_6.

55. (E) Since the temperature (in kelvins) increases from 303 K to 333 K, the volume of the gas should increase with pressure held constant. The correct fraction is $\frac{333}{303}$.

56. (D) One mole of dissolved substance (which does not ionize) causes a 1.86 degree drop (Celsius) in the freezing point of a 1 molal solution. Since 1,500 grams of water were used, the solution has $\frac{684}{342}$ or 2 moles in 1,500 grams of water. Then

$$\frac{2\,\text{moles}}{1{,}500\,\text{g}} = \frac{x\,\text{moles}}{1{,}000\,\text{g}}$$
$$x = \frac{2{,}000}{1{,}500} = 1.33\,\text{moles} / 1{,}000\,\text{g}$$

So the freezing point is depressed $1.33 \times -1.86° = -2.48°C$.

57. (B) The pH is $-\log[H^+]$. A 0.003 molar solution of H_2SO_4 ionizes in a dilute solution to release two H^+ ions per molecule of H_2SO_4. Therefore the molar concentration of H^+ ion is 2×0.003 mole/liter or 0.006 mole/liter. Substituting this in the formula, you have:

$$pH = -\log[0.006] = -\log[6 \times 10^{-3}]$$

The log of a number is the exponent to which the base 10 is raised to express that number in exponential form:

$1 = 10^0$ so the log $1 = 0$ and
$10 = 10^1$ so the log $10 = 1$

The log of 6 must be 10 to a decimal exponent less than 1 and probably close to 0.7. Then
$-\log [6 \times 10^{-3}] = -\log (10^{0.7} \times 10^{-3}) = -[0.7 + (-3)] = -[-2.3] = 2.3$ (approx.)

58. (B) If the solution is to be 5% sodium hydroxide, then 5% of 100 g is 5 grams. % is always by weight unless otherwise specified.

59. (C) Since this equation is exothermic, higher temperatures will decrease the reaction to the right and increase the reaction to the left.

60. (B) The reaction is:

$$H_2SO_4 + 2NaOH \rightarrow Na_2SO_4 + H_2O$$
$$\underset{\substack{\text{1 mole}\\\text{acid}}}{} \quad \underset{\substack{\text{2 moles}\\\text{base}}}{}$$

1 mole acid = $\frac{1}{2}$ (moles) of base
Since

molarity × volume (L) = moles,
then

$$M_a V_a = \tfrac{1}{2} M_b V_b$$
$$(1\ M)(x\ L) = \tfrac{1}{2}(1\ M)(0.05\ L)$$
$$x\ L = \tfrac{1}{2}(0.05L)$$
$$x\ L = 0.025\ L \text{ or } 25\ mL$$

61. (E) The reaction is:
150 g

$$CaCO_3 + 2HCl \rightarrow CaCl_2 + H_2O + CO_2$$

The gram-molecular mass of calcium is 100 g. Then 150 g = $\frac{150}{100}$ or 1.5 moles of calcium carbonate. According to the equation, 1 mole of calcium carbonate yields 1 mole of carbon dioxide, so 1.5 moles of calcium carbonate would yield 1.5 moles of carbon dioxide.

The gram-molecular mass of CO_2 = 44 g:
1.5 moles of CO_2 = 1.5 × 44 g = 66 g CO_2

62. (E) The other choices are wrong because:
(A) is lighter than air
(B) reacts with air
(C) is lighter than air
(D) needs heat to be evolved

63. (A) The other choices are wrong because:
(B) needs no heat
(C) needs a condenser for collection
(D) is lighter than air
(E) is lighter than air

64. (C) To insure that all the water vapor collected in the U-tube comes from the reaction, the first drying tube is placed in the path of the hydrogen to absorb any evaporated water.

65. (C) Calcium chloride is deliquescent, and its weight gain of water indicates that water was formed from the reaction.

66. (D)

$CaCl_2$ and U-tube after	= 57.32 g
$CaCl_2$ and U-tube before	= 39.32 g
Mass of water formed	= 18.00 g
CuO and boat before	= 30.23 g
CuO and boat after	= 14.23 g
Mass of oxygen reacted	= 16.00 g
So, mass of water	= 18 g
mass of oxygen	= 16 g
Mass of hydrogen	= 2 g

The ratio of mass of water to mass of hydrogen is 18:2 or 9:1.

67. (D) 1K = 39, 1Al = 27, 2(SO_4) = 2(32 + 16 × 4) = 192, and 12H_2O = 12(2 + 16) = 216. This totals 474 g.

68. (D)
$$\underset{108\,g}{4Al} + \underset{67.2\,L}{3O_2} \rightarrow 2Al_2O_3$$
with x g and 44.8 L

$$\frac{x}{108} = \frac{44.8}{67.2}, \text{ so } x = 72 \text{ g}.$$

69. (B) Metal oxides are generally basic anhydrides.

70. (A) Since A is marked +, the oxidation (or loss of electrons) will occur on this pole.

71. (C) "Salt bridge" is the correct terminology.

72. (C) Starch solution is used to test for the presence of I_2. Since this half-cell produces I_2, the identifying blue-black color due to the I_2 reacting with the starch will appear.

73. (D)
$$\underset{213\,g}{2NaClO_3} \rightarrow 2NaCl + \underset{x\,L}{3O_2}$$

$$\frac{213}{213} = \frac{x}{67.2}, \text{ so } x = 67.2 \text{ L}.$$

74. (C) The balanced equation has the coefficients 2, 3, 1, and 6: $2Al(OH)_3 + 3H_2SO_4 \rightarrow Al_2(SO_4)_3 + 6H_2O$.

75. (E) The reaction is $NaClO_3 \rightarrow NaCl + \frac{3}{2}O_2$.
$$\Delta H_{reaction} = \Delta H_{f(products)} - H_{f(reactants)}$$
$$\Delta H_{reaction} = (-98.2 + 0) - (-85.7)$$
$$\Delta H_{reaction} = -12.5 \text{ kcal}$$

76. (D) II and III are identical; isotopes differ only in the number of neutrons in the nucleus and this affects the atomic mass only.

77. (E) I, II, and III would affect this reaction.

78. (D) This setup depends on water displacement of an insoluble gas.

79. (B) The coefficients give the molar relations, so 2 moles of HCl give off 1 mole of CO_2. Given 3 moles of HCl, you have

$$\frac{3}{2} = \frac{x}{1}, \text{ then } x = \frac{3}{2} = 1.5 \text{ moles.}$$

80. (E) $K_{sp} = [B_2^{2+}][SO_4^{2-}] = (3.9 \times 10^{-5})$
$(3.9 \times 10^{-5}) = 1.5 \times 10^{-9}$
These two will be the same since $BaSO_4 \rightarrow Ba^{2+} + SO_4^{2-}$

81. (A) According to the gas laws, only I will cause an increase in the volume of a confined gas.

82. (B) The introduction of the "common ion" SO_4^{2-} at 0.1 molar forces the equilibrium to shift to the left and reduce the Ba^{2+} concentration.

83. (D) In 1 mole of $Al_2(CO_3)_3$, nine oxygens (three carbonates with three oxygen atoms each) are in each molecule, or 9 moles of oxygen atoms are in 1 mole of $Al_2(CO_3)_3$. Since only 0.5 mole is given, there would be $\frac{1}{2}(9)$ or 4.5 moles of oxygen atoms. In 4.5 moles of oxygen, there would be:

$$4.5 \text{ moles O atoms} \times \frac{6.0 \times 10^{23} \text{ atoms}}{1 \text{ mole O atoms}}$$
27.0×10^{23} atoms or 2.7×10^{24} atoms

84. (D) $K_a = 1.8 \times 10^{-5} = \frac{[H_3O^+][Ac^-]}{[HAc]} = \frac{(x)(x)}{.1-x}$.
Since acetic acid is a weak acid, the 0.1 molar concentration is much greater than the amount ionized (x), so $0.1 - x$ is essentially equal to 0.1.

Therefore $1.8 \times 10^{-5} = \frac{(x)(x)}{0.1} = \frac{x^2}{0.1}$

and $x = 1.3 \times 10^{-3}$.

85. (C) Percent dissociation =

$$\frac{\text{moles/liter of } H_3O^+}{\text{original concentration}} \times 100. \text{ Per-}$$

cent dissociation = $\frac{1 \times 10^{-3}}{1 \times 10^{-1}} \times 100 =$ 1%.

DIAGNOSING YOUR NEEDS

After taking Practice Test 2, check your answers against the correct ones. Then fill in the chart below.

In the space under each question number, place a check if you answered that question correctly.

> • **EXAMPLE:**

If your answer to question 5 was correct, place a check in the appropriate box.

Next, total the check marks for each section and insert the number in the designated block. Now do the arithmetic indicated and insert your percent for each area.

SUBJECT AREA	(✔) QUESTIONS ANSWERED CORRECTLY

I. Atomic Theory and Structure, including periodic relationships

19	20	21	23	101	115	44	76

☐ No. of checks ÷ 8 × 100 = _____%

II. Nuclear Reactions

2	48	49

☐ No. of checks ÷ 3 × 100 = _____%

III. Chemical Bonding and Molecular Structure

4	8	9	10	11	110	111	45	46

☐ No. of checks ÷ 9 × 100 = _____%

IV. States of Matter and Kinetic Molecular Theory of Gases

3	12	13	14	43	55	81

☐ No. of checks ÷ 7 × 100 = _____%

V. Solutions, including concentration units, solubility, and colligative properties

105	112	56	60

☐ No. of checks ÷ 4 × 100 = _____%

VI. Acids and Bases

50	52	57	69	84	85

☐ No. of checks ÷ 6 × 100 = _____%

VII. Oxidation-Reduction and Electrochemistry

114	116	47	51	70	71

☐ No. of checks ÷ 6 × 100 = _____%

SUBJECT AREA	(✔)QUESTIONS ANSWERED CORRECTLY											
VIII. Stoichiometry	103	53	54	58	61	66	67	68	73	74	79	83
☐ No. of checks ÷ 12 × 100 = _____ %												

SUBJECT AREA			
IX. Reaction Rates		59	77
☐ No. of checks ÷ 2 × 100 = _____ %			

SUBJECT AREA					
X. Equilibrium	106	108	109	80	82
☐ No. of checks ÷ 5 × 100 = _____ %					

SUBJECT AREA						
XI. Thermodynamics: energy changes in chemical reactions, randomness, and criteria for spontaneity	5	6	7	102	107	75
☐ No. of checks ÷ 6 × 100 = _____ %						

SUBJECT AREA						
XII. Descriptive Chemistry: physical and chemical properties of elements and their familiar compounds; organic chemistry; periodic properties	1	15	16	17	18	22
	104	113	40	41	42	72
☐ No. of checks ÷ 12 × 100 = _____ %						

SUBJECT AREA					
XIII. Laboratory: equipment, procedures, observations, safety, calculations, and interpretation of results	62	63	64	65	78
☐ No. of checks ÷ 5 × 100 = _____ %					

PLANNING YOUR STUDY

The percentages give you an idea of how you have done on the various major areas of the test. Because of the limited number of questions on some parts, these percentages may not be as reliable as the percentages for parts with larger numbers of questions. However, you should now have at least a rough idea of the areas in which you have done well and those in which you need more study.

Start your study with the areas in which you are the weakest. The corresponding chapters are indicated below:

Subject Area	Chapters to Review
I. Atomic Theory and Structure, including periodic relationships	2
II. Nuclear Reactions	15
III. Chemical Bonding and Molecular Structure	3, 4
IV. States of Matter and Kinetic Molecular Theory of Gases	1
V. Solutions, including concentration units, solubility, and colligative properties	7
VI. Acids and Bases	11
VII. Oxidation-reduction and Electrochemistry	12
VIII. Stoichiometry	5, 6
IX. Reaction Rates	9
X. Equilibrium	10
XI. Thermodynamics, including energy changes in chemical reactions, randomness, and criteria for spontaneity	8
XII. Descriptive Chemistry: physical and chemical properties of elements and their familiar compounds; organic chemistry; periodic properties	1, 2, 13, 14
XIII. Laboratory: equipment, procedures, observations, safety, calculations, and interpretation of results.	All lab diagrams, 16

ANSWER SHEET FOR PRACTICE TEST 3

Determine the correct answer for each question. Then, using a No. 2 pencil, blacken completely the oval containing the letter of your choice.

1. Ⓐ Ⓑ Ⓒ Ⓓ Ⓔ
2. Ⓐ Ⓑ Ⓒ Ⓓ Ⓔ
3. Ⓐ Ⓑ Ⓒ Ⓓ Ⓔ
4. Ⓐ Ⓑ Ⓒ Ⓓ Ⓔ
5. Ⓐ Ⓑ Ⓒ Ⓓ Ⓔ
6. Ⓐ Ⓑ Ⓒ Ⓓ Ⓔ
7. Ⓐ Ⓑ Ⓒ Ⓓ Ⓔ
8. Ⓐ Ⓑ Ⓒ Ⓓ Ⓔ
9. Ⓐ Ⓑ Ⓒ Ⓓ Ⓔ
10. Ⓐ Ⓑ Ⓒ Ⓓ Ⓔ
11. Ⓐ Ⓑ Ⓒ Ⓓ Ⓔ
12. Ⓐ Ⓑ Ⓒ Ⓓ Ⓔ
13. Ⓐ Ⓑ Ⓒ Ⓓ Ⓔ
14. Ⓐ Ⓑ Ⓒ Ⓓ Ⓔ
15. Ⓐ Ⓑ Ⓒ Ⓓ Ⓔ
16. Ⓐ Ⓑ Ⓒ Ⓓ Ⓔ
17. Ⓐ Ⓑ Ⓒ Ⓓ Ⓔ
18. Ⓐ Ⓑ Ⓒ Ⓓ Ⓔ
19. Ⓐ Ⓑ Ⓒ Ⓓ Ⓔ
20. Ⓐ Ⓑ Ⓒ Ⓓ Ⓔ
21. Ⓐ Ⓑ Ⓒ Ⓓ Ⓔ
22. Ⓐ Ⓑ Ⓒ Ⓓ Ⓔ
23. Ⓐ Ⓑ Ⓒ Ⓓ Ⓔ

ON THE ACTUAL CHEMISTRY TEST, THE FOLLOWING TYPE OF QUESTION MUST BE ANSWERED ON A SPECIAL SECTION (LABELED "CHEMISTRY") AT THE LOWER LEFT-HAND CORNER OF PAGE 2 OF YOUR ANSWER SHEET. THESE QUESTIONS WILL BE NUMBERED BEGINNING WITH 101 AND MUST BE ANSWERED ACCORDING TO THE FOLLOWING DIRECTIONS.

CHEMISTRY* Fill in oval CE only if II is a correct explanation of I.

	I	II	CE*
101.	Ⓣ Ⓕ	Ⓣ Ⓕ	◯
102.	Ⓣ Ⓕ	Ⓣ Ⓕ	◯
103.	Ⓣ Ⓕ	Ⓣ Ⓕ	◯
104.	Ⓣ Ⓕ	Ⓣ Ⓕ	◯
105.	Ⓣ Ⓕ	Ⓣ Ⓕ	◯
106.	Ⓣ Ⓕ	Ⓣ Ⓕ	◯
107.	Ⓣ Ⓕ	Ⓣ Ⓕ	◯
108.	Ⓣ Ⓕ	Ⓣ Ⓕ	◯
109.	Ⓣ Ⓕ	Ⓣ Ⓕ	◯
110.	Ⓣ Ⓕ	Ⓣ Ⓕ	◯
111.	Ⓣ Ⓕ	Ⓣ Ⓕ	◯
112.	Ⓣ Ⓕ	Ⓣ Ⓕ	◯
113.	Ⓣ Ⓕ	Ⓣ Ⓕ	◯
114.	Ⓣ Ⓕ	Ⓣ Ⓕ	◯
115.	Ⓣ Ⓕ	Ⓣ Ⓕ	◯
116.	Ⓣ Ⓕ	Ⓣ Ⓕ	◯

ON THE ACTUAL CHEMISTRY TEST, THE REMAINING QUESTIONS MUST BE ANSWERED BY RETURNING TO THE SECTION OF YOUR ANSWER SHEET YOU STARTED FOR CHEMISTRY.

40. Ⓐ Ⓑ Ⓒ Ⓓ Ⓔ
41. Ⓐ Ⓑ Ⓒ Ⓓ Ⓔ
42. Ⓐ Ⓑ Ⓒ Ⓓ Ⓔ
43. Ⓐ Ⓑ Ⓒ Ⓓ Ⓔ
44. Ⓐ Ⓑ Ⓒ Ⓓ Ⓔ
45. Ⓐ Ⓑ Ⓒ Ⓓ Ⓔ
46. Ⓐ Ⓑ Ⓒ Ⓓ Ⓔ
47. Ⓐ Ⓑ Ⓒ Ⓓ Ⓔ
48. Ⓐ Ⓑ Ⓒ Ⓓ Ⓔ
49. Ⓐ Ⓑ Ⓒ Ⓓ Ⓔ
50. Ⓐ Ⓑ Ⓒ Ⓓ Ⓔ
51. Ⓐ Ⓑ Ⓒ Ⓓ Ⓔ
52. Ⓐ Ⓑ Ⓒ Ⓓ Ⓔ
53. Ⓐ Ⓑ Ⓒ Ⓓ Ⓔ
54. Ⓐ Ⓑ Ⓒ Ⓓ Ⓔ
55. Ⓐ Ⓑ Ⓒ Ⓓ Ⓔ

56. Ⓐ Ⓑ Ⓒ Ⓓ Ⓔ
57. Ⓐ Ⓑ Ⓒ Ⓓ Ⓔ
58. Ⓐ Ⓑ Ⓒ Ⓓ Ⓔ
59. Ⓐ Ⓑ Ⓒ Ⓓ Ⓔ
60. Ⓐ Ⓑ Ⓒ Ⓓ Ⓔ
61. Ⓐ Ⓑ Ⓒ Ⓓ Ⓔ
62. Ⓐ Ⓑ Ⓒ Ⓓ Ⓔ
63. Ⓐ Ⓑ Ⓒ Ⓓ Ⓔ
64. Ⓐ Ⓑ Ⓒ Ⓓ Ⓔ
65. Ⓐ Ⓑ Ⓒ Ⓓ Ⓔ
66. Ⓐ Ⓑ Ⓒ Ⓓ Ⓔ
67. Ⓐ Ⓑ Ⓒ Ⓓ Ⓔ
68. Ⓐ Ⓑ Ⓒ Ⓓ Ⓔ
69. Ⓐ Ⓑ Ⓒ Ⓓ Ⓔ
70. Ⓐ Ⓑ Ⓒ Ⓓ Ⓔ

71. Ⓐ Ⓑ Ⓒ Ⓓ Ⓔ
72. Ⓐ Ⓑ Ⓒ Ⓓ Ⓔ
73. Ⓐ Ⓑ Ⓒ Ⓓ Ⓔ
74. Ⓐ Ⓑ Ⓒ Ⓓ Ⓔ
75. Ⓐ Ⓑ Ⓒ Ⓓ Ⓔ
76. Ⓐ Ⓑ Ⓒ Ⓓ Ⓔ
77. Ⓐ Ⓑ Ⓒ Ⓓ Ⓔ
78. Ⓐ Ⓑ Ⓒ Ⓓ Ⓔ
79. Ⓐ Ⓑ Ⓒ Ⓓ Ⓔ
80. Ⓐ Ⓑ Ⓒ Ⓓ Ⓔ
81. Ⓐ Ⓑ Ⓒ Ⓓ Ⓔ
82. Ⓐ Ⓑ Ⓒ Ⓓ Ⓔ
83. Ⓐ Ⓑ Ⓒ Ⓓ Ⓔ
84. Ⓐ Ⓑ Ⓒ Ⓓ Ⓔ
85. Ⓐ Ⓑ Ⓒ Ⓓ Ⓔ

Note: For all questions involving solutions and/or chemical equations, assume that the system is in water unless otherwise stated.

Part A

Directions: Each set of lettered choices below refers to the numbered statements or formulas immediately following it. Select the one lettered choice that best fits each statement or formula and then fill in the corresponding oval on the answer sheet. A choice may be used once, more than once, or not at all in each set.

Questions 1–4 refer to the following diagram:

1. The activation energy of the forward reaction
2. The activation energy of the reverse reaction
3. The heat of the reaction for the forward reaction
4. The potential energy of the reactants

(A) A
(B) B
(C) C
(D) D
(E) E

Questions 5–7 refer to the following diagram:

5. To plate silver on the spoon, the position to which the wire from the spoon must be connected

6. The position of the anode

7. The position from which silver that is plated out emerges

Questions 8–11 refer to the following graph:

8. The solid area

9. The liquid area

10. The gaseous area

11. The area that can have both liquid and gas

Questions 12–14

(A) 1
(B) 2
(C) 3
(D) 4
(E) 5

12. When the following equation: $Cu + HNO_3 \rightarrow Cu(NO_3)_2 + H_2O + NO$ is balanced, the coefficient, in the lowest whole number, of NO

13. When the following equation: $KMnO_4 + H_2SO_3 \rightarrow K_2SO_4 + MnSO_4 + H_2SO_4 + H_2O$ is balanced, the coefficient, in the lowest whole number, of H_2SO_3

14. When the following equation: $Br_2 + SO_2 + H_2O \rightarrow H_2SO_4 + HBr$ is balanced, the coefficient, in the lowest whole number, of HBr

Questions 15–18

    (A) Ionic substance
    (B) Polar covalent substance
    (C) Nonpolar covalent substance
    (D) Amorphous substance
    (E) Metallic network

15. KCl

16. HCl(g)

17. CH_4

18. Li

Questions 19–23

    (A) Brownian movement
    (B) Litmus paper reaction
    (C) Phenolphthalein reaction
    (D) Dehydration
    (E) Deliquescent

19. The reason why a blue crystal of $CuSO_4 \cdot 5H_2O$ turns white when heated

20. The zigzag path of colloidal particles in light

21. The pink color in a basic solution

22. The pink color in an acid solution

23. The adsorbtion of water to the surface of a crystal

Part B

ON THE ACTUAL CHEMISTRY TEST, THE FOLLOWING TYPE OF QUESTION MUST BE ANSWERED ON A SPECIAL SECTION (LABELED "CHEMISTRY") AT THE LOWER LEFT-HAND CORNER OF PAGE 2 OF YOUR ANSWER SHEET. THESE QUESTIONS WILL BE NUMBERED BEGINNING WITH 101 AND MUST BE ANSWERED ACCORDING TO THE FOLLOWING DIRECTIONS.

Directions: Each question below consists of two statements, I in the left-hand column and II in the right-hand column. For each question, determine whether statement I is true or false <u>and</u> if statement II is true or false and fill in the corresponding T or F ovals on your answer sheet. <u>Fill in oval CE only if statement II is a correct explanation of statement I.</u>

Sample Answer Grid:

CHEMISTRY * Fill in oval CE only if II is a correct explanation of I.

	I	II	CE*
101.	ⓉⒻ	ⓉⒻ	◯

101. Elements in the upper right of the periodic table are active nonmetals   BECAUSE   they have larger ionic radii than atomic radii.

102. A synthesis reaction with a high heat of formation will not occur until some heat is added   BECAUSE   exothermic reactions need enough activation energy to get them started.

103. Transition elements in a particular period often have the same oxidation number   BECAUSE   they have a complete outer energy level.

104. When a crystal is added to a super-saturated solution, of itself, the crystal does not change   BECAUSE   the supersaturated solution is holding more solute than its normal solubility.

105. Equilibrium is a static condition   BECAUSE   at equilibrium, the forward reaction rate equals the reverse reaction rate.

106. The ionic bond is the strongest bond   BECAUSE   it has an electrostatic attraction due to the loss and gain of electron(s).

107. Pressure applied to a gaseous system in equilibrium favors the forward reaction   BECAUSE   the product side of an equilibrium would show an increase in gas pressure.

108. If the forward reaction of an equilibrium is exothermic, adding heat to the system favors the reverse reaction   BECAUSE   additional heat causes a stress on the system, and the system moves in the direction that releases the stress.

109. The ions of an element in the halogen family have a larger radius than the atom   BECAUSE   more positive charges in the nucleus than electron charges in the energy levels will cause the radius of the ion to be different than that of the atom.

110. The most electronegative elements in the periodic chart are found among nonmetals   BECAUSE   electronegativity is a measure of the ability of an atom to draw valence electrons to itself.

111. Basic anhydrides react in water to form bases   BECAUSE   nonmetallic oxides react with water to form a solution that has an excess of hydroxide ions.

112. There are 3 moles of atoms in 18 grams of water   BECAUSE   there are 6×10^{23} atoms in 1 mole.

113. Benzene is a good electrolyte   BECAUSE   a good electrolyte has charged ions that carry the electric current.

114. Normal butyl alcohol and 2-butanol are isomers   BECAUSE   isomers vary in the number of neutrons in the nucleus of the atom.

115. The reaction of $CaCO_3$ and HCl goes to completion   BECAUSE   reactions that form a precipitate go to completion.

116. A large number of alpha particles were deflected in the Rutherford experiment   BECAUSE   alpha particles that came close to the nucleus of the gold atoms were deflected.

Part C

Directions: Each of the questions or incomplete statements below is followed by five suggested answers or completions. Select the one that is best in each case and then fill in the corresponding oval on the answer sheet.

40. What are the simplest coefficients that balance this equation?

    ... C_4H_{10} + ... O_2 → ... CO_2 + ... H_2O

    (A) 1, 6, 4, 2
    (B) 2, 13, 8, 10
    (C) 1, 6, 1, 5
    (D) 3, 10, 16, 20
    (E) 4, 26, 16, 20

41. How many atoms are present in the formula $KAl(SO_4)_2$?
    (A) 7
    (B) 9
    (C) 11
    (D) 12
    (E) 13

42. All of the following are compounds EXCEPT
    (A) copper sulfate
    (B) carbon dioxide
    (C) washing soda
    (D) air
    (E) lime

43. What volume of gas, in liters, would 1.5 moles of hydrogen occupy at STP?
    (A) 11.2
    (B) 22.4
    (C) 33.6
    (D) 44.8
    (E) 67.2

44. What is the maximum number of electrons held in the d orbitals?
    (A) 2
    (B) 6
    (C) 8
    (D) 10
    (E) 14

45. If an element has an atomic number of 11, it will combine most readily with an element that has an atomic number of
    (A) 12
    (B) 14
    (C) 15
    (D) 16
    (E) 17

46. An example of a physical property is
    (A) rusting
    (B) decay
    (C) souring
    (D) low melting point
    (E) high heat of formation

47. A gas at STP that contains 6.02×10^{23} atoms and forms diatomic molecules will occupy
    (A) 11.2 liters
    (B) 22.4 liters
    (C) 33.6 liters
    (D) 67.2 liters
    (E) 1.06 quarts

48. When excited electrons cascade to lower energy levels in an atom,
    (A) visible light is always emitted
    (B) the potential energy of the atom increases
    (C) the electrons always fall back to the first energy level
    (D) the electrons fall indiscriminately to all levels
    (E) the electrons fall back to the lowest unfilled energy level

49. The spectroscope uses the concept that
    (A) charged particles are evenly deflected in a magnetic field
    (B) charged particles are deflected in a magnetic field inversely to the mass of the particles
    (C) particles of heavier mass are deflected in a magnetic field to a greater degree than lighter particles
    (D) particles are evenly deflected in a magnetic field

50. The bond that includes an upper and a lower sharing of electron orbitals is called
    (A) a pi bond
    (B) a sigma bond
    (C) a hydrogen bond
    (D) a covalent bond
    (E) an ionic bond

51. What is the boiling point of water at the top of Pike's Peak?
    (A) It is 100°C.
    (B) It is > 100°C since the pressure is less than at ground level.
    (C) It is < 100°C since the pressure is less than at ground level.
    (D) It is > 100°C since the pressure is greater than at ground level.
    (E) It is < 100°C since the pressure is greater than at ground level.

52. The atomic structure of the alkane series contains the hybrid orbitals designated as
    (A) sp
    (B) sp^2
    (C) sp^3
    (D) sp^3d^2
    (E) sp^4d^3

53. Which of the following is (are) true for this reaction?

$$Cu + 4HNO_3 \rightarrow Cu(NO_3)_2 + 2H_2O + 2NO_2 \uparrow$$

 I. It is an oxidation-reduction reaction.
 II. Copper is oxidized.
 III. The oxidation number of nitrogen goes from +5 to +4.
  (A) I, II, and III
  (B) I and II only
  (C) II and III only
  (D) I only
  (E) III only

54. Which of the following properties can be attributed to water?

 I. It is attracted by a magnet.
 II. It is a very good conductor of electricity.
 III. It has its polar covalent bonds with hydrogen on opposite sides of the oxygen atom, so that the molecule is linear.
  (A) I, II, and III
  (B) I and II only
  (C) II and III only
  (D) I only
  (E) III only

55. Which of the following define(s) an acid?

 I. It is a good proton donor.
 II. It is a good electron pair acceptor.
 III. It has an excess of H^+ in solution.
  (A) I, II, and III
  (B) I and II only
  (C) II and III only
  (D) I only
  (E) III only

56. A nuclear reactor must include which of the following parts?

 I. Electric generator
 II. Fissionable fuel elements
 III. Moderator
  (A) I, II, and III
  (B) I and II only
  (C) II and III only
  (D) I only
  (E) III only

57. Which of the following salts will hydrolyze in water to form basic solutions?

 I. NaCl
 II. $CuSO_4$
 III. K_3PO_4
  (A) I, II, and III
  (B) I and II only
  (C) II and III only
  (D) I only
  (E) III only

58. The dissolving of 1 mole of NaCl in 1,000 grams of water will change the boiling point of the water to
    (A) 100.51°C
    (B) 101.02°C
    (C) 101.53°C
    (D) 101.86°C
    (E) 103.62°C

59. What is the structure associated with the BF_3 molecule?
    (A) linear
    (B) triangle
    (C) tetrahedron
    (D) trigonal pyramidal
    (E) bent or V shaped

Questions 60–61 refer to the following setup:

60. What letter designates an error in this laboratory setup?
    (A) A
    (B) B
    (C) C
    (D) D
    (E) E

61. If the reaction in question 60 created a gas, where would the contents of the flask be expelled under these conditions?
    (A) A
    (B) B
    (C) C
    (D) D
    (E) E

62. The most active nonmetal has
    (A) a high electronegativity
    (B) a low electronegativity
    (C) a medium electronegativity
    (D) large atomic radii
    (E) a deliquescent property

63. In the reaction $Fe + 2S \rightarrow FeS_2$, which is true?
    (A) $Fe^0 + 2e^- \rightarrow Fe^{2+}$
    (B) $Fe^0 - 2e^- \rightarrow Fe^{2+}$
    (C) $Fe^{2+} - 2e^- \rightarrow Fe^0$
    (D) $S^0 - 2e^- \rightarrow S^{2-}$
    (E) $S^{-2} + 2e^- \rightarrow S^0$

64. What is the pH of a solution with a hydroxide ion concentration of 0.00001 mole/liter?
    (A) -5
    (B) -1
    (C) 5
    (D) 9
    (E) 14

65. Electrolysis of a dilute solution of sodium chloride results in the cathode product
    (A) sodium
    (B) hydrogen
    (C) chlorine
    (D) oxygen
    (E) peroxide

66. $.....C_2H_4(g) + ...O_2(g) \rightarrow .....CO_2(g) + ...H_2O(l)$

    If the equation for the above reaction is balanced, what is the appropriate coefficient for oxygen gas?
    (A) 1
    (B) 2
    (C) 3
    (D) 4
    (E) 5

67. Five liters of gas at STP have a mass of 12.5 grams. What is the molecular mass of the gas?
    (A) 12.5 g
    (B) 25.0 g
    (C) 47.5 g
    (D) 56.0 g
    (E) 125 g

68. A compound whose molecular mass is 90 grams contains 40.0% carbon, 6.67% hydrogen, and 53.33% oxygen. What is the true formula of the compound?
    (A) $C_2H_2O_4$
    (B) CH_2O_4
    (C) C_3H_6O
    (D) C_3HO_3
    (E) $C_3H_6O_3$

69. How many moles of CaO are needed to react with an excess of water to form 370 grams of calcium hydroxide?
    (A) 1
    (B) 2
    (C) 3
    (D) 4
    (E) 5

70. To what volume, in milliliters, must 50.0 milliliters of 3.50 M H_2SO_4 be diluted in order to make 2 M H_2SO_4?
    (A) 25
    (B) 60.1
    (C) 87.5
    (D) 93.2
    (E) 101

71. If K_e is small, it indicates that equilibrium occurs
    (A) at a low product concentration
    (B) at a high product concentration
    (C) after considerable time
    (D) with the help of a catalyst
    (E) with no forward reaction

72. A student measured 10.0 milliliters of an HCl solution into a beaker and titrated it with a standard NaOH solution that was 0.09 M. The initial NaOH burette reading was 34.7 milliliters while the final reading showed 49.2 milliliters.

    How many grams of HCl were present in 100 milliliters of the HCl solution?
    (A) 0.13
    (B) 0.47
    (C) 4.7
    (D) 14.5
    (E) 36.5

73. A student made the following observations in the laboratory:
    (a) Sodium metal reacted vigorously with water while a strip of magnesium did not seem to react at all.
    (b) The magnesium strip reacted with dilute hydrochloric acid faster than an iron strip.
    (c) A copper rivet suspended in silver nitrate solution was covered with silver-colored stalactites in several days, and the resulting solution had a blue color.
    (d) Iron filings dropped into the blue solution were coated with an orange color.

    The order of *decreasing* strength as reducing agents is:
    (A) Na, Mg, Fe, Ag, Cu
    (B) Mg, Na, Fe, Cu, Ag
    (C) Ag, Cu, Fe, Mg, Na
    (D) Na, Fe, Mg, Cu, Ag
    (E) Na, Mg, Fe, Cu, Ag

74. A student placed water, sodium chloride, potassium dichromate, sand, chalk, and hydrogen sulfide into a distilling flask and proceeded to distill. What ingredient besides water would be found in the distillate?
    (A) sodium chloride
    (B) chalk
    (C) sand
    (D) hydrogen sulfide
    (E) chrome sulfate

75. Which of these statements is NOT correct?
    (A) In an exothermic reaction, ΔH is negative and the enthalpy decreases.
    (B) In an endothermic reaction, ΔH is positive and the enthalpy increases.
    (C) In a reaction where ΔG is negative, the forward reaction is spontaneous.
    (D) In a reaction where ΔG is positive, ΔS may also be positive.
    (E) In a reaction where ΔH is positive and ΔS is negative, the forward reaction is spontaneous.

76. A student used a steam-jacketed eudiometer and filled it with 32 milliliters of oxygen and 4 milliliters of hydrogen over mercury. How much of which gas would be left uncombined after sparking the mixture?
    (A) none of either
    (B) 3 mL H_2
    (C) 24 mL O_2
    (D) 28 mL O_2
    (E) 30 mL O_2

77. What would be the *total* volume, in milliliters, of gases in question 76 after sparking?
    (A) 16
    (B) 24
    (C) 34
    (D) 36
    (E) 40

78. How can the addition of a catalyst affect an exothermic reaction?
    I. Speed up the reaction.
    II. Slow down the reaction.
    III. Increase the amount of product formed.
        (A) I, II, and III
        (B) I and II only
        (C) II and III only
        (D) I only
        (E) II only

79. In which period is the most electronegative element found?
    (A) 1
    (B) 2
    (C) 3
    (D) 4
    (E) 5

80. What could be the equilibrium constant for this reaction: aA + bB $\rightleftharpoons$ cC + dD, if A and D are solids?
    (A) $\dfrac{[C]^c[D]^d}{[A]^a[B]^b}$
    (B) $\dfrac{[A]^a[B]^b}{[C]^c[D]^d}$
    (C) $\dfrac{[C]^c}{[B]^b}$
    (D) $\dfrac{[C]^c[D]^d}{[A]^a}$
    (E) $[A]^a[B]^b[C]^c[D]^d$

81. Which of the following is NOT a commonly occurring sulfur compound?
    (A) H_2S
    (B) H_2SO_4
    (C) SO_2
    (D) Ag_2S
    (E) SO_3

82. Which of these statements is the best expression for the sp^3 hybridization of carbon electrons?
    (A) The new orbitals are one s orbital and three p orbitals.
    (B) The s electron is promoted to the p orbitals.
    (C) The s orbital is deformed into a p orbital.
    (D) Four new and equivalent orbitals are formed.
    (E) The s orbital electron loses energy to fall back into a partially filled p orbital.

83. The bonding that explains the variation of the boiling point of water from the boiling points of similarly structured molecules is
    (A) hydrogen bonding
    (B) van der Waals forces
    (C) covalent bonding
    (D) ionic bonding
    (E) coordinate covalent bonding

84. If K for the reaction $H_2 + I_2 \leftrightarrows 2HI$ is equal to 45.9 at 450°C, and 1 mole of H_2 and 1 mole of I_2 are introduced into a 1-liter box at that temperature, what would be the concentration of H_2 at equilibrium?
    (A) 0.114 mole/liter
    (B) 0.228 mole/liter
    (C) 0.456 mole/liter
    (D) 0.516 mole/liter
    (E) 0.772 mole/liter

85. What is the gram-molecular mass of a nonionizing solid if 10 grams of this solid, dissolved in 100 grams of water, formed a solution that froze at −1.22°C?
    (A) 0.65 g
    (B) 6.5 g
    (C) 130 g
    (D) 154 g
    (E) 265 g

STOP

IF YOU FINISH BEFORE ONE HOUR IS UP, YOU MAY GO BACK TO CHECK YOUR WORK OR COMPLETE UNANSWERED QUESTIONS.

Answers and
Explanations to Test

3

1. (B) The activation energy is the energy needed to begin the reaction.

2. (D) For the reverse reaction to occur, the sum of B + C is needed. This is shown by D.

3. (C) The heat of the reaction is the heat liberated between the level of potential energy of the reactants and that of the products. This is the quantity C on the diagram.

4. (A) The potential energy of the reactants is the total of the original potential energies of the reactants shown by A.

5. (B) The spoon must be made the cathode to attract the Ag^+ ions.

6. (D) The silver plate is the anode.

7. (E) The solution of Ag^+ provides the silver for plating.

8. (A)
9. (B) } In a phase diagram, the zones are as shown below.
10. (C)

11. (E) The boundry of liquid and gas is shown on the diagram.

12–14. The balanced equations with half-reactions are as follows:

12. (B)
$$3Cu^0 \rightarrow 3Cu^{2+} + 2e^-$$
$$\underline{2NO_3^- + 8H^+ + 3e^- \rightarrow 2NO + 4H_2O}$$
$$3Cu + 8HNO_3 \rightarrow 3Cu(NO_3)_2 + 4H_2O + 2NO$$

13. (E)
$$2(MnO_4^{1-} + 8H^+ + 5e^- \rightarrow Mn^{2+} + 4H_2O)$$
$$\underline{5(SO_3^{2-} + H_2O \rightarrow SO_4^{2-} + 2H^+ + 2e^-)}$$
$$2KMnO_4 + 5H_2SO_3 \rightarrow K_2SO_4 + 2MnSO_4 + 2H_2SO_4 + 3H_2O$$

14. (B)
$$Br_2 + 2e^- \rightarrow 2Br^{1-}$$
$$\underline{SO_2 + 2H_2O \rightarrow SO_4^{2-} + 4H^+ + 2e^-}$$
$$Br_2 + SO_2 + 2H_2O \rightarrow H_2SO_4 + 2HBr$$

15. (A) KCl is ionic because it is the product of a very active metal combining with a very active nonmetal.

16. (B) The electronegativity difference between H and Cl is between 0.5 and 1.7. This indicates an unequal sharing of electrons, which results in a polar covalent bond.

17. (C) Because these polar bonds are symmetrically arranged in the methane molecule, the molecule is nonpolar covalent.

18. (E) Lithium (Li) is a metal.

19. (D) When hydrated copper sulfate is heated, the crystal crumples as the water is forced out of the structure, and a white powder is the result.

20. (A) Brownian movement is due to molecular collisions with colloidal particles, which knock the particles about in a zigzag path noted by the reflected light from these particles.

21. (C) The indicator phenolphthalein turns pink in a basic solution.

22. (B) Litmus paper turns pink in an acid solution.

23. (E) A substance that is deliquescent draws water to its surface. At times it can draw enough water to form a water solution.

101. (T, T) The assertion is true because the nonmetals have a tendency to gain electrons readily. The reason is a true statement but does not explain the assertion.

102. (T, T, CE) The assertion is explained by the reason. The graphic display of this is:

Reaction continuum

103. (T, F) The assertion is true but the reasoning is false. Transition elements have incomplete inner energy levels that are being filled with the additional electrons, thus leaving the outer energy level the same in most cases. This gives these elements common oxidation numbers.

104. (F, T) The assertion is false and the reason is true. A supersaturated solution is holding more than its normal solubility, and the addition of a crystal causes a crystallization to occur.

105. (F, T) Equilibrium is a dynamic condition because of the reason stated. The assertion is false; the reason, true.

106. (T, T, CE) Both statements are true, and the reason explains why ionic bonding is the strongest.

107. (F, F) An equilibrium system must have a gaseous reactant or product for pressure to affect the equilibrium. Then increased pressure will cause the reaction to go in the direction that reduces the concentration of gaseous substances.

108. (T, T, CE) The assertion is explained by the reason; both are true.

109. (T, T) While both assertion and reason are true, the reason does not apply because halogen family ions have a nucleus with less positive charges than the negative charges of the electrons in their energy levels.

110. (T, T, CE) The assertion is true; the reason is true and explains why nonmetals have the highest electronegativity.

111. (T, T, CE) The assertion is explained by the reason; both are true.

112. (T, T) There are three moles of atoms in 18 grams of water because 18 grams is 1 mole of water molecules and each molecule has three atoms. The reason does not explain the assertion but is also true.

113. (F, T) Benzene is a nonionizing substance and therefore a nonelectrolyte. The reason is a true statement.

114. (T, F) Isomers have the same empirical formula but vary in their structural formulas.

115. (T, F) The reaction does go to completion, but a gaseous product is formed, not a precipitate. $2HCl + CaCO_3 \rightarrow CaCl_2 + H_2O + CO_2(g)$

116. (F, T) In the Rutherford experiment relatively few alpha particles were deflected, indicating a great deal of empty space in the atom. The reason is a true statement.

40. (B) The correct coefficients are 2, 13, 8, and 10.

41. (D) $1K + 1Al + 2S + 8O = 12$ total

42. (D) Air is a mixture; all others are compounds. Washing soda (C) is sodium carbonate, and lime (E) is calcium oxide.

43. (C) One mole of a gas at STP occupies 22.4 L, so 1.5 moles × 22.4 L = 33.6 L.

44. (D) The maximum number of electrons in each kind of orbital is:
$s = $ 2 in one orbital
$p = $ 6 in three orbitals
$d = $ 10 in five orbitals
$f = $ 14 in seven orbitals

45. (E) An element with atomic number 11 will have an electron structure of 2, 8, 1. This gives it an oxidation number of +1, which would combine most readily with an oxidation number of −1. The atomic number 17 has an oxidation number of −1.

46. (D) The only physical property named in the list is low melting point.

47. (A) If the gas is diatomic, then 6.02×10^{23} atoms will form $6.02 \times 10^{23}/2$ molecules. At STP, 6.02×10^{23} molecules occupy 22.4 liters. So half that amount would occupy 11.2 liters.

48. (E) Cascading excited electrons can fall only to the lowest energy level that is unfilled.

49. (B) The spectroscope uses a magnetic field to separate isotopes by bending their path. The lighter ones are bent farther than the heavier ones.

50. (A) The pi bond is a bond between two p orbitals, like this:

Upper portion

Lower portion

51. (C) At Pike's Peak (alt. approx. 14,000 ft.) the pressure is lower than at ground level; therefore the vapor pressure at a lower temperature will equal the outside pressure and boiling will occur.

52. (C) The alkanes use the sp^3 hybrid orbitals.

53. (A) I, II, and III are correct.

54. (D) Only I is correct.

55. (A) I, II, and III are acid definitions.

56. (C) I is not necessary for the reactor, but often nuclear energy is used to operate an electric generator. The others, II and III, are necessary for fuel and neutron-speed control, respectively.

57. (E) III is a salt from a strong base and a weak acid, which hydrolyzes to form a basic solution with water.

58. (B) Since the boiling point is increased by 0.51°C for each mole of particles, 1 mole of $NaCl \rightarrow Na^+ + Cl^-$ gives 2 moles of particles. Therefore the boiling point will be 1.02° higher or 101.02°C.

59. (B) The VSEPR model shows BF_3 is trigonal planar and so is related to the triangle shape.

60. (C) The delivery tube is below the fluid level in the flask and will cause liquid to be forced up the thistle tube when gas is evolved in the reaction.

61. (A) At first the fluid will be expelled up the thistle tube by the gas generated and exerting pressure in the reaction flask. When the level of the fluid falls below the end of the thistle tube, the gas will then be released through the thistle tube.

62. (A) The most active nonmetal has a high attraction for another electron—thus high electronegativity.

63. (B) Fe loses electrons to form the Fe^{2+} ion.

64. (D) The K_w of water = $[H^+][OH^-]$ = 10^{-14}. If $[OH^-] = 10^{-5}$ mole/liter, then $[H^+] = 10^{-14}/10^{-5} = 10^{-9}$
pH = $-\log [H^+]$ (by definition)
pH = $-[-9]$
pH = 9

65. (B) When dilute NaCl solution is electrolyzed, hydrogen is given off at the cathode, chlorine is given off at the anode, and sodium hydroxide is left in the container.

66. (E) The correctly balanced equation is
$2C_2H_2(g) + 5O_2(g) \rightarrow 4CO_2(g) + 2H_2O(\ell)$

67. (D) One mole of a gas at STP occupies 22.4 liters.

If $\dfrac{5 \text{ liters}}{12.5 \text{ g}}$, then

$\dfrac{22.4 \text{ liters}}{\text{gram-molecular mass}}$

$\dfrac{5}{12.5} = \dfrac{22.4}{x}$

$x = 56$ g

68. (E) To find the simple or empirical formula, divide each % by the element's atomic mass.

Carbon
$40 \div 12 = 3.333$

Hydrogen
$6.67 \div 1 = 6.67$

Oxygen
$53.33 \div 16 = 3.33$

Next, divide each quotient by the smallest quotient in an attempt to get small whole numbers.
$3.33 \div 3.33 = 1$ Carbon
$6.67 \div 3.33 = 2$ Hydrogen
$3.33 \div 3.33 = 1$ Oxygen
The simplest formula is CH_2O, which has a molecular mass of 30. The true molecular mass is given as 90, which is three times the simplest. Therefore the true formula is $C_3H_6O_3$.

69. (E) The reaction is
$$\overset{370 \text{ g}}{CaO} + H_2O \rightarrow Ca(OH)_2$$
$Ca(OH)_2$ molecular wt. = 74
So 370 g ÷ 74 = 5 moles of $Ca(OH)_2$ wanted. The reaction shows 1 mole of CaO produces 1 mole of $Ca(OH)_2$, so the answer is 5 moles.

70. (C) In dilution problems, this formula can be used:

$$M_{before} \times V_{before} = M_{after} \times V_{after}$$

Substituting gives:

$$3.5 \times 50 = 2 \times (?x)$$

$x = 87.5$ mL, new volume after dilution

71. (A) For the K_e to be small, the numerator, which is made up of the concentration(s) of the product(s) at equilibrium, must be smaller than the denominator. This generally means that equilibrium is reached rather rapidly.

72. (B) The amount of NaOH used is

49.2 − 34.7 = 14.5 mL

Using $M_1 = V_1 = M_2 \times V_2$ gives

0.09 M × 14.5 mL = M_2 × 10 mL

$M_2 = 0.13$ M

To find the grams of HCl present:
Multiply the $M \times$ molar mass.

$$\frac{0.13 \text{ moles}}{1,000 \text{ mL}} \times \frac{36.5 \text{ g}}{1 \text{ mole}} = 4.7 \text{ g HCl}/1,000 \text{ mL}$$

If there are 4.7 g in 1,000 mL, then there will be 4.7 ÷ 10 = 0.47 g of HCl in 100 mL of 0.13 M.

73. (E) The reactions recorded indicated that the ease of losing electrons is greater in sodium than magnesium, greater in magnesium than iron, greater in iron than copper, and finally greater in copper than silver.

74. (D) Distillation removes only dissolved solids from the distillate. The volatile gases, like H_2S, will be carried into the distillate.

75. (E) All the first four statements are correct.

The Gibbs free energy equation is:

$$\Delta G = \Delta H - T\Delta S$$

In choice (E), if ΔH is positive and ΔS is negative, then ΔG will definitely be positive, which means that the forward reaction will not occur spontaneously.

76. (E) H_2 to O_2 ratio by volume is 2:1 in the formation of water. Therefore, 4 mL H_2 will react with 2 mL of O_2 to make 4 mL of steam.

$$2H_2(g) + O_2(g) \rightarrow 2H_2O(g)$$

This leaves 30 mL of O_2 uncombined.

77. (C) There will be 30 mL of O_2 + 4 mL of steam = 34 mL total.

78. (B) By definition, a catalyst can be used to speed up or slow down a reaction.

79. (B) The most electronegative element is fluorine (F), found in period 2.

80. (C) Solids are incorporated into the K value and therefore do not appear on the right side of the equation.

81. (E) SO_3 is not easily formed. The commercial process uses a catalyst.

82. (D) When hybridization forms the sp^3 orbitals, four entirely new orbitals different from the former s and p orbitals result.

83. (A) Hydrogen bonding between water molecules causes the boiling point to be higher than would be expected.

84. (B) At the beginning of the reaction

$[H_2]$ = 1 mole/liter

$[I_2]$ = 1 mole liter

$[HI]$ = 0

At equilibrium

$$(H_2 + I_2 \leftrightarrows 2HI)$$

(Let x = moles/liter of H_2 and I_2 in HI form)

$[H_2] = (1 - x)$ mole/liter

$[I_2] = (1 - x)$ mole/liter

$[HI] = 2x$ mole/liter

$$K = \frac{[HI]^2}{[H_2][I_2]}$$

$$= \frac{(2x)^2}{(1-x)(1-x)} = 45.9$$

Taking the square root of both sides gives

$$\frac{2x}{1-x} = 6.8$$

$x = 0.772$ mole/liter

Therefore $[H_2] = 1 - x = 1 - .0772 = 0.228$ mole/liter.

85. (D) 10 g/100 g water = 100 g/1,000 g water. The freezing point depression, 1.22, is divided by 1.86, which is the depression caused by 1 mole in 1,000 g of water, to find how many moles are dissolved.

1.22/1.86 = 0.65 mole percent

So 100 g = 0.65 mole. Then 1 mole =

$$\frac{100 \text{ g}}{0.65} = 153.8 \text{ or } 154 \text{ g.}$$

DIAGNOSING YOUR NEEDS

After taking Practice Test 3, check your answers against the correct ones. Then fill in the chart below.

In the space under each question number, place a check if you answered that question correctly.

• **EXAMPLE:**

If your answer to question 5 was correct, place a check in the appropriate box.

Next, total the check marks for each section and insert the number in the designated block. Now do the arithmetic indicated and insert your percent for each area.

SUBJECT AREA   (✔) QUESTIONS ANSWERED
                                                        CORRECTLY

I. **Atomic Theory and Structure,** including periodic relationships	110	112	116	44	48	50	82

☐ No. of checks ÷ 7 × 100 = _____%

II. **Nuclear Reactions**	49	56

☐ No. of checks ÷ 2 × 100 = _____%

III. **Chemical Bonding and Molecular Structure**	15	16	17	18	106	41	45	54	59	83

☐ No. of checks ÷ 10 × 100 = _____%

IV. **States of Matter and Kinetic Molecular Theory of Gases**	8	9	10	11	43	47	51

☐ No. of checks ÷ 7 × 100 = _____%

V. **Solutions,** including concentration units, solubility, and colligative properties	19	20	104	58	70

☐ No. of checks ÷ 5 × 100 = _____%

VI. **Acids and Bases**	21	22	113	115	55	57	64

☐ No. of checks ÷ 7 × 100 = _____%

VII. **Oxidation-Reduction and Electrochemistry**	5	6	7	53	65	63

☐ No. of checks ÷ 6 × 100 = _____%

SUBJECT AREA

(✔) QUESTIONS ANSWERED CORRECTLY

VIII. Stoichiometry		12	13	14	40	66	67	68	69	76	77	85

☐ No. of checks ÷ 11 × 100 = _____%

IX. Reaction Rates						1	108

☐ No. of checks ÷ 2 × 100 = _____%

X. Equilibrium		105	107	71	80	84

☐ No. of checks ÷ 5 × 100 = _____%

XI. Thermodynamics: energy changes in chemical reactions, randomness, and criteria for spontaneity		2	3	4	102	75	78

☐ No. of checks ÷ 6 × 100 = _____%

XII. Descriptive Chemistry: physical and chemical properties of elements and their familiar compounds; organic chemistry; periodic properties		23	101	103	109	111	114
		42	46	52	62	79	81

☐ No. of checks ÷ 12 × 100 = _____%

XIII. Laboratory: equipment, procedures, observations, safety, calculations, and interpretation of results		60	61	72	73	74

☐ No. of checks ÷ 5 × 100 = _____%

PLANNING YOUR STUDY

The percentages give you an idea of how you have done on the various major areas of the test. Because of the limited number of questions on some parts, these percentages may not be as reliable as the percentages for parts with larger numbers of questions. However, you should now have at least a rough idea of the areas in which you have done well and those in which you need more study.

Start your study with the areas in which you are the weakest. The corresponding chapters are indicated below:

Subject Area	Chapters to Review
I. Atomic Theory and Structure, including periodic relationships	2, 15
II. Nuclear Reactions	15
III. Chemical Bonding and Molecular Structure	3, 4
IV. States of Matter and Kinetic Molecular Theory of Gases	1
V. Solutions, including concentration units, solubility, and colligative properties	7
VI. Acids and Bases	11
VII. Oxidation-reduction and Electrochemistry	12
VIII. Stoichiometry	5, 6
IX. Reaction Rates	9
X. Equilibrium	10
XI. Thermodynamics, including energy changes in chemical reactions, randomness, and criteria for spontaneity	8
XII. Descriptive Chemistry: physical and chemical properties of elements and their familiar compounds; organic chemistry; periodic properties	1, 2, 13, 14
XIII. Laboratory: equipment, procedures, observations, safety, calculations, and interpretation of results.	All lab diagrams, 16

ANSWER SHEET FOR PRACTICE TEST 4

Determine the correct answer for each question. Then, using a No. 2 pencil, blacken completely the oval containing the letter of your choice.

1. (A) (B) (C) (D) (E)
2. (A) (B) (C) (D) (E)
3. (A) (B) (C) (D) (E)
4. (A) (B) (C) (D) (E)
5. (A) (B) (C) (D) (E)
6. (A) (B) (C) (D) (E)
7. (A) (B) (C) (D) (E)
8. (A) (B) (C) (D) (E)
9. (A) (B) (C) (D) (E)
10. (A) (B) (C) (D) (E)
11. (A) (B) (C) (D) (E)
12. (A) (B) (C) (D) (E)
13. (A) (B) (C) (D) (E)
14. (A) (B) (C) (D) (E)
15. (A) (B) (C) (D) (E)
16. (A) (B) (C) (D) (E)
17. (A) (B) (C) (D) (E)
18. (A) (B) (C) (D) (E)
19. (A) (B) (C) (D) (E)
20. (A) (B) (C) (D) (E)
21. (A) (B) (C) (D) (E)
22. (A) (B) (C) (D) (E)
23. (A) (B) (C) (D) (E)

ON THE ACTUAL CHEMISTRY TEST, THE FOLLOWING TYPE OF QUESTION MUST BE ANSWERED ON A SPECIAL SECTION (LABELED "CHEMISTRY") AT THE LOWER LEFT-HAND CORNER OF PAGE 2 OF YOUR ANSWER SHEET. THESE QUESTIONS WILL BE NUMBERED BEGINNING WITH 101 AND MUST BE ANSWERED ACCORDING TO THE FOLLOWING DIRECTIONS.

CHEMISTRY* Fill in oval CE only if II is a correct explanation of I.

	I	II	CE*
101.	(T) (F)	(T) (F)	◯
102.	(T) (F)	(T) (F)	◯
103.	(T) (F)	(T) (F)	◯
104.	(T) (F)	(T) (F)	◯
105.	(T) (F)	(T) (F)	◯
106.	(T) (F)	(T) (F)	◯
107.	(T) (F)	(T) (F)	◯
108.	(T) (F)	(T) (F)	◯
109.	(T) (F)	(T) (F)	◯
110.	(T) (F)	(T) (F)	◯
111.	(T) (F)	(T) (F)	◯
112.	(T) (F)	(T) (F)	◯
113.	(T) (F)	(T) (F)	◯
114.	(T) (F)	(T) (F)	◯
115.	(T) (F)	(T) (F)	◯
116.	(T) (F)	(T) (F)	◯

ON THE ACTUAL CHEMISTRY TEST, THE REMAINING QUESTIONS MUST BE ANSWERED BY RETURNING TO THE SECTION OF YOUR ANSWER SHEET YOU STARTED FOR CHEMISTRY.

40. (A) (B) (C) (D) (E)
41. (A) (B) (C) (D) (E)
42. (A) (B) (C) (D) (E)
43. (A) (B) (C) (D) (E)
44. (A) (B) (C) (D) (E)
45. (A) (B) (C) (D) (E)
46. (A) (B) (C) (D) (E)
47. (A) (B) (C) (D) (E)
48. (A) (B) (C) (D) (E)
49. (A) (B) (C) (D) (E)
50. (A) (B) (C) (D) (E)
51. (A) (B) (C) (D) (E)
52. (A) (B) (C) (D) (E)
53. (A) (B) (C) (D) (E)
54. (A) (B) (C) (D) (E)
55. (A) (B) (C) (D) (E)

56. (A) (B) (C) (D) (E)
57. (A) (B) (C) (D) (E)
58. (A) (B) (C) (D) (E)
59. (A) (B) (C) (D) (E)
60. (A) (B) (C) (D) (E)
61. (A) (B) (C) (D) (E)
62. (A) (B) (C) (D) (E)
63. (A) (B) (C) (D) (E)
64. (A) (B) (C) (D) (E)
65. (A) (B) (C) (D) (E)
66. (A) (B) (C) (D) (E)
67. (A) (B) (C) (D) (E)
68. (A) (B) (C) (D) (E)
69. (A) (B) (C) (D) (E)
70. (A) (B) (C) (D) (E)

71. (A) (B) (C) (D) (E)
72. (A) (B) (C) (D) (E)
73. (A) (B) (C) (D) (E)
74. (A) (B) (C) (D) (E)
75. (A) (B) (C) (D) (E)
76. (A) (B) (C) (D) (E)
77. (A) (B) (C) (D) (E)
78. (A) (B) (C) (D) (E)
79. (A) (B) (C) (D) (E)
80. (A) (B) (C) (D) (E)
81. (A) (B) (C) (D) (E)
82. (A) (B) (C) (D) (E)
83. (A) (B) (C) (D) (E)
84. (A) (B) (C) (D) (E)
85. (A) (B) (C) (D) (E)

<u>Note</u>: For all questions involving solutions and/or chemical equations, assume that the system is in water unless otherwise stated.

Part A

<u>Directions</u>: Each set of lettered choices below refers to the numbered statements or formulas immediately following it. Select the one lettered choice that best fits each statement or formula and then fill in the corresponding oval on the answer sheet. A choice may be used once, more than once, or not at all in each set.

<u>Questions 1–3</u> refer to the following graphs:

(A)   (B)   (C)   (D)   (E)

1. The graph that best shows the relationship of gas volume to temperature, with pressure held constant

2. The graph that best shows the relationship of gas volume to pressure, with temperature held constant

3. The graph that best shows the relationship of the number of grams of solute that is soluble in 100 grams of H_2O at varying temperatures if the solubility begins at a small quantity and increases in direct proportion to the increase in temperature

<u>Questions 4–7</u>

      (A) A molecule
      (B) A mixture of compounds
      (C) An isotope
      (D) An isomer
      (E) An acid salt

4. The simplest unit of water that retains its properties

5. A commercial cake mix

6. An atom with the same number of protons as another atom of the same element but a different number of neutrons

7. Classification of $NaHCO_3$

Questions 8–10

        (A) 1
        (B) 7
        (C) 9
        (D) 10
        (E) 14

8. The atomic number of an atom with an electron dot arrangement similiar to $\ddot{\underset{\cdot\cdot}{I}}$

9. The number of atoms represented in the formula $Al(OH)_3$

10. The number that represents the most acid pH

Questions 11–14

        (A) Density
        (B) Equilibrium constant
        (C) Molecular mass
        (D) Freezing point
        (E) Molarity

11. Can be expressed in moles per liter of solution

12. Can be expressed in grams per liter of a gas

13. Will NOT be affected by changes in temperature and pressure

14. At STP, can be used to determine the molecular mass of a pure gas

Questions 15–18

        (A)   Buffer
        (B)   Indicator
        (C)   Arrhenius acid
        (D)   Arrhenius base
        (E)   Neutral condition

15. Resists a rapid change of pH

16. Exhibits different colors in acidic and basic solutions

17. At 25°C, the aqueous solution has a pH < 7.

18. At 25°C, the aqueous solution has a pH > 7.

Questions 19–23

        (A) H_2
        (B) NH_3
        (C) CO_2
        (D) HCl
        (E) O_2

19. A gas produced by the reaction of zinc with hydrochloric acid

20. A gas combustion product that is heavier than air

21. A gas produced by the heating of potassium chlorate

22. A gas that is slightly soluble in water and gives a weakly acid solution

23. A gas that is very soluble in water and gives a weakly basic solution

Part B

ON THE ACTUAL CHEMISTRY TEST, THE FOLLOWING TYPE OF QUESTION MUST BE ANSWERED ON A SPECIAL SECTION (LABELED "CHEMISTRY") AT THE LOWER LEFT-HAND CORNER OF PAGE 2 OF YOUR ANSWER SHEET. THESE QUESTIONS WILL BE NUMBERED BEGINNING WITH 101 AND MUST BE ANSWERED ACCORDING TO THE FOLLOWING DIRECTIONS.

<u>Directions</u>: Each question below consists of two statements, I in the left-hand column and II in the right-hand column. For each question, determine whether statement I is true or false <u>and</u> if statement II is true or false and fill in the corresponding T or F ovals on your answer sheet. <u>Fill in oval CE only if statement II is a correct explanation of statement I.</u>

Sample Answer Grid:

CHEMISTRY * Fill in oval CE only if II is a correct explanation of I.

	I	II	CE*
101.	(T) (F)	(T) (F)	◯

101. According to the kinetic molecular gas theory, particles are in random motion   BECAUSE   the molecules undergo inelastic collisions.

102. The electron can have wave properties as well as corpuscular properties   BECAUSE   the design of the experiment determines which is verified.

103. The alkanes are considered a homologous series   BECAUSE   homologous series have the same functional group but differ in formula by the addition of a fixed group of atoms.

104. When an atom of an active metal becomes an ion, the radius of the ion is less than that of the atom   BECAUSE   the nucleus has less positive charge than the electron "cloud."

105. When the heat of formation for a compound is negative, the ΔH is negative   BECAUSE   a negative heat of formation indicates that the reaction is exothermic with a negative enthalpy.

106. Water is a polar substance   BECAUSE   the sharing of the bonding electrons is equal.

107. A catalyst accelerates a chemical reaction   BECAUSE   a catalyst lowers the activation energy of the reaction.

108. Copper is an oxidizing agent in the reaction with silver nitrate solution   BECAUSE   copper loses electrons in a reaction with silver ions.

109. The rate of diffusion (or effusion) of hydrogen gas compared with that of helium gas is 1:4   BECAUSE   the rate of diffusion (or effusion) of gases varies inversely as the square root of the molecular weights.

110. A gas heated from 10°C to 100°C at constant pressure will increase in volume   BECAUSE   Charles's Law states that, if the pressure remains constant, the volume varies directly as the absolute temperature.

111. The Gibbs free energy equation can be used to predict the solubility of a solute   BECAUSE   the solubility of most salts increases as temperature increases.

112. The complete electrolysis of 45 grams of water will yield 40 grams of H_2 and 5 grams of O_2 BECAUSE   water is composed of hydrogen and oxygen in a ratio of 8:1 by mass.

113. 320 calories of heat will melt 4 grams of water at 0°C   BECAUSE   the heat of fusion of water is 80 calories per gram.

114. When 2 liters of oxygen gas react with 2 liters of hydrogen completely, the limiting factor is the volume of the oxygen.   BECAUSE   the coefficients of gaseous reactions give the volume relationship of the reacting gases.

115. Water is a good solvent   BECAUSE   water shows hydrogen bonding between oxygen atoms.

116. Ammonia gas, NH_3, has a smaller density than argon gas, Ar, at STP   BECAUSE   the density of a gas at STP is found by dividing the atomic mass by 22.4 liters.

Part C

Directions: Each of the questions or incomplete statements below is followed by five suggested answers or completions. Select the one that is best in each case and then fill in the corresponding oval on the answer sheet.

40. In this graphic representation of a chemical reaction, which arrow depicts the activation energy?
    (A) A
    (B) B
    (C) C
    (D) D
    (E) E

41. How many liters (STP) of O_2 can be produced by completely decomposing 1 mole of $KClO_3$?
    (A) 11.2
    (B) 22.4
    (C) 33.6
    (D) 44.8
    (E) 77.2

42. Which of the following is the correct structural representation of sodium?
    (A) $\begin{array}{c} 11\,p \\ 11\,n \end{array}$)2   )8   )8   )5

(B) $\begin{array}{c} 11\,p \\ 12\,n \end{array}$)2   )8   )11   )2

(C) $\begin{array}{c} 23\,p \\ 23\,n \end{array}$)2   )8   )1

(D) $\begin{array}{c} 23\,p \\ 23\,n \end{array}$)2   )8   )11   )2

(E) $\begin{array}{c} 11\,p \\ 12\,n \end{array}$)2   )8   )1

43. Which of the following statements is true?
   (A) A catalyst cannot lower the activation energy.
   (B) A catalyst can lower the activation energy.
   (C) A catalyst affects only the activation energy of the forward reaction.
   (D) A catalyst affects only the activation energy of the reverse reaction.
   (E) A catalyst is permanently changed after the activiation energy is achieved.

44. If the molecular mass of NH_3 is 17, what is the density of this compound at STP?
   (A) 0.25 g/L
   (B) 0.76 g/L
   (C) 1.52 g/L
   (D) 3.04 g/L
   (E) 9.11 g/L

45. Which bond(s) is (are) ionic?
   I. H—Cl(g)
   II. S—Cl(g)
   III. Cs—F(s)
   (A) I, II, and III
   (B) I and II only
   (C) II and III only
   (D) I only
   (E) III only

46. Aromatic hydrocarbons are represented by which of the following?

   (A) I, II, and III
   (B) I and II only
   (C) II and III only
   (D) I only
   (E) III only

47. According to placement in the periodic table, which statement(s) regarding the first ionization energies of certain elements should be true?
    I. Li has a higher value than Na.
    II. K has a higher value than Cs.
    III. Na has a higher value than Al.
        (A) I, II, and III
        (B) I and II only
        (C) II and III only
        (D) I only
        (E) III only

48. Correctly expressed half-reactions include which of the following?
    I. $CrO_4^{2-} + 8H^+ + 6e^- \rightarrow Cr^{3+} + 4H_2O$
    II. $I^- + 6OH^- \rightarrow IO_3^- + 3H_2O + 6e^-$
    III. $MnO_4^- + 2H_2O + 3e^- \rightarrow MnO_2 + 4OH^-$
        (A) I, II, and III
        (B) I and II only
        (C) II and III only
        (D) I only
        (E) III only

What is the apparent oxidation number of the underlined element in the compound

49. Na<u>N</u>O$_3$?
    (A) +1
    (B) +2
    (C) +3
    (D) +4
    (E) +5

50. <u>Ca</u>SO$_4$?
    (A) +1
    (B) −1
    (C) +2
    (D) −2
    (E) +3

51. <u>N</u>H$_3$?
    (A) +2
    (B) −2
    (C) +3
    (D) −3
    (E) +5

52. An atom with an orbital notation of $1s^2\ 2s^2\ 2p^6\ 3s^2\ 3p^4$ will probably exhibit which oxidation state?
    (A) +2
    (B) −2
    (C) +3
    (D) −3
    (E) +5

53. In the Lewis dot structure shown, what is the predictable oxidation number? x :
    (A) +1
    (B) −1
    (C) +2
    (D) −2
    (E) +3

Questions 54–57 refer to the following apparatus, assembled by a student:

Trough   Flask   Bottle   Glass bend and tubing   Glass plate

54. The apparatus assembled is used to prepare a gas by a reaction that takes place when a solid
    (A) is heated
    (B) is exposed
    (C) reacts with a liquid
    (D) is heated in a vacuum
    (E) is decomposed by a catalyst

55. This apparatus suggests that the student plans to collect a gas that
    (A) supports combustion
    (B) is heavier than air
    (C) is not flammable
    (D) is insoluble in water
    (E) reacts with water

56. If the student planned to prepare hydrogen, what would also be needed?
    (A) mercuric oxide
    (B) acid plus zinc
    (C) potassium chlorate
    (D) carbon disulfide
    (E) benzene

57. If this apparatus were to be modified to prepare oxygen by the decomposition of mercuric oxide, which piece would be replaced?
    (A) delivery tube
    (B) collecting bottle
    (C) trough of water
    (D) flask with the thistle tube
    (E) glass plate

58. What occurs when a reaction is at equilibrium and more reactant is added to the container?
    (A) The equilibrium remains unchanged.
    (B) The forward reaction rate increases.
    (C) The reverse reaction rate increases.
    (D) The forward reaction rate decreases.
    (E) The reverse reaction rate decreases.

59. How much heat energy is released when 8 grams of hydrogen are burned? The thermal equation is: $2H_2 + O_2 \rightarrow 2H_2O + 136.64$ kcal.
    (A) 68.32 kcal
    (B) 102.48 kcal
    (C) 136.64 kcal
    (D) 273.28 kcal
    (E) 546.56 kcal

60. Would a spontaneous reaction occur between zinc *ions* and gold *atoms*?

    $Zn^{2+} + 2e^- \leftrightarrows Zn°$ $E° = -0.76$ volt
    $Au^{3+} + 3e^- \leftrightarrows Au°$ $E° = +1.42$ volt

    (A) yes—Reaction potential 2.18 volts
    (B) no—Reaction potential −2.18 volts
    (C) yes—Reaction potential 0.66 volt
    (D) no—Reaction potential −0.66 volt
    (E) yes—Reaction potential 0.56 volt

61. Four moles of electrons ($4 \times 6.02 \times 10^{23}$ electrons) would electroplate how many grams of silver from a silver nitrate solution?
    (A) 108
    (B) 216
    (C) 324
    (D) 432
    (E) 540

62. A 5 M solution of HCl would have how many moles of H^+ ion in 1 liter?
    (A) 0.5
    (B) 1.0
    (C) 2.0
    (D) 2.5
    (E) 5.0

63. What is the K_{sp} for silver acetate if a saturated solution contains 2×10^{-3} mole of silver ion/liter of solution?
    (A) 2×10^{-3}
    (B) 2×10^{-6}
    (C) 4×10^{-3}
    (D) 4×10^{-6}
    (E) 4×10^6

64. The following data were obtained for H_2O and H_2S:

	Formula Mass	Freezing Point (°C)	Boiling Point (°C)
H_2O	18	0	100
H_2S	34	−83	−60

What is the best explanation for the variation of physical properties between these two compounds?
(A) The H_2S has stronger bonds between molecules.
(B) The H_2O has a great deal of hydrogen bonding.
(C) The bond angles differ by about 15°.
(D) The formula mass is of prime importance.
(E) The oxygen atom has a smaller radius and thus cannot bump into other molecules as often as the sulfur.

65. What is the pH of a solution that has 0.00001 mole of H^+/liter of solution?
(A) 2
(B) 3
(C) 4
(D) 5
(E) 9

66. How many grams of sulfur are present in 1 mole of H_2SO_4?
(A) 2
(B) 32
(C) 49
(D) 64
(E) 98

67. What is the approximate mass, in grams, of 1 liter of nitrous oxide, N_2O, at STP?
(A) 1
(B) 2
(C) 11.2
(D) 22
(E) 44

68. If the simplest formula of a substance is CH_2 and its molecular mass is 56, what is its true formula?
(A) CH_2
(B) C_2H_4
(C) C_3H_4
(D) C_4H_8
(E) C_5H_{10}

Questions 69–70 refer to the following diagrams of two methods of collecting gases:

69. Method 1 is BEST suited to collect
    (A) a gas heavier than air
    (B) a gas lighter than air
    (C) a gas that is insoluble in water
    (D) a gas that is soluble in water
    (E) a gas that has a distinct color

70. Which of these gases because of its density and solubility should be collected by Method 2?
    (A) NH_3
    (B) H_2
    (C) HCl
    (D) CO_2
    (E) He

71. What is the gram-formula mass of $CaCO_3$?
    (A) 68 g
    (B) 75 g
    (C) 82 g
    (D) 100 g
    (E) 116 g

72. What volume, in liters, will be occupied at STP by 4 grams of H_2?
    (A) 11.2
    (B) 22.4
    (C) 33.6
    (D) 44.8
    (E) 56.0

73. What mass, in grams, of KOH is needed to neutralize 20 grams of sulfuric acid?
    (A) 22.8
    (B) 45.6
    (C) 68.4
    (D) 110
    (E) 205

74. What volume, in liters, of ammonia is produced when 22.4 liters of nitrogen are made to combine with a sufficient quantity of hydrogen under appropriate conditions?
    (A) 11.2
    (B) 22.4
    (C) 44.8
    (D) 67.0
    (E) 78.2

75. What volume, in liters, of SO_2 will result from the complete burning of 64 grams of sulfur?
    (A) 2
    (B) 11.2
    (C) 44.8
    (D) 126
    (E) 158

76. The amount of energy to melt 5 grams of ice at 0°C would also heat 1 gram of water at 4°C to what condition?
    (A) water at 90°C
    (B) water at 100°C
    (C) steam at 100°C
    (D) 0.2 g water and 0.8 g steam at 100°C
    (E) 0.44 g water and 0.56 g steam at 100°C

77. How many moles of electrons are needed to electroplate a deposit of 0.5 mole of silver from a silver nitrate solution?
    (A) 0.5
    (B) 1
    (C) 27
    (D) 54
    (E) 108

78. All of the following statements about carbon dioxide are true EXCEPT:
    (A) It can be prepared by the action of acid on $CaCO_3$.
    (B) It is used in fire extinguishers.
    (C) It dissolves in water at room temperature.
    (D) It sublimes rather than melts at 20°C and 1 atm pressure.
    (E) It is a product of photosynthesis in plants.

79. Three moles of H_2 and 3 moles and I_2 are introduced into a liter box at a temperature of 490°C. What will the K_e expression be for this reaction? ($K_e = 45.9$)

    (A) $K_e = \dfrac{[H_2][I_2]}{[HI]}$

    (B) $K_e = \dfrac{[HI]}{[H_2][I_2]}$

    (C) $K_e = \dfrac{2x}{(x)(x)}$

    (D) $K_e = \dfrac{(2x)^2}{(3-x)^2}$

    (E) $K_e = \dfrac{(3-x)^2}{(2x)^2}$

80. What is the concentration, in moles/liter, of H_2 at equilibrium in question 79?

    (A) 2.3

    (B) 0.68

    (C) 1.6

    (D) 2.1

    (E) 4.64

81. Which of the following correctly completes this nuclear reaction: $^{14}_{7}N + ^{4}_{2}He \rightarrow \cdots + ^{1}_{1}H$?

    (A) $^{17}_{8}O$

    (B) $^{16}_{9}O$

    (C) $^{17}_{8}N$

    (D) $^{17}_{7}N$

    (E) $^{16}_{8}O$

82. How many grams of NaCl will be needed to make 100 cm^3 of 2 M solution?

    (A) 5.85

    (B) 11.7

    (C) 29.2

    (D) 58.5

    (E) 117

83. How many grams of H_2SO_4 are in 1,000 grams of a 10% solution? (1 mole of H_2SO_4 = 98 g)

    (A) 1.0

    (B) 9.8

    (C) 10

    (D) 98

    (E) 100

84. If 1 mole of ethyl alcohol in 1,000 grams of water depresses the freezing point by 1.86° Celsius, what will be the freezing point of a solution of 1 mole of ethyl alcohol in 500 grams of water?

    (A) −0.93°C

    (B) −1.86°C

    (C) −2.79°C

    (D) −3.72°C

    (E) −5.58°C

85. Which substance may be used as fuel in a nuclear reactor?

    (A) ^{226}Ra

    (B) ^{235}U

    (C) ^{220}Fr

    (D) ^{204}Pb

    (E) ^{14}C

STOP

IF YOU FINISH BEFORE ONE HOUR IS UP, YOU MAY GO BACK TO CHECK YOUR WORK OR COMPLETE UNANSWERED QUESTIONS.

Answers and
Explanations to Test

4

1. (A) The volume of a gas increases as temperature increases provided that pressure remains constant. This is a direct proportion. Heating a balloon is a good example of this.

2. (B) The volume of a gas decreases as the pressure is increased provided that the temperature is held constant. This is shown by the inversely proportional curve in (B). Pressure increase on a closed cylinder is a good example.

3. (E) The chart shows that there is a starting quantity in solution, and as you move to the right a slight positive slope indicates a directly proportional change.

4. (A) This is the definition of any molecule.

5. (B) A commercial cake mix is a mixture of ingredients.

6. (C) This is the definition of an isotope.

7. (E) An acid salt contains one or more H atoms in the salt formula separating a positive ion and the hydrogen-bearing negative ion. For example, Na_2SO_4 is a *normal* salt and $NaHSO_4$ is an *acid* salt because of the presence of H in the hydrogen sulfate ion.

8. (C) An atom with atomic number 9 would have a 2,7 electron configuration, which matches the outer energy level of iodine.

9. (B) Three $(OH)^-$ each have 2 atoms = 6 atoms plus one Al = 7.

10. (A) pH from 1 to 6 is acid, 7 neutral, 8 to 14 basic. Most acid is 1.

11. (E) Molarity is defined as moles of solute/liter of solution.

12. (A) Gas densities can be expressed in grams/liter.

13. (C) Molecular mass is not affected by pressure and temperature.

14. (A) If the density of a gas is known, the mass of 1 L can be multiplied by 22.4 to find the molecular mass because 1 mole occupies 22.4 L at STP.

15. (A) Buffers resist changes in pH.

16. (B) Color change is the function of indicators.

17. (C) On the pH scale, from 1 to 6 is acid and 7 is neutral.

18. (D) On the pH scale, from 8 to 14 is basic.

19. (A) $Zn + 2HCl \rightarrow ZnCl_2 + H_2(g)$ is the reaction that occurs.

20. (C) Only CO_2, with a molecular mass of 44, is heavier than air, for which the molecular mass is 29.

21. (E) $2KClO_3 \rightarrow 2KCl + 3O_2(g)$ is the reaction that occurs.

22. (C) CO_2 is slightly soluble in water, forming carbonic acid, H_2CO_3, which is a weak acid.

23. (B) NH_3 is very soluble in water, and forms a solution of the weak ammonium hydroxide base.

101. (T, F) The assertion is true, but the particles undergo elastic collisions.

102. (T, T, CE) Assertion and reason are true; the electron can be treated as either an electromagnetic wave or a bundle of negative charge.

103. (T, T) A homologous series increases each member by a constant number of carbons and hydrogens. Examples are the alkane, alkene, and alkyne series, which each increase the chain by a CH_2 group. The reason is true but does not explain the assertion.

104. (T, F) The nuclear charge is greater than the electron cloud. The reason is false.

105. (T, T, CE) A negative heat of formation indicates that the reaction is exothermic and the enthalpy is negative.

106. (T, F) Water is a polar molecule because there is unequal, not equal, sharing of bonding electrons.

107. (T, T, CE) This is a function of a catalyst—to either speed up or slow down a reaction without permanent change to itself. Assertion and reason are true.

108. (F, T) The Cu is losing electrons and thus being oxidized; the assertion is false. It is furnishing electrons and thus is a reducing agent; the reason is true.

109. (F, T) $H_2 = 2$, $He = 4$ (molecular mass); then inversely $\sqrt{4}:\sqrt{2} = 2:\sqrt{2}$ is the rate of diffusion of hydrogen to helium. The assertion is false; the reason, true.

110. (T, T, CE) Since the volume is being heated at constant pressure, it expands. The temperatures are converted to kelvins (K) by adding 273° to the Celsius readings. The fraction must be $\frac{373}{273}$ and this will increase the volume.

111. (F, T) Gibbs free energy is useful in indicating the conditions under which a chemical reaction will occur. It is not related to solubility. It is true that, generally speaking, solubility of a solute increases with an increase in the temperature of the solvent.

112. (F, F) Water is $\frac{1}{9}$ hydrogen and $\frac{8}{9}$ oxygen by weight. Both assertion and reason are false.

113. (T, T, CE) Four grams of ice would require 4×80 cal/g or 320 cal to melt the ice.

114. (F, T) The reaction is: $2H_2 + O_2 \rightarrow 2H_2O$. The coefficients of this gaseous reaction show that 2 liters of hydrogen react with 1 liter of oxygen. This would leave 1 liter of unreacted oxygen. The limiting factor is the hydrogen.

115. (T, F) The reason why water is a good solvent is false.

116. (T, F) The density of a gas at STP is found by dividing the molecular mass by 22.4 liters. NH_3 has a gram molecular mass of $N = 14 + 3H = 3$ or a total of 17 g. The gram molecular mass of Ar is 40 g. The density of each can be found by dividing by 22.4 liters, but obviously, the density of the ammonia will be smaller.

40. (C) The energy necessary to get the reaction started, which is the activation energy, is shown at C.

41. (C) $2KClO_3 \rightarrow 2KCl + 3O_2$ (g) shows that 2 moles of $KClO_3$ yield 3 moles of O_2. 1 mole will produce $\frac{1}{2} \times 3 = 1.5$ moles

$$1.5 \text{ moles} \times \frac{22.4 \text{ L}}{1 \text{ mole}} = 33.6 \text{ L}$$

42. (E) The atomic number gives the number of protons in the nucleus and the total number of electrons. The atomic mass indicates the total number of protons and neutrons in the nucleus—for Na, 23.

43. (B) A catalyst can speed up a reaction by lowering the activation energy needed to start the reaction and then keep it going.

44. (B) Density = $\dfrac{\text{Mass}}{\text{Volume}}$. For gases this is expressed as grams per liter. Since 1 gram-molecular mass of a gas occupies 22.4 L, 17 g/22.4 L = 0.76 g/L.

45. (E) Choice III is made up of elements from extreme sides of the periodic table and will therefore form ionic bonds.

46. (E) Only III is a ring hydrocarbon of the aromatic series.

47. (B) Since Li is higher in Group 1 than Na, and K is higher than Cs, they have smaller radii and hence higher ionization energies. Al is to the right of Na and therefore has a higher ionization energy.

48. (C) Only II and III are correctly balanced. To be correct, I should only have $3e^-$.

49. (E) ⎫ These answers are based on the
50. (C) ⎬ fact that the total of the assigned oxidation numbers times their occurrence for all the atoms in a
51. (D) ⎭ compound is zero.

52. (B) This orbital notation shows 6 electrons in the third energy level. The atom would like to gain $2e^-$ to fill the $3p$ and thereby gain a –2 charge.

53. (C) With this structure, the atom would tend to lose these electrons and get a +2 charge.

54. (C) Assembled, the apparatus would look like this:

and could be used to prepare a gas by reacting a solid with a liquid.

55. (D) The setup depends on the property of insolubility of the gas collected over water.

56. (B) This apparatus would need an acid to cause the reaction.

57. (D) A test tube would replace the thistle tube and flask portion:

58. (B) The equilibrium shifts in the direction that tends to relieve the stress and thus regain equilibrium.

59. (D) The thermal reaction shows 2 moles of hydrogen reacting or 4 grams. Therefore 8 grams would release twice the amount of energy; 2×136.64 kcal = 273.28 kcal.

60. (B) The reaction potential calculation would be:

$$Zn^{2+} + 2e^- \rightarrow Zn^0 \qquad E^0 = -0.76V$$
$$\underline{Au^0 \rightarrow Au^{3+} + 3e^- \qquad E^0 = -1.42V}$$
$$Zn^{2+} + Au^0 \rightarrow Zn^0 + Au^{3+} \quad E^0 = -2.18V$$

61. (D) Since $Ag^+ + 1e^- \rightarrow Ag^0$, 1 mole of electrons yields 1 mole of silver; 1 mole silver = 6.02×10^{23} atoms
4×108 g/mole = 432 g

62. (E) $5 M = \dfrac{5 \text{ moles}}{\text{liter}}$, and since HCl ionizes completely there would be 5 moles of H^+ and 5 moles of Cl^- ions.

63. (D) $K_{sp} = [Ag^+][C_2H_3O_2^-]$
$\qquad = [2 \times 10^{-3}][2 \times 10^{-3}]$
(Since
$AgC_2H_3O_2 \rightleftharpoons Ag^+ + C_2H_3O_2^-$

the silver ion and acetate ion concentrations are equal.)
$K_{sp} = 4 \times 10^{-6}$

64. (B) It is the explanation for the observed high boiling point and high freezing point of water compared with hydrogen sulfide.

65. (D) pH = $-\log [H^+]$
$\qquad = -\log [10^{-5}] = -[-5] = 5$

66. (B) 1 mole H_2SO_4 contains 1 molar mass of sulfur, that is, 32 g.

67. (B) N_2O = 44g/mole
$\qquad\qquad (2 \times 14 + 16 = 44)$
1 mole of a gas occupies 22.4 liters. So 44 g/22.4 liters = 1.99 g/liter.

68. (D) $CH_2 = 14$
$(12 + 2 = 14$ molecular mass)
$56 \div 14 = 4$
So $4 \times CH_2 = C_4H_8$

69. (C) Only insoluble gases can be collected this way.

70. (C) HCl is very soluble in water and heavier than air, so it is suited to the No. 2 collection method.

71. (D) Ca = 40
$\quad$ C = 12
$\quad \underline{30 = 48}$
$\qquad\quad 100$ g

72. (D) The gram-molecular mass of H_2 is 2 g. 4 g is 2 moles, and each mole occupies 22.4 L. $2 \times 22.4 = 44.8$ L.

73. (A)
$$\overset{x\,g}{} \overset{20\,g}{}$$
$$2KOH + H_2SO_4 \rightarrow K_2SO_4 + 2H_2O$$
$$\underset{112\,g}{} \underset{98\,g}{}$$
So $\dfrac{x\,g}{112\,g} = \dfrac{20\,g}{98\,g}$.
Then $x = 22.8$ g.

74. (C)
$$\overset{22.4\,L}{} \qquad \overset{x\,L}{}$$
$$N_2 + 3H_2 \rightarrow 2NH_3$$
$$\underset{1\,L}{} \qquad \underset{2\,L}{}$$
So $\dfrac{22.4\,L}{1\,L} = \dfrac{x\,L}{2\,L}$.
Then $x = 44.8$ L.

75. (C)
$$\overset{64\,g}{} \qquad \overset{x\,L}{}$$
$$S + O_2 \rightarrow SO_2$$
$$\underset{32\,g}{} \qquad \underset{22.4\,L}{}$$
So $\dfrac{64\,g}{32\,g} = \dfrac{x\,L}{22.4\,L}$.
Then $x = 44.8$ L.

76. (E) 5 g ice to water = 5×80 cal = 400 cal. 1 g at 4° can go to 100°C as water and absorb 1 cal per °C. So 400 cal − (100° − 4°) = 400 − 96 = 304 cal. 304

cal can change $\frac{304 \text{ cal}}{540 \text{ cal/g}}$ or 0.56 g of water to steam. So 0.56 g steam and 0.44 g water would remain.

77. (A) Since Ag^+ gains $1e^-$ to become $Ag°$, 0.5 mole would require 0.5 mole of electrons.

78. (E) CO_2 is a reactant in photosynthesis, not a product. The reaction is
$$6CO_2 + 6H_2O \rightarrow C_6H_{12}O_6 + 6O_2(g)$$
$$\text{simple sugar}$$
$$\text{or}$$
$$6CO_2 + 5H_2O \rightarrow C_6H_{12}O_5 + 6O_2(g)$$
$$\text{cellulose}$$

79. (D) $K_e = \dfrac{[HI]^2}{[H_2][I_2]}$

Let x = moles of H_2 and also I_2 that combine to form HI.
Then at equilibrium $[H_2] = 3 - x$, $[I_2] = 3 - x$, $[HI] = 2x$.

So $K_e = \dfrac{(2x)^2}{(3-x)(3-x)}$.

80. (B) $45.9 = \dfrac{(2x)^2}{(3-x)^2}$

Taking the square root of both sides gives $6.77 = \dfrac{2x}{3-x}$.
Then $x = 2.3$ or $[H_2] = 3 - x$ or .68.

81. (A) This is Rutherford's famous artificial transmutation experiment done in 1919.

82. (B) $2\,M = \dfrac{2 \text{ moles}}{1,000 \text{ mL}}$
2 moles of NaCl $= 2 \times 58.5$ g
$= 117.0$ g
$2\,M = 117$ g/1,000 mL,
so $\dfrac{117 \text{ g}}{1,000 \text{ mL}} = \dfrac{x \text{ g}}{100 \text{ mL}}$,
$x = 11.7$ g

83. (E) Percent is by mass. So 10% is 0.1 × 1,000 g or 100 g.

84. (D) First find the molality. 1 mole in 500 g = 2 moles in 1,000 g. Then 2 × 1.86°C = 3.72°C drop from 0°C or –3.72°C.

85. (B) Uranium is the chief fuel of nuclear reactors. The most abundant isotope of uranium found in the earth is $^{238}_{92}U$. However ^{235}U is needed to run a reactor.

DIAGNOSING YOUR NEEDS

After taking Practice Test 4, check your answers against the correct ones. Then fill in the chart below.

In the space under each question number, place a check if you answered that question correctly.

- **EXAMPLE:**

If your answer to question 5 was correct, place a check in the appropriate box.

Next, total the check marks for each section and insert the number in the designated block. Now do the arithmetic indicated, and insert your percent for each area.

SUBJECT AREA   **(✔) QUESTIONS ANSWERED CORRECTLY**

I. Atomic Theory and Structure, including periodic relationships	6	8	102	107	42	47	52	53

☐ No. of checks ÷ 8 × 100 = _____ %

II. Nuclear Reactions	81	85

☐ No. of checks ÷ 2 × 100 = _____ %

III. Chemical Bonding and Molecular Structure	4	9	106	45	49	50	51	62	64

☐ No. of checks ÷ 9 × 100 = _____ %

IV. States of Matter and Kinetic Molecular Theory of Gases	1	2	14	101	109	110	44

☐ No. of checks ÷ 7 × 100 = _____ %

V. Solutions, including concentration units, solubility, and colligative properties	3	11	82	83	84

☐ No. of checks ÷ 5 × 100 = _____ %

VI. Acids and Bases	7	10	15	16	17	18

☐ No. of checks ÷ 6 × 100 = _____ %

VII. Oxidation-Reduction and Electrochemistry	108	112	48	60	61	77

☐ No. of checks ÷ 6 × 100 = _____ %

SUBJECT AREA | (✔)QUESTIONS ANSWERED CORRECTLY

VIII. Stoichiometry	113	114	116	41	66	67	68	71	72	73	74	75

☐ No. of checks ÷ 12 × 100 = _____%

IX. Reaction Rates	107	43

☐ No. of checks ÷ 2 × 100 = _____%

X. Equilibrium	58	63	65	79	80

☐ No. of checks ÷ 5 × 100 = _____%

XI. Thermodynamics: energy changes in chemical reactions, randomness, and criteria for spontaneity	105	111	40	59	76

☐ No. of checks ÷ 5 × 100 = _____%

XII. Descriptive Chemistry: physical and chemical properties of elements and their familiar compounds; organic chemistry; periodic properties	5	12	13	19	20	21
	22	23	103	115	46	78

☐ No. of checks ÷ 12 × 100 = _____%

XIII. Laboratory: equipment, procedures, observations, safety, calculations, and interpretation of results	54	55	56	57	69	70

☐ No. of checks ÷ 6 × 100 = _____%

PLANNING YOUR STUDY

The percentages give you an idea of how you have done on the various major areas of the test. Because of the limited number of questions on some parts, these percentages may not be as reliable as the percentages for parts with larger numbers of questions. However, you should now have at least a rough idea of the areas in which you have done well and those in which you need more study.

Start your study with the areas in which you are the weakest. The corresponding chapters are indicated below:

Subject Area	Chapters to Review
I. Atomic Theory and Structure, including periodic relationships	2
II. Nuclear Reactions	15
III. Chemical Bonding and Molecular Structure	3, 4
IV. States of Matter and Kinetic Molecular Theory of Gases	1
V. Solutions, including concentration units, solubility, and colligative properties	7
VI. Acids and Bases	11
VII. Oxidation-reduction and Electrochemistry	12
VIII. Stoichiometry	5, 6
IX. Reaction Rates	9
X. Equilibrium	10
XI. Thermodynamics, including energy changes in chemical reactions, randomness, and criteria for spontaneity	8
XII. Descriptive Chemistry: physical and chemical properties of elements and their familiar compounds; organic chemistry; periodic properties	1, 2, 13, 14
XIII. Laboratory: equipment, procedures, observations, safety, calculations, and interpretation of results.	All lab diagrams, 16

MODERN PERIODIC

TABLE

s-block

18

2	0
He	
Helium	
4.00260	
$1s^2$	

— p-block —

GROUP

13	14	15	16	17	18
5 +3 **B** Boron 10.81 [He]$2s^2 2p^1$	6 −4 +2 +4 **C** Carbon 12.011 [He]$2s^2 2p^2$	7 −3 −2 −1 +1 +2 +3 +4 +5 **N** Nitrogen 14.0067 [He]$2s^2 2p^3$	8 −2 **O** Oxygen 15.9994 [He]$2s^2 2p^4$	9 −1 **F** Fluorine 18.998403 [He]$2s^2 2p^5$	10 0 **Ne** Neon 20.1797 [He]$2s^2 2p^6$
13 +3 **Al** Aluminum 26.98154 [Ne]$3s^2 3p^1$	14 −4 +2 +4 **Si** Silicon 28.0855 [Ne]$3s^2 3p^2$	15 −3 +3 +5 **P** Phosphorus 30.97376 [Ne]$3s^2 3p^3$	16 −2 +4 +6 **S** Sulfur 32.066 [Ne]$3s^2 3p^4$	17 −1 +3 +5 +7 **Cl** Chlorine 35.453 [Ne]$3s^2 3p^5$	18 0 **Ar** Argon 39.948 [Ne]$3s^2 3p^6$

10	11	12						
28 +2 +3 **Ni** Nickel 58.69 [Ar]$3d^8 4s^2$	29 +1 +2 **Cu** Copper 63.546 [Ar]$3d^{10} 4s^1$	30 +2 **Zn** Zinc 65.39 [Ar]$3d^{10} 4s^2$	31 +3 **Ga** Gallium 69.72 [Ar]$3d^{10} 4s^2 4p^1$	32 +2 +4 **Ge** Germanium 72.61 [Ar]$3d^{10} 4s^2 4p^2$	33 +3 +5 **As** Arsenic 74.9216 [Ar]$3d^{10} 4s^2 4p^3$	34 −2 +4 +6 **Se** Selenium 78.96 [Ar]$3d^{10} 4s^2 4p^4$	35 −1 +5 **Br** Bromine 79.904 [Ar]$3d^{10} 4s^2 4p^5$	36 0 +2 **Kr** Krypton 83.80 [Ar]$3d^{10} 4s^2 4p^6$
46 +2 +4 **Pd** Palladium 106.42 [Kr]$4d^{10}$	47 +1 **Ag** Silver 107.8682 [Kr]$4d^{10} 5s^1$	48 +2 **Cd** Cadmium 112.41 [Kr]$4d^{10} 5s^2$	49 +3 **In** Indium 114.82 [Kr]$4d^{10} 5s^2 5p^1$	50 +4 **Sn** Tin 118.710 [Kr]$4d^{10} 5s^2 5p^2$	51 −3 +3 +5 **Sb** Antimony 121.757 [Kr]$4d^{10} 5s^2 5p^3$	52 −2 +4 +6 **Te** Tellurium 127.60 [Kr]$4d^{10} 5s^2 5p^4$	53 −1 +1 +5 +7 **I** Iodine 126.9045 [Kr]$4d^{10} 5s^2 5p^5$	54 0 +2 +4 +6 **Xe** Xenon 131.29 [Kr]$4d^{10} 5s^2 5p^6$
78 +2 +4 **Pt** Platinum 195.08 [Xe]$4f^{14} 5d^9 6s^1$	79 +1 +3 **Au** Gold 196.9665 [Xe]$4f^{14} 5d^{10} 6s^1$	80 +1 +2 **Hg** Mercury 200.59 [Xe]$4f^{14} 5d^{10} 6s^2$	81 +1 +3 **Tl** Thallium 204.383 [Xe]$4f^{14} 5d^{10} 6s^2 6p^1$	82 +2 +4 **Pb** Lead 207.2 [Xe]$4f^{14} 5d^{10} 6s^2 6p^2$	83 +3 +5 **Bi** Bismuth 208.9804 [Xe]$4f^{14} 5d^{10} 6s^2 6p^3$	84 +2 +4 **Po** Polonium (209) [Xe]$4f^{14} 5d^{10} 6s^2 6p^4$	85 **At** Astatine (210) [Xe]$4f^{14} 5d^{10} 6s^2 6p^5$	86 0 **Rn** Radon (222) [Xe]$4f^{14} 5d^{10} 6s^2 6p^6$
110 **Uun** Ununnilium (269) [Rn]$5f^{14} 6d^8 7s^2$	111 **Uuu** Unununium (272) [Rn]$5f^{14} 6d^9 7s^2$	112 **Uub** Ununbium (277) [Rn]$5f^{14} 6d^{10} 7s^2$						

— f-block —

63 +2 +3 **Eu** Europium 151.96 [Xe]$4f^7 6s^2$	64 +3 **Gd** Gadolinium 157.25 [Xe]$4f^7 5d^1 6s^2$	65 +3 **Tb** Terbium 158.9254 [Xe]$4f^9 6s^2$	66 +3 **Dy** Dysprosium 162.50 [Xe]$4f^{10} 6s^2$	67 +3 **Ho** Holmium 164.9304 [Xe]$4f^{11} 6s^2$	68 +3 **Er** Erbium 167.26 [Xe]$4f^{12} 6s^2$	69 +3 **Tm** Thulium 168.9342 [Xe]$4f^{13} 6s^2$	70 +2 +3 **Yb** Ytterbium 173.04 [Xe]$4f^{14} 6s^2$	71 +3 **Lu** Lutetium 174.967 [Xe]$4f^{14} 5d^1 6s^2$
95 +3 +4 +5 +6 **Am** Americium (243) [Rn]$5f^7 7s^2$	96 +3 **Cm** Curium (247) [Rn]$5f^7 6d^1 7s^2$	97 +3 +4 **Bk** Berkelium (247) [Rn]$5f^9 7s^2$	98 +3 **Cf** Californium (251) [Rn]$5f^{10} 7s^2$	99 **Es** Einsteinium (252) [Rn]$5f^{11} 7s^2$	100 **Fm** Fermium (257) [Rn]$5f^{12} 7s^2$	101 **Md** Mendelevium (258) [Rn]$5f^{13} 7s^2$	102 **No** Nobelium (259) [Rn]$5f^{14} 7s^2$	103 **Lr** Lawrencium (262) [Rn]$5f^{14} 6d^1 7s^2$

Tables for Reference

(A) DENSITY AND BOILING POINTS OF SOME COMMON GASES

Name		Density grams/liter at STP*	Boiling Point (at 1 atm) K
Air	—	1.29	—
Ammonia	NH_3	0.771	240
Carbon dioxide	CO_2	1.98	195
Carbon monoxide	CO	1.25	82
Chlorine	Cl_2	3.21	238
Hydrogen	H_2	0.0899	20
Hydrogen chloride	HCl	1.64	188
Hydrogen sulfide	H_2S	1.54	212
Methane	CH_4	0.716	109
Nitrogen	N_2	1.25	77
Nitrogen (II) oxide	NO	1.34	121
Oxygen	O_2	1.43	90
Sulfur dioxide	SO_2	2.92	263
*STP is defined as 273 K and 1 atm			

(B) SOLUBILITY CURVES

(C) TABLE OF SOLUBILITIES IN WATER

i—nearly insoluble
ss—slightly soluble
s—soluble
d—decomposes
n—not isolated

	acetate	bromide	carbonate	chloride	chromate	hydroxide	iodide	nitrate	phosphate	sulfate	sulfide
Aluminum	ss	s	n	s	n	i	s	s	i	s	d
Ammonium	s	s	s	s	s	s	s	s	s	s	s
Barium	s	s	i	s	i	s	s	s	i	i	d
Calcium	s	s	i	s	s	ss	s	s	i	ss	d
Copper II	s	s	i	s	i	i	n	s	i	s	i
Iron II	s	s	i	s	n	i	s	s	i	s	i
Iron III	s	s	n	s	i	i	n	s	i	ss	d
Lead	s	ss	i	ss	i	i	ss	s	i	i	i
Magnesium	s	s	i	s	s	i	s	s	i	s	d
Mercury I	ss	i	i	i	ss	n	i	s	i	ss	i
Mercury II	s	ss	i	s	ss	i	i	s	i	d	i
Potassium	s	s	s	s	s	s	s	s	s	s	s
Silver	ss	i	i	i	ss	n	i	s	i	ss	i
Sodium	s	s	s	s	s	s	s	s	s	s	s
Zinc	s	s	i	s	s	i	s	s	i	s	i

(D) SELECTED POLYATOMIC IONS

Hg_2^{2+}	dimercury (I)	CrO_4^{2-}	chromate
NH_4^+	ammonium	$Cr_2O_7^{2-}$	dichromate
$C_2H_3O_2^-$	} acetate	MnO_4^-	permanganate
CH_3COO^-		MnO_4^{2-}	manganate
CN^-	cyanide	NO_2^-	nitrite
CO_3^{2-}	carbonate	NO_3^-	nitrate
HCO_3^-	hydrogen carbonate	OH^-	hydroxide
		PO_4^{3-}	phosphate
$C_2O_4^{2-}$	oxalate	SCN^-	thiocyanate
ClO^-	hypochlorite	SO_3^{2-}	sulfite
ClO_2^-	chlorite	SO_4^{2-}	sulfate
ClO_3^-	chlorate	HSO_4^-	hydrogen sulfate
ClO_4^-	perchlorate	$S_2O_3^{2-}$	thiosulfate

E

STANDARD ENERGIES OF FORMATION
OF COMPOUNDS AT 1 atm AND 298 K

Compound	Heat (Enthalpy) of Formation* (ΔH_f°)		Free Energy of Formation (ΔG_f°)	
	kJ/mol	kcal/mol	kJ/mol	kcal/mol
Aluminum oxide $Al_2O_3(s)$	−1676.5	−400.5	−1583.1	−378.2
Ammonia $NH_3(g)$	−46.0	−11.0	−16.3	−3.9
Barium sulfate $BaSO_4(s)$	−1473.9	−352.1	−1363.0	−325.6
Calcium hydroxide $Ca(OH)_2(s)$	−986.6	−235.7	−899.2	−214.8
Carbon dioxide $CO_2(g)$	−393.9	−94.1	−394.7	−94.3
Carbon monoxide $CO(g)$	−110.5	−26.4	−137.3	−32.8
Copper(II) sulfate $CuSO_4(s)$	−771.9	−184.4	−662.2	−158.2
Ethane $C_2H_6(g)$	−84.6	−20.2	−33.1	−7.9
Ethene (ethylene) $C_2H_4(g)$	52.3	12.5	68.2	16.3
Ethyne (acetylene) $C_2H_2(g)$	226.9	54.2	209.3	50.0
Hydrogen fluoride $HF(g)$	−271.3	−64.8	−273.3	−65.3
Hydrogen iodide $HI(g)$	26.4	6.3	1.7	0.4
Iodine chloride $ICl(g)$	18.0	4.3	−5.4	−1.3
Lead(II) oxide $PbO(s)$	−215.6	−51.5	−188.4	−45.0
Methane $CH_4(g)$	−74.9	−17.9	−50.7	−12.1
Magnesium oxide $MgO(s)$	−601.9	−143.8	−569.7	−136.1
Nitrogen(II) oxide $NO(g)$	90.4	21.6	86.7	20.7
Nitrogen(IV) oxide $NO_2(g)$	33.1	7.9	51.5	12.3
Potassium chloride $KCl(s)$	−437.0	−104.4	−409.4	−97.8
Sodium chloride $NaCl(s)$	−411.5	−98.3	−384.3	−91.8
Sulfur dioxide $SO_2(g)$	−296.8	−70.9	−300.1	−71.7
Water $H_2O(g)$	−242.0	−57.8	−228.6	−54.6
Water $H_2O(\ell)$	−285.9	−68.3	−237.3	−56.7

*Minus sign indicates an exothermic reaction.
Sample equations:

$$2Al(s) + \frac{3}{2}O_2(g) \rightarrow Al_2O_3(s) + 400.5 \text{ kcal}$$

$$2Al(s) + \frac{3}{2}O_2(g) \rightarrow Al_2O_3(s) \quad \Delta H = -400.5 \text{ kcal/mol}$$

F

SELECTED RADIOISOTOPES

Nuclide	Half-Life	Decay Mode
^{198}Au	2.69 d	β^-
^{14}C	5730 y	β^-
^{60}Co	5.26 y	β^-
^{137}Cs	30.23 y	β^-
^{220}Fr	27.5 s	α
^{3}H	12.26 y	β^-
^{131}I	8.07 d	β^-
^{37}K	1.23 s	β^+
^{42}K	12.4 h	β^-
^{85}Kr	10.76 y	β^-
^{85m}Kr*	4.39 h	γ
^{16}N	7.2 s	β^-
^{32}P	14.3 d	β^-
^{239}Pu	2.44×10^4 y	α
^{226}Ra	1600 y	α
^{222}Rn	3.82 d	α
^{90}Sr	28.1 y	β^-
^{99}Tc	2.13×10^5 y	β^-
^{99m}Tc*	6.01 h	γ
^{232}Th	1.4×10^{10} y	α
^{233}U	1.62×10^5 y	α
^{235}U	7.1×10^8 y	α
^{238}U	4.51×10^9 y	α

y = years; d = days; h = hours; s = seconds
*m = meta stable or excited state of the same nucleus. Gamma decay from such a state is called an isomeric transition (IT).
Nuclear isomers are different energy states of the same nucleus, each having a different measurable lifetime.

G

HEATS OF REACTION AT 1 atm AND 298 K

Reaction	ΔH	
	kJ	kcal
$CH_4(g) + 2O_2(g) \rightarrow CO_2(g) + 2H_2O(\ell)$	−890.8	−212.8
$C_3H_8(g) + 5O_2(g) \rightarrow 3CO_2(g) + 4H_2O(\ell)$	−2221.1	−530.6
$CH_3OH(\ell) + \frac{3}{2}O_2(g) \rightarrow CO_2(g) + 2H_2O(\ell)$	−726.7	−173.6
$C_6H_{12}O_6(s) + 6O_2(g) \rightarrow 6CO_2(g) + 6H_2O(\ell)$	−2804.2	−669.9
$CO(g) + \frac{1}{2}O_2(g) \rightarrow CO_2(g)$	−283.4	−67.7
$C_8H_{18}(\ell) + \frac{25}{2}O_2(g) \rightarrow 8CO_2(g) + 9H_2O(\ell)$	−5453.1	−1302.7
$KNO_3(s) \xrightarrow{H_2O} K^+(aq) + NO_3^-(aq)$	34.7	8.3
$NaOH(s) \xrightarrow{H_2O} Na^+(aq) + OH^-(aq)$	−44.4	−10.6
$NH_4Cl(s) \xrightarrow{H_2O} NH_4^+(aq) + Cl^-(aq)$	14.7	3.5
$NH_4NO_3(s) \xrightarrow{H_2O} NH_4^+(aq) + NO_3^-(aq)$	25.5	6.1
$NaCl(s) \xrightarrow{H_2O} Na^+(aq) + Cl^-(aq)$	3.8	0.9
$KClO_3(s) \xrightarrow{H_2O} K^+(aq) + ClO_3^-(aq)$	41.4	9.9
$LiBr(s) \xrightarrow{H_2O} Li^+(aq) + Br^-(aq)$	−49.0	−11.7
$H^+(aq) + OH^-(aq) \rightarrow H_2O(\ell)$	−57.8	−13.8

H

SYMBOLS USED IN
NUCLEAR CHEMISTRY

alpha particle	$^4_2$He	α
beta particle (electron)	$^0_{-1}$e	β^-
gamma radiation		γ
neutron	$^1_0$n	n
proton	$^1_1$H	p
deuteron	$^2_1$H	
triton	$^3_1$H	
positron	$^0_{+1}$e	β^+

Ⓘ

IONIZATION ENERGIES AND ELECTRONEGATIVITIES

313 — First Ionization Energy (kcal/mol of atoms)
2.2 — Electronegativity*

1	2		13	14	15	16	17	18
H 313 2.2								He 567
Li 125 1.0	Be 215 1.5		B 191 2.0	C 260 2.6	N 336 3.1	O 314 3.5	F 402 4.0	Ne 497
Na 119 0.9	Mg 176 1.2		Al 138 1.5	Si 188 1.9	P 242 2.2	S 239 2.6	Cl 300 3.2	Ar 363
K 100 0.8	Ca 141 1.0		Ga 138 1.6	Ge 182 1.9	As 226 2.0	Se 225 2.5	Br 273 2.9	Kr 323
Rb 96 0.8	Sr 131 1.0		In 133 1.7	Sn 169 1.8	Sb 199 2.1	Te 208 2.3	I 241 2.7	Xe 280
Cs 90 0.7	Ba 120 0.9		Tl 141 1.8	Pb 171 1.8	Bi 168 1.9	Po 194 2.0	At 2.2	Rn 248
Fr 0.7	Ra 122 0.9							

*Arbitrary scale based on fluorine = 4.0

Ⓙ

RELATIVE STRENGTHS OF ACIDS IN AQUEOUS SOLUTION AT 1 atm AND 298 K

Conjugate Pairs ACID BASE	K_a
$HI = H^+ + I^-$	very large
$HBr = H^+ + Br^-$	very large
$HCl = H^+ + Cl^-$	very large
$HNO_3 = H^+ + NO_3^-$	very large
$H_2SO_4 = H^+ + HSO_4^-$	large
$H_2O + SO_2 = H^+ + HSO_3^-$	1.5×10^{-2}
$HSO_4^- = H^+ + SO_4^{2-}$	1.2×10^{-2}
$H_3PO_4 = H^+ + H_2PO_4^-$	7.5×10^{-3}
$Fe(H_2O)_6^{3+} = H^+ + Fe(H_2O)_5(OH)^{2+}$	8.9×10^{-4}
$HNO_2 = H^+ + NO_2^-$	4.6×10^{-4}
$HF = H^+ + F^-$	3.5×10^{-4}
$Cr(H_2O)_6^{3+} = H^+ + Cr(H_2O)_5(OH)^{2+}$	1.0×10^{-4}
$CH_3COOH = H^+ + CH_3COO^-$	1.8×10^{-5}
$Al(H_2O)_6^{3+} = H^+ + Al(H_2O)_5(OH)^{2+}$	1.1×10^{-5}
$H_2O + CO_2 = H^+ + HCO_3^-$	4.3×10^{-7}
$HSO_3^- = H^+ + SO_3^{2-}$	1.1×10^{-7}
$H_2S = H^+ + HS^-$	9.5×10^{-8}
$H_2PO_4^- = H^+ + HPO_4^{2-}$	6.2×10^{-8}
$NH_4^+ = H^+ + NH_3$	5.7×10^{-10}
$HCO_3^- = H^+ + CO_3^{2-}$	5.6×10^{-11}
$HPO_4^{2-} = H^+ + PO_4^{3-}$	2.2×10^{-13}
$HS^- = H^+ + S^{2-}$	1.3×10^{-14}
$H_2O = H^+ + OH^-$	1.0×10^{-14}
$OH^- = H^+ + O^{2-}$	$< 10^{-36}$
$NH_3 = H^+ + NH_2^-$	very small

Note: $H^+(aq) = H_3O^+$
Sample equation: $HI + H_2O = H_3O^+ + I^-$

Ⓚ

CONSTANTS FOR VARIOUS EQUILIBRIA AT 1 atm AND 298 K

$H_2O(\ell) = H^+(aq) + OH^-(aq)$	$K_w = 1.0 \times 10^{-14}$
$H_2O(\ell) + H_2O(\ell) = H_3O^+(aq) + OH^-(aq)$	$K_w = 1.0 \times 10^{-14}$
$CH_3COO^-(aq) + H_2O(\ell) = CH_3COOH(aq) + OH^-(aq)$	$K_b = 5.6 \times 10^{-10}$
$Na^+F^-(aq) + H_2O(\ell) = Na^+(OH)^- + HF(aq)$	$K_b = 1.5 \times 10^{-11}$
$NH_3(aq) + H_2O(\ell) = NH_4^+(aq) + OH^-(aq)$	$K_b = 1.8 \times 10^{-5}$
$CO_3^{2-}(aq) + H_2O(\ell) = HCO_3^-(aq) + OH^-(aq)$	$K_b = 1.8 \times 10^{-4}$
$Ag(NH_3)_2^+(aq) = Ag^+(aq) + 2NH_3(aq)$	$K_{eq} = 8.9 \times 10^{-8}$
$N_2(g) + 3H_2(g) = 2NH_3(g)$	$K_{eq} = 6.7 \times 10^5$
$H_2(g) + I_2(g) = 2HI(g)$	$K_{eq} = 3.5 \times 10^{-1}$

Compound	K_{sp}	Compound	K_{sp}
AgBr	5.0×10^{-13}	Li_2CO_3	2.5×10^{-2}
AgCl	1.8×10^{-10}	$PbCl_2$	1.6×10^{-5}
Ag_2CrO_4	1.1×10^{-12}	$PbCO_3$	7.4×10^{-14}
AgI	8.3×10^{-17}	$PbCrO_4$	2.8×10^{-13}
$BaSO_4$	1.1×10^{-10}	PbI_2	7.1×10^{-9}
$CaSO_4$	9.1×10^{-6}	$ZnCO_3$	1.4×10^{-11}

(L)

STANDARD ELECTRODE REDUCTION POTENTIALS

| Ionic Concentrations 1 M Water at 298 K, 1 atm ||
Half-Reaction	$E°$ (volts)
$F_2(g) + 2e^- \rightarrow 2F^-$	+2.87
$8H^+ + MnO_4^- + 5e^- \rightarrow Mn^{2+} + 4H_2O$	+1.51
$Au^{3+} + 3e^- \rightarrow Au(s)$	+1.50
$Cl_2(g) + 2e^- \rightarrow 2Cl^-$	+1.36
$14H^+ + Cr_2O_7^{2-} + 6e^- \rightarrow 2Cr^{3+} + 7H_2O$	+1.23
$4H^+ + O_2(g) + 4e^- \rightarrow 2H_2O$	+1.23
$4H^+ + MnO_2(s) + 2e^- \rightarrow Mn^{2+} + 2H_2O$	+1.22
$Br_2(\ell) + 2e^- \rightarrow 2Br^-$	+1.09
$Hg^{2+} + 2e^- \rightarrow Hg(\ell)$	+0.85
$Ag^+ + e^- \rightarrow Ag(s)$	+0.80
$Hg_2^{2+} + 2e^- \rightarrow 2Hg(\ell)$	+0.80
$Fe^{3+} + e^- \rightarrow Fe^{2+}$	+0.77
$I_2(s) + 2e^- \rightarrow 2I^-$	+0.54
$Cu^+ + e^- \rightarrow Cu(s)$	+0.52
$Cu^{2+} + 2e^- \rightarrow Cu(s)$	+0.34
$4H^+ + SO_4^{2-} + 2e^- \rightarrow SO_2(aq) + 2H_2O$	+0.17
$Sn^{4+} + 2e^- \rightarrow Sn^{2+}$	+0.15
$2H^+ + 2e^- \rightarrow H_2(g)$	0.00
$Pb^{2+} + 2e^- \rightarrow Pb(s)$	−0.13
$Sn^{2+} + 2e^- \rightarrow Sn(s)$	−0.14
$Ni^{2+} + 2e^- \rightarrow Ni(s)$	−0.26
$Co^{2+} + 2e^- \rightarrow Co(s)$	−0.28
$Fe^{2+} + 2e^- \rightarrow Fe(s)$	−0.45
$Cr^{3+} + 3e^- \rightarrow Cr(s)$	−0.74
$Zn^{2+} + 2e^- \rightarrow Zn(s)$	−0.76
$2H_2O + 2e^- \rightarrow 2OH^- + H_2(g)$	−0.83
$Mn^{2+} + 2e^- \rightarrow Mn(s)$	−1.19
$Al^3 + 3e^- \rightarrow Al(s)$	−1.66
$Mg^{2+} + 2e^- \rightarrow Mg(s)$	−2.37
$Na^+ + e^- \rightarrow Na(s)$	−2.71
$Ca^{2+} + 2e^- \rightarrow Ca(s)$	−2.87
$Sr^{2+} + 2e^- \rightarrow Sr(s)$	−2.89
$Ba^{2+} + 2e^- \rightarrow Ba(s)$	−2.91
$Cs^+ + e^- \rightarrow Cs(s)$	−2.92
$K^+ + e^- \rightarrow K(s)$	−2.93
$Rb^+ + e^- \rightarrow Rb(s)$	−2.98
$Li^+ + e^- \rightarrow Li(s)$	−3.04

(M)

PHYSICAL CONSTANTS AND CONVERSION FACTORS

Name	Symbol	Value(s)	Units
Angstrom unit	Å	1×10^{-10} m	meter
Avogadro number	N_A	6.02×10^{23} per mol	
Charge of electron	e	1.60×10^{-19} C	coulomb
Electron-volt	eV	1.60×10^{-19} J	joule
Speed of light	c	3.00×10^8 m/s	meters/second
Planck's constant	h	6.63×10^{-34} J·s	joule-second
		1.58×10^{-37} kcal·s	kilocalorie-second
Universal gas constant	R	0.0821 L · atm/mol·K	liter-atmosphere/mole-kelvin
		1.98 cal/mol·K	calories/mole-kelvin
		8.31 J/mol·K	joules/mole-kelvin
Atomic mass unit	μ	1.66×10^{-24} g	gram
Volume standard, liter	L	1×10^3 cm³ = 1 dm³	cubic centimeters, cubic decimeter
Standard pressure, atmosphere	atm	101.3 kPa	kilopascals
		760 mmHg	millimeters of mercury
		760 torr	torr
Heat equivalent, kilocalorie	kcal	4.18×10^3 J	joules

Physical Constants for H₂O		
Molal freezing point depression	....	1.86°C/m
Molal boiling point elevation	....	0.52°C/m
Heat of fusion	....	79.72 cal/g
Heat of vaporization	....	539.4 cal/g

STANDARD UNITS

Symbol	Name	Quantity
m	meter	length
kg	kilogram	mass
Pa	pascal	pressure
K	kelvin	thermodynamic temperature
mol	mole	amount of substance
J	joule	energy, work, quantity of heat
s	second	time
C	coulomb	quantity of electricity
V	volt	electric potential, potential difference
L	liter	volume

(N)

VAPOR PRESSURE OF WATER

°C	torr (mmHg)	°C	torr (mmHg)
0	4.6	26	25.2
5	6.5	27	26.7
10	9.2	28	28.3
15	12.8	29	30.0
16	13.6	30	31.8
17	14.5	40	55.3
18	15.5	50	92.5
19	16.5	60	149.4
20	17.5	70	233.7
21	18.7	80	355.1
22	19.8	90	525.8
23	21.1	100	760.0
24	22.4	105	906.1
25	23.8	110	1074.6

Selected Prefixes		
Factor	Prefix	Symbol
10^{12}	tera	T
10^9	giga	G
10^6	mega	M
10^3	kilo	k
10^{-1}	deci	d
10^{-2}	centi	c
10^{-3}	milli	m
10^{-6}	micro	μ
10^{-9}	nano	n
10^{-12}	pico	p

O

RADII OF ATOMS

KEY

Symbol →	**F**
Covalent Radius, Å →	0.64
Atomic Radius in Metals, Å →	(−)
van der Waals Radius, Å →	1.35

A dash (−) indicates data are not available.

Symbol	Covalent Radius, Å	Atomic Radius in Metals, Å	van der Waals Radius, Å
H	0.37	(−)	1.2
He	(−)	(−)	1.22
Li	1.23	1.52	(−)
Be	0.89	1.13	(−)
B	0.88	(−)	2.08
C	0.77	(−)	1.85
N	0.70	(−)	1.54
O	0.66	(−)	1.40
F	0.64	(−)	1.35
Ne	(−)	(−)	1.60
Na	1.57	1.54	2.31
Mg	1.36	1.60	(−)
Al	1.25	1.43	(−)
Si	1.17	(−)	2.0
P	1.10	(−)	1.90
S	1.04	(−)	1.85
Cl	0.99	(−)	1.81
Ar	(−)	(−)	1.91
K	2.03	2.27	2.31
Ca	1.74	1.97	(−)
Sc	1.44	1.61	(−)
Ti	1.32	1.45	(−)
V	1.22	1.32	(−)
Cr	1.17	1.25	(−)
Mn	1.17	1.24	(−)
Fe	1.17	1.24	(−)
Co	1.16	1.25	(−)
Ni	1.15	1.25	(−)
Cu	1.17	1.28	(−)
Zn	1.25	1.33	(−)
Ga	1.25	1.22	(−)
Ge	1.22	1.23	(−)
As	1.21	(−)	2.0
Se	1.17	(−)	2.0
Br	1.14	(−)	1.95
Kr	1.89	(−)	1.98
Rb	2.16	2.48	2.44
Sr	1.92	2.15	(−)
Y	1.62	1.81	(−)
Zr	1.45	1.60	(−)
Nb	1.34	1.43	(−)
Mo	1.29	1.36	(−)
Tc	(−)	1.36	(−)
Ru	1.24	1.33	(−)
Rh	1.25	1.35	(−)
Pd	1.28	1.38	(−)
Ag	1.34	1.44	(−)
Cd	1.41	1.49	(−)
In	1.50	1.63	(−)
Sn	1.40	1.41	(−)
Sb	1.41	(−)	2.2
Te	1.37	(−)	2.20
I	1.33	(−)	2.15
Xe	2.09	(−)	(−)
Cs	2.35	2.65	2.62
Ba	1.98	2.17	(−)
La-Lu			
Hf	1.44	1.56	(−)
Ta	1.34	1.43	(−)
W	1.30	1.37	(−)
Re	1.28	1.37	(−)
Os	1.26	1.34	(−)
Ir	1.26	1.36	(−)
Pt	1.29	1.38	(−)
Au	1.34	1.44	(−)
Hg	1.44	1.60	(−)
Tl	1.55	1.70	(−)
Pb	1.54	1.75	(−)
Bi	1.52	1.55	(−)
Po	1.53	1.67	(−)
At	(−)	(−)	(−)
Rn	2.14	(−)	(−)
Fr	(−)	2.7	(−)
Ra	(−)	2.20	(−)
Ac-Lr			

Lanthanides (La–Lu)

	La	Ce	Pr	Nd	Pm	Sm	Eu	Gd	Tb	Dy	Ho	Er	Tm	Yb	Lu
Covalent	1.69	1.65	1.65	1.64	(−)	1.66	1.85	1.61	1.59	1.59	1.58	1.57	1.56	1.70	1.56
Metal	1.88	1.83	1.83	1.82	1.81	1.80	2.04	1.80	1.78	1.77	1.77	1.76	1.75	1.94	1.73
van der Waals	(−)	(−)	(−)	(−)	(−)	(−)	(−)	(−)	(−)	(−)	(−)	(−)	(−)	(−)	(−)

Actinides (Ac–Lr)

	Ac	Th	Pa	U	Np	Pu	Am	Cm	Bk	Cf	Es	Fm	Md	No	Lr
Covalent	(−)	(−)	(−)	(−)	(−)	(−)	(−)	(−)	(−)	(−)	(−)	(−)	(−)	(−)	(−)
Metal	1.88	1.80	1.61	1.39	1.31	1.51	1.84	(−)	(−)	(−)	(−)	(−)	(−)	(−)	(−)
van der Waals	(−)	(−)	(−)	(−)	(−)	(−)	(−)	(−)	(−)	(−)	(−)	(−)	(−)	(−)	(−)

THE CHEMICAL ELEMENTS

(Atomic masses in this table are based on the atomic mass of carbon-12 being exactly 12.)

Name	Symbol	Atomic Number	Atomic Mass	Name	Symbol	Atomic Number	Atomic Mass
Actinium	Ac	89	(227)	Mercury	Hg	80	200.59
Aluminum	Al	13	26.98	Molybdenum	Mo	42	95.94
Americium	Am	95	(243)	Neilsbohrium	Ns	107	(262)
Antimony	Sb	51	121.75	Neodymium	Nd	60	144.24
Argon	Ar	18	39.95	Neon	Ne	10	20.18
Arsenic	As	33	74.92	Neptunium	Np	93	237.05
Astatine	At	85	(210)	Nickel	Ni	28	58.71
Barium	Ba	56	137.34	Niobium	Nb	41	92.90
Berkelium	Bk	97	(247)	Nitrogen	N	7	14.01
Beryllium	Be	4	9.01	Nobelium	No	102	(259)
Bismuth	Bi	83	208.98	Osmium	Os	76	190.2
Boron	B	5	10.81	Oxygen	O	8	16.00
Bromine	Br	35	79.90	Palladium	Pd	46	106.4
Cadmium	Cd	48	112.40	Phosphorus	P	15	30.97
Caesium	Cs	55	132.91	Platinum	Pt	78	195.09
Calcium	Ca	20	40.08	Plutonium	Pu	94	(244)
Californium	Cf	98	(251)	Polonium	Po	84	(210)
Carbon	C	6	12.01	Potassium	K	19	39.10
Cerium	Ce	58	140.12	Praseodymium	Pr	59	140.90
Chlorine	Cl	17	35.45	Promethium	Pm	61	(145)
Chromium	Cr	24	52.00	Protactinium	Pa	91	231.04
Cobalt	Co	27	58.93	Radium	Ra	88	(226)
Copper	Cu	29	63.55	Radon	Rn	86	(222)
Curium	Cm	96	(247)	Rhenium	Re	75	186.2
Dysprosium	Dy	66	162.50	Rhodium	Rh	45	102.91
Einsteinium	Es	99	(254)	Rubidium	Rb	37	85.47
Erbium	Er	68	167.26	Ruthenium	Ru	44	101.07
Europium	Eu	63	151.96	Rutherfordium	Rf	104	(261)
Fermium	Fm	100	(257)	Samarium	Sm	62	150.35
Fluorine	F	9	19.00	Scandium	Sc	21	44.95
Francium	Fr	87	(223)	Selenium	Se	34	78.96
Gadolinium	Gd	64	157.25	Silicon	Si	14	28.09
Gallium	Ga	31	69.72	Silver	Ag	47	107.89
Germanium	Ge	32	72.59	Sodium	Na	11	22.99
Gold	Au	79	196.97	Strontium	Sr	38	87.62
Hafnium	Hf	72	178.49	Sulfur	S	16	32.06
Hahnium	Ha	105	(262)	Tantalum	Ta	73	180.95
Hassia	Hs	108	(265)	Technetium	Tc	43	(99)
Helium	He	2	4.00	Tellurium	Te	52	127.60
Holmium	Ho	67	164.93	Terbium	Tb	65	158.92
Hydrogen	H	1	1.008	Thallium	Tl	81	204.37
Indium	In	49	114.82	Thorium	Th	90	232.03
Iodine	I	53	126.90	Thulium	Tm	69	168.93
Iridium	Ir	77	192.2	Tin	Sn	50	118.69
Iron	Fe	26	55.85	Titanium	Ti	22	47.90
Krypton	Kr	36	83.80	Tungsten	W	74	183.85
Lanthanum	La	57	138.91	Unnilhexium	Unh	106	(263)
Lawrencium	Lr	103	(261)	Uranium	U	92	238.03
Lead	Pb	82	207.19	Vanadium	V	23	50.94
Lithium	Li	3	6.94	Xenon	Xe	54	131.30
Lutetium	Lu	71	174.97	Ytterbium	Yb	70	173.04
Magnesium	Mg	12	24.31	Yttrium	Y	39	88.91
Manganese	Mn	25	54.94	Zinc	Zn	30	65.37
Meitnerium	Mt	109	(266)	Zirconium	Zr	40	91.22
Mendelevium	Md	101	(258)				

*A number in parentheses is the mass number of the most stable isotope.

Glossary of Common Terms

(See Index for additional references)

absolute temperature   Temperature measured on the absolute scale, which has its origin at absolute zero. *See also* **Kelvin scale.**

absorption   The process of taking up by capillary, osmotic, chemical, or solvent action, as a sponge absorbs water.

acid   A water solution that has an excess of hydrogen ions; an acid turns litmus paper pink or red, has a sour taste, and neutralizes bases to form salts.

acid anhydride   A nonmetallic oxide that, when placed in water, reacts to form an acid solution.

acid salt   A salt formed by replacing part of the hydrogen ions of a dibasic or tribasic acid with metallic ions. Examples: $NaHSO_4$, NaH_2PO_4.

actinide series   The series of radioactive elements starting with actinium, No. 89, and ending with lawrencium, No. 103.

activated charcoal   A specially treated and finely divided form of carbon, which possesses a high degree of adsorption.

activation energy   The minimum energy necessary to start a reaction.

adsorption   The adhesion (in an extremely thin layer) of the molecules of gases, of dissolved substances, or of liquids to the surfaces of solid or liquid bodies with which they come into contact.

alcohol   An organic hydroxyl compound formed by replacing one or more hydrogen atoms of a hydrocarbon with an equal number of hydroxyl (OH) groups.

aldehyde   An organic compound formed by dehydrating oxidized alcohol; contains the characteristic –CHO group.

alkali   Usually, a strong base, such as sodium hydroxide or potassium hydroxide.

alkaline   Referring to any substance that has basic properties.

alkyl   A substitutent obtained from a saturated hydrocarbon by removing one hydrogen atom.

Examples: methyl (CH_3^-), ethyl ($C_2H_5^-$).

allotropic forms   Forms of the same element that differ in their crystalline structures.

alloy   A substance composed of two or more metals, which are intimately mixed; usually made by melting the metals together.

alpha particles   Positively charged helium nuclei.

alum   A double sulfate of a monovalent metal and a trivalent metal, such as K_2SO_4 $Al_2(SO_4)_3$ $24H_2O$; also, a common name for commercial aluminum sulfate.

amalgam   An alloy of mercury and another metal.

amine   A compound such as CH_3NH_2, derived from ammonia by substituting one or more hydrocarbon radicals for hydrogen atoms.

amino acid   One of the "building blocks" of proteins; contains one or more NH_2^- groups that have replaced the same number of hydrogen atoms in an organic acid.

amorphous   Having no definite crystalline structure.

amphoteric   Referring to a hydroxide that may have either acidic or basic properties, depending on the substance with which it reacts.

analysis   The breaking down of a compound into two or more simpler substances.

angstrom   A unit of length used in atomic measurements; equal to 10^{-8} cm (abbreviation: Å).

anhydride   A compound derived from another compound by the removal of water; it will combine with water to form an acid (acid anhydride) or a base (basic anhydride).

anhydrous   Containing no water.

anion   An ion or particle that has a negative charge and thus is attracted to a positively charged anode.

anode   The electrode in an electrolytic cell that has a positive charge and attracts negative ions.

antichlor   A substance used to remove the excess of chlorine in the bleaching process.

aromatic compounds   A compound whose basic structure contains the benzene ring; it usually has an odor.

atmosphere   The layer of gases surrounding the earth; also, a unit of pressure (1 atm = approx. 760 mm or torr).

atom   The smallest particle of an element that retains the properties of that element and can enter into a chemical reaction.

atomic energy   *See* **nuclear energy,** a more accurate term.

atomic mass (relative atomic mass or atomic weight)   The average mean value of the isotopic masses of the atoms of an element. It indicates the relative mass of the element as compared with that of carbon-12, which is assigned a mass of exactly 12 atomic mass units.

atomic mass unit   One twelfth of the mass of a carbon-12 atom; equivalent to 1.660531×10^{-27} kilogram (abbreviation: amu).

atomic number   The number that indicates the order of an element in the periodic system; numerically equal to the number of protons in the nucleus of the atom, or the number of negative electrons located outside the nucleus of the atom.

atomic radius   One-half the distance between adjacent nuclei in the crystalline or solid phase of an element; the distance from the atomic nucleus to the valence electrons.

atomic weight   *See* **atomic mass.**

Avogadro's hypothesis   *See under* **laws.**

Avogadro's number   The number of molecules in 1 gram-molecular volume of a substance, or the number of atoms in 1 gram-atomic mass of an element; equal to 6.022169×10^{23}. *See also* **mole.**

barometer   An instrument, invented by Torricelli in 1643, used for measuring atmospheric pressure.

base   A water solution that contains an excess of hydroxide ions; a proton acceptor; a base turns litmus paper blue and neutralizes acids to form salts.

basic anhydride   A metallic oxide that forms a base when placed in water.

beta particles   High-speed, negatively charged electrons emitted in radiation.

binary   Referring to a compound composed of two elements, such as H_2O.

boiling point   The temperature at which the vapor pressure of a liquid equals the atmospheric pressure.

bond energy   The energy needed to break a chemical bond and form a neutral atom.

bonding   The union of atoms to form compounds or molecules by filling their outer shells of electrons. This can be done through giving and taking electrons (ionic) or by sharing electrons (covalent).

Boyle's law   *See under* **laws**.

brass   An alloy of copper and zinc.

breeder reactor   A nuclear reactor in which more fissionable material is produced than is used up during operation.

British thermal unit   A unit of heat; the amount of heat necessary to raise the temperature of 1 pound of water 1 degree on the Fahrenheit scale (abbreviation: Btu).

Brownian movement   Continuous zigzagging movement of colloidal particles in a dispersing medium, as viewed through an ultramicroscope.

buffer   A substance that, when added to a solution, makes changing the pH of the solution more difficult.

calorie   A unit of heat; the amount of heat needed to raise the temperature of 1 gram of water 1 degree on the Celsius scale.

calorimeter   An instrument used to measure the amount of heat liberated or absorbed during a change.

carbonated water   Water containing dissolved carbon dioxide.

carbon dating   The use of radioactive carbon-14 to estimate the ages of ancient materials, such as archeological or paleontological specimens.

catalyst   A substance that speeds up or slows down a reaction without being permanently changed itself.

cathode   The electrode in an electrolytic cell that is negatively charged and attracts positive ions.

cathode rays   Streams of electrons given off by the cathode of a vacuum tube.

cation   An ion that has a positive charge.

Celsius scale   A temperature scale divided into 100 equal divisions and based on water freezing at 0° and boiling at 100°. Synonymous with centigrade.

chain reaction   A reaction produced during nuclear fission when at least one neutron from each fission produces another fission, so that the process becomes self-sustaining without additional external energy.

Charles's law   *See under* **laws**.

chemical change   A change that alters the atomic structures of the substances involved and results in different properties.

chemical property   A property that determines how a substance will behave in a chemical reaction.

chemistry   The science concerned with the compositions of substances and the changes that they undergo.

colligative properties   Properties of a solution that depend primarily on the concentration, not the type, of particles present.

colloids   Particles larger than those found in a solution but smaller than those in a suspension.

Combining volumes   *See Gay-Lussac's, under* **laws**.

combustion   A chemical action in which both heat and light are given off.

compound   A substance composed of elements chemically united in definite proportions by weight.

condensation   (a) A change from gaseous to liquid state; (b) the union of like or unlike molecules with the elimination of water, hydrogen chloride, or alcohol.

Conservation of energy   *See under* **laws**.

Conservation of matter   *See under* **laws**.

control rod   In a nuclear reactor, a rod of a certain metal such as cadmium, which controls the speed of the chain reaction by absorbing neutrons.

coordinate covalence   Covalence in which both electrons in a pair come from the same atom.

covalent bonding   Bonding accomplished through the sharing of electrons so that atoms can fill their outer shells.

critical mass   The smallest amount of fissionable material that will sustain a chain reaction.

critical temperature   The temperature above which no gas can be liquefied, regardless of the pressure applied.

crystalline   Having a definite molecular or ionic structure.

crystallization   The process of forming definitely shaped crystals when water is evaporated from a solution of the substance.

cyclotron   A device used to accelerate charged particles to high energies for bombarding the nuclei of atoms.

Dalton's law of partial pressures   *See under* **laws**.

decomposition   The breaking down of a compound into simpler substances or into its constituent elements.

definite composition   *See under* **laws**.

dehydrate   To take water from a substance.

dehydrating agent   A substance able to withdraw water from another substance, thereby drying it.

deliquescence   The absorption by a substance of water from the air, so that the substance becomes wet.

denatured alcohol   Ethyl alcohol that has been "poisoned" in order to produce (by avoiding federal tax) a cheaper alcohol for industrial purposes.

density   The mass per unit volume of a substance; the mathematical formula is $D = m/V$, where D = density, m = mass, and V = volume.

destructive distillation   The process of heating an organic substance, such as coal, in the absence of air to break it down into solid and volatile products.

deuterium   An isotope of hydrogen, sometimes called heavy hydrogen, with an atomic weight of 2.

dew point   The highest temperature at which water vapor condenses out of the air.

dialysis   The process of separation of a solution by diffusion through a semipermeable membrane.

diffusion   The process whereby gases or liquids intermingle freely of their own accord.

dipole-dipole attraction   A relatively weak force of attraction between polar molecules; a component of van der Waals forces.

displacement   A change by which an element takes the place of another element in a compound.

dissociation (ionic)   The separation of the ions of an electrovalent compound due to the action of a solvent.

distillation   The process of first vaporizing a liquid and then condensing the vapor back into a liquid, leaving behind the nonvolatile impurities.

double bond   A bond between atoms involving two electron pairs. In organic chemistry: unsaturated.

double displacement   A reaction in which two chemical substances exchange ions with the formation of two new compounds.

dry ice   Solid carbon dioxide.

ductile   Capable of being drawn into thin wire.

effervescence   The rapid escape of excess gas that has been dissolved in a liquid.

efflorescence   The loss by a substance of its water of hydration on exposure to air at ordinary temperatures.

effusion   The flow of a gas through a small aperture.

Einstein equation   The equation $E = mc^2$, which relates mass to energy; E is energy in ergs, m is mass in grams, and c is velocity of light, 3×10^{10} cm/s.

electrode   A terminal of an electrolytic cell.

electrode potential   The difference in potential between an electrode and the solution in which it is immersed.

electrolysis   The process of separating the ions in a compound by means of electrically charged poles.

electrolysis (electrolytic) cell   A cell in which electrolysis is carried out.

electrolyte   A liquid that will conduct an electric current.

electron   A negatively charged particle found outside the nucleus of the atom; it has a mass of 9.109×10^{-28} g.

electron dot symbol   *See* **Lewis dot symbol.**

electronegativity   The numerical expression of the relative strength with which the atoms of an element attract valence electrons to themselves; the higher the number, the greater the attraction.

electron volt   A unit for expressing the kinetic energy of subatomic particles; the energy acquired by an electron when it is accelerated by a potential difference of 1 volt; equals 1.6×10^{-12} erg or 23.1 kcal/mole (abbreviation: eV).

electroplating   Depositing a thin layer of (usually) a metallic element on the surface of another metal by electrolysis.

electrovalence   The bond of attraction between negatively and positively charged ions.

element   One of the more than 100 "building blocks" of which all matter is composed. An element consists of atoms of only one kind and cannot be decomposed further by ordinary chemical means.

empirical formula   A formula that shows only the simplest ratio of the numbers and kinds of atoms, such as CH_4.

emulsifying agent   A colloidal substance that forms a film about the particles of two immiscible liquids, so that one liquid remains suspended in the other.

emulsion   A suspension of fine particles or droplets of one liquid in another, the two liquids being immiscible in each other; the droplets are surrounded by a colloidal (emulsifying) agent.

endothermic   Referring to a chemical reaction that results in an overall absorption of heat from its surroundings.

energy   The capacity to do work. In every chemical change energy is either given off or taken in. Forms of energy are heat, light, motion, sound, and electrical, chemical, and nuclear energy.

enthalpy   The heat content of a chemical system.

entropy   The measure of the randomness that exists in a system.

equation   A shorthand method of showing the changes that take place in a chemical reaction.

equilibrium   The point in a reversible reaction at which the forward reaction is occurring at the same rate as the opposing reaction.

erg   A unit of energy or work done by a force of 1 dyne (1/980 gram of force) acting through a distance of 1 centimeter; equals 2.4×10^{-11} kcal.

ester   An organic salt formed by the reaction of an alcohol with an organic (or inorganic) acid.

esterification   A chemical reaction between an alcohol and an acid, in which an ester is formed.

ether   An organic compound containing the –O– group.

eudiometer   A graduated glass tube into which gases are placed and subjected to an electric spark; used to measure the individual volumes of combining gases.

evaporation   The process in which molecules of a liquid (or a solid) leave the surface in the form of vapor.

exothermic   Referring to a chemical reaction that results in the giving off of heat to its surroundings.

Fahrenheit scale   The temperature scale that has 32° as the freezing point of water and 212° as the boiling point.

fallout   The residual radioactivity from an atmospheric nuclear test, which eventually settles on the surface of the earth.

Faraday's law   *See under* **laws.**

filtration   The process by which suspended matter is removed from a liquid by passing the liquid through a porous material.

First Law of Thermodynamics   *See under* **laws.**

fission   A nuclear reaction that releases energy because of the splitting of large nuclei into smaller ones.

fixation of nitrogen   Any process for converting atmospheric nitrogen into compounds, such as ammonia and nitric acid.

flame   The glowing mass of gas and luminous particles produced by the burning of a gaseous substance.

flammable   Capable of being easily set on fire; combustible (same as *inflammable*).

fluorescence   Emission by a substance of electromagnetic radiation, usually visible, as the immediate result of (and only during) absorption of energy from another source.

fluoridation   Addition of small amounts of fluoride (usually NaF) to drinking water to help prevent tooth decay.

flux   In metallurgy: a substance that helps to melt and remove the solid impurities as slag. In soldering: a substance that cleans the surface of the metal to be soldered. In nucleonics: the concentration of nuclear particles or rays.

formula   An expression that uses the symbols for elements and subscripts to show the basic makeup of a substance.

formula mass   The sum of the atomic masses of all the atoms (or ions) contained in a formula.

fractional crystallization   The separation of the components in a mixture of dissolved solids by evaporation according to individual solubilities.

fractional distillation   The separation of the components in a mixture of liquids having different boiling points by vaporization.

freezing point   The specific temperature at which a given liquid and its solid form are in equilibrium.

fuel   Any substance used to furnish heat by combustion. *See also* **nuclear fuel**.

fuel cell   A device for converting an ordinary fuel such as hydrogen or methane directly into electricity.

functional group   A group of atoms that characterizes certain types of organic compounds, such as –OH for alcohols, and that reacts more or less independently.

fusion   A nuclear reaction that releases energy because of the union of smaller nuclei to form larger ones.

fusion melting   Changing a solid to the liquid state by heating.

galvanizing   Applying a coating of zinc to iron or steel to protect the latter from rusting.

gamma rays   A type of radiation consisting of high-energy waves that can pass through most materials.

gas   A phase of matter that has neither definite shape nor definite volume.

Gay-Lussac's law   *See under* **laws**.

glass   An amorphous, usually translucent substance consisting of a mixture of silicates. Ordinary glass is made by fusing together silica and sodium carbonate and lime; the various forms of glass contain many other silicates.

Graham's law   *See under* **laws**.

gram   A unit of weight in the metric system; the weight of 1 milliliter of water at 4°C (abbreviation: g).

gram-atomic mass   The atomic mass, in grams, of an element.

gram-formula weight   The formula weight, in grams, of a substance.

gram-molecular mass   The molecular mass, in grams, of a substance.

gram-molecular volume   The volume occupied by 1 gram-molecular mass of a gaseous substance at standard conditions; equals 22.4 L.

group   A vertical column of elements in the periodic table that generally have similar properties.

half-life   The time required for half of the mass of a radioactive substance to disintegrate.

half-reaction   One of the two parts, either the reduction part or the oxidation part, of a redox reaction.

halogen   Any of the five nonmetallic elements (fluorine, chlorine, bromine, iodine, astatine) that form part of Group 17 (Group VII) of the periodic table.

heat   A form of molecular energy; it passes from a warmer body to a cooler one.

heat capacity (specific heat)   The quantity of heat, in calories, needed to raise the temperature of 1 gram of a substance 1 degree on the Celsius scale.

heat of formation   The quantity of heat either given off or absorbed in the formation of 1 mole of a substance from its elements.

heat of fusion   The amount of heat, in calories, required to melt 1 gram of a solid; for water, 80 cal.

heat of vaporization   The quantity of heat needed to vaporize 1 gram of a liquid at constant temperature and pressure; for water at 100°C, 540 cal.

heavy water (deuterium oxide, D_2O)   Water in which the hydrogen atoms are replaced by atoms of the isotope of hydrogen, deuterium.

Henry's law   *See under* **laws**.

homogeneous   Uniform; having every portion exactly like every other portion.

homologous   Alike in structure; referring to series of organic compounds, such as hydrocarbons, in which each member differs from the next by the addition of the same group.

humidity   The amount of moisture in the air.

hybridization   The combination of two or more orbitals to form new orbitals.

hydrate   A compound that has water molecules included in its crystalline makeup.

hydride   Any binary compound containing hydrogen, such as HCl.

hydrogenation   A process in which hydrogen is made to combine with another substance, usually organic, in the presence of a catalyst.

hydrogen bond   A weak chemical linkage between the hydrogen of one polar molecule and the oppositely charged portion of a closely adjacent molecule.

hydrolysis   Of carbohydrates: the action of water in the presence of a catalyst upon one carbohydrate to form simpler carbohydrates. Of salts: a reaction involving the splitting of water into its ions by the formation of a weak acid, a weak base, or both.

hydronium ion   A hydrated ion, $H_2O \cdot H^+$ or H_3O^+.

hydroponics   Growing plants without the use of soil, as in nutrient solution or in sand irrigated with nutrient solution.

hydroxyl   Referring to the –OH radical.

hygroscopic   Referring to the ability of a substance to draw water vapor from the atmosphere to itself and become wet.

hypothesis   A possible explanation of the nature of an action or phenomenon; a hypothesis is not as completely developed as a theory.

Ideal Gas law   *See under* **laws**.

immiscible   Referring to the inability of two liquids to mix.

indicator   A dye that shows one color in the presence of the hydrogen ion (acid) and a different color in the presence of the hydroxyl ion (base).

inert   Referring to an atom having a complete outer energy level of electrons and consequently not reactive.

inert gas structure   The outer energy level electron configuration characteristic of the inert gases—two electrons for helium; eight electrons for all others.

inertia   The property of matter whereby it remains at rest or, if in motion, remains in motion in a straight line unless acted upon by an outside force.

ion   An atom or a group of combined atoms that carries one or more electric charges.

Examples: NH_4^+, OH^-.

ionic bonding   The bonding of ions due to their opposite charges.

ionic equation   An equation showing a reaction among ions.

ionization   The process in which ions are formed from neutral atoms.

ionization equation   An equation showing the ions set free from an electrolyte.

isomerization   The rearrangement of atoms in a molecule to form isomers.

isomers   Two or more compounds having the same percentage composition but different arrangements of atoms in their molecules and hence different properties.

isotopes   Two or more forms of an element that differ only in the number of neutrons in the nucleus and hence in their mass numbers.

IUPAC   International Union of Pure and Applied Chemistry, an organization that establishes standard rules for naming compounds.

Kelvin scale   A temperature scale based on water freezing at 273 and boiling at 373 Kelvin units; its origin is absolute zero. Synonymous with *absolute scale*.

kernel (atomic)   The nucleus and all the electron shells of an atom except the outer one; usually designated by the symbol for the atom.

ketone   An organic compound containing the –CO– group.

kilocalorie   A unit of heat; the amount of heat needed to raise the temperature of 1 kilogram of water 1 degree on the Celsius scale.

kindling temperature   The temperature to which a given substance must be raised before it ignites.

kinetic molecular theory   The theory that all molecules are in motion; this motion is most rapid in gases, less rapid in liquids, and very slow in solids.

lanthanide series   The "rare earth" series of elements starting with lanthanum, No. 57, and ending with lutetium, No. 71.

law (in science)   A generalized statement about the uniform behavior in natural processes.

laws

Avogadro's   Equal volumes of gases under identical conditions of temperature and pressure contain equal numbers of particles (atoms, molecules, ions, or electrons).

Boyle's   The volume of a confined gas is inversely proportional to the pressure to which it is subjected, provided that the temperature remains the same.

Charles's   The volume of a confined gas is directly proportional to the absolute temperature, provided that the pressure remains the same.

Combining volumes   See *Gay-Lussac's* under **laws**.

Conservation of energy   Energy can be neither created nor destroyed, so that the energy of the universe is constant.

Conservation of matter   Matter can be neither created nor destroyed (or weight remains constant in an ordinary chemical change).

Dalton's   When a gas is made up of a mixture of different gases, the pressure of the mixture is equal to the sum of the partial pressures of the components.

Definite composition   A compound is composed of two or more elements chemically combined in a definite ratio by weight.

Faraday's   During electrolysis, the weight of any element liberated is proportional (1) to the quantity of electricity passing through the cell, and (2) to the equivalent weight of the element.

First Law of Thermodynamics   The total energy of the universe is constant and cannot be created or destroyed.

Gay-Lussac's   The ratio between the combining volumes of gases and the product, if gaseous, can be expressed in small whole numbers.

Graham's   The rate of diffusion (or effusion) of a gas is inversely proportional to the square root of its molecular mass.

Henry's   The solubility of a gas (unless the gas is very soluble) is directly proportional to the pressure applied to the gas.

Ideal Gas   Any gas that obeys the gas laws perfectly. No such gas actually exists.

Multiple Proportions   When any two elements, A and B, combine to form more than one compound, the different weights of B that unite with a fixed weight of A bear a small whole-number ratio to each other.

Periodic   The chemical properties of elements vary periodically with their atomic numbers.

Second Law of Thermodynamics   Heat cannot, of itself, pass from a cold body to a hot body.

Le Châtelier's principle   If a stress is placed on a system in equilibrium, the system will react in the direction that relieves the stress.

leptons Elementary particles that are believed to make up the electron and neutrino.

Lewis dot symbol The chemical symbol (kernel) for an atom, surrounded by dots to represent its outer level electrons. Examples: K ·, Sr :.

liquid A phase of matter that has a definite volume but takes the shape of the container.

liquid air Air that has been cooled and compressed until it liquefies.

litmus An organic substance, obtained from the lichen plant and used as an indicator; it turns red in acidic solution and blue in basic solution.

London force The weakest of the van der Waal forces between molecules. These weak, attractive forces become apparent only when the molecules approach one another closely (usually at low temperatures and high pressure). They are due to the way the positive charges of one molecule attract the negative charges of another molecule because of the charge distribution at any one instant.

luminous Emitting a steady, suffused light.

malleable Capable of being hammered or pounded into thin sheets.

manometer A U-tube (containing mercury or some other liquid) used to measure the pressure of a confined gas.

mass The quantity of matter that a substance possesses; it can be measured by its resistance to a change in position or motion, and is not related to the force of gravity.

mass number The nearest whole number to the combined atomic mass of the individual atoms of an isotope when that mass is expressed in atomic mass units.

mass spectograph A device for determining the masses of electrically charged particles by separating them into distinct streams by means of magnetic deflection.

matter A substance that occupies space, has mass, and cannot be created or destroyed easily.

melting The change in phase of a substance from solid to liquid.

melting point The specific temperature at which a given solid changes to a liquid.

meson Any unstable, elementary nuclear particle having a mass between that of an electron and that of a proton.

metal (a) An element whose oxide combines with water to form a base; (b) an element that readily loses electrons and acquires a positive valence.

metallurgy The process involved in obtaining a metal from its ores.

meter The basic unit of length in the metric system; defined as 1,650,763.73 times the wavelength of krypton-86 when excited to give off an orange-red spectral line.

MeV A unit for expressing the kinetic energy of subatomic particles; equals 10^6 electron volts.

micron One thousandth of a millimeter (abbreviation: μ).

millibar A unit of pressure used in meteorology; equals 1,000 dynes per square centimeter.

millimicron One millionth of a millimeter (abbreviation: mμ).

mineral An inorganic substance of definite composition found in nature.

miscible Referring to the ability of two liquids to mix with one another.

mixture A substance composed of two or more components, each of which retains its own properties.

moderator A substance such as graphite, paraffin, or heavy water used in a nuclear reactor to slow down neutrons.

molal solution A solution containing 1 mole of solute in 1,000 grams of solvent (indicated by m).

molar mass The mass arrived at by the addition of the atomic masses of the units that make up a molecule of an element or compound. Expressed in grams, the molar mass of a gaseous substance at STP occupies a molar volume equal to 22.4 liters.

molar solution A solution containing 1 mole of solute in 1,000 ml of solution (indicated by M).

mole A unit of quantity that consists of 6.02×10^{23} particles.

molecular mass The sum of the atomic masses of all the atoms in a molecule of a substance.

molecular theory *See* **kinetic-molecular theory**.

molecule The smallest particle of a substance that retains the physical and chemical properties of that substance.

monobasic acid An acid having only one hydrogen atom that can be replaced by a metal or a positive radical.

mordant A chemical, such as aluminum sulfate, used for fixing colors on textiles.

multiple proportions *See under* **laws**.

nascent (atomic) Referring to an element in the atomic form as it has just been liberated in a chemical reaction.

neutralization The union of the hydrogen ion of an acid and the hydroxyl ion of a base to form water.

neutron A subatomic particle found in the nucleus of the atom; it has no charge and has the same mass as the proton.

neutron capture A nuclear reaction in which a neutron attaches itself to a nucleus; a gamma ray is usually emitted simultaneously.

nitriding A process in which ammonia or a cyanide is used to produce case-hardened steel; a nitride is formed instead of a carbide.

nitrogen fixation Any process by which atmospheric nitrogen is converted into a compound such as ammonia or nitric acid.

noble gas A gaseous element that has a complete outer level of electrons; any of a group of rare gases (helium, neon, argon, krypton, xenon, and radon) that exhibit great stability and very low reaction rates.

nonelectrolyte A substance whose solution does not conduct a current of electricity.

nonmetal (a) An element whose oxide reacts with water to form an acid; (b) an element that takes on electrons and acquires a negative valence.

nonpolar compound A compound in whose molecules the atoms are arranged symmetrically so that the electric charges are uniformly distributed.

normal salt   A salt in which all the hydrogen of the acid has been displaced by a metal.

normal solution   A solution that contains 1 gram of H^+ (or its equivalent: 17 grams of OH^-, 23 grams of Na^+, 20 grams of Ca^{2+}, and so forth) in 1 liter of solution (indicated by N).

nuclear energy   The energy released by spontaneously or artificially produced fission, fusion, or disintegration of the nuclei of atoms.

nuclear fuel   A substance that is consumed during nuclear fission or fusion.

nuclear reaction   Any reaction involving a change in nuclear structure.

nuclear reactor   A device in which a controlled chain reaction of fissionable material can be produced.

nucleonics   The science that deals with the constituents and all the changes in the atomic nucleus.

nucleus   The center of the atom, which contains protons and neutrons.

nuclide   A species of atom characterized by the constitution of its nucleus.

octane number   A conventional rating for gasoline based on its behavior as compared with that of isooctane as 100.

orbital   A subdivision of a nuclear shell; it may contain none, one, or two electrons.

ore   A natural mineral substance from which an element, usually a metal, may be obtained with profit.

organic acid   An organic compound that contains the –COOH group.

organic chemistry   The branch of chemistry dealing with carbon compounds, usually those found in nature.

oxidation   The chemical process by which oxygen is attached to a substance; the process of losing electrons.

oxidation number (state)   A positive or negative number representing the charge that an ion has or an atom appears to have when its electrons are counted according to arbitrarily accepted rules: (1) electrons shared by two unlike atoms are counted with the more electronegative atom; (2) electrons shared by two like atoms are divided equally between the atoms.

oxidation potential   An electrode potential associated with the oxidation half-reaction.

oxidizing agent   A substance that (a) gives up its oxygen readily, (b) removes hydrogen from a compound, (c) takes electrons from an element.

ozone   An allotropic and very active form of oxygen, having the formula O_3.

paraffin series   The methane series of hydrocarbons.

pascal   The SI unit of pressure equal to 1 newton per square meter.

pasteurization   Partial sterilization of a substance, such as milk, by heating to approximately 65°C for $1/2$ hour.

Pauli exclusion principle   Each electron orbital of an atom can be filled by only two electrons, each with an opposite spin.

period   A horizontal row of elements in the periodic table.

Periodic law   *See under* **laws**.

petroleum (meaning "oil from stone")   A complex mixture of gaseous, liquid, and solid hydrocarbons obtained from the earth.

pH   A numerical expression of the hydrogen or hydronium ion concentration in a solution; defined as $-\log [H^+]$, where $[H^+]$ is the concentration of hydrogen ions, in moles per liter.

phenolphthalein   An organic indicator; it is colorless in acid solution and red in the presence of OH^- ions.

photosynthesis   The reaction taking place in all green plants that produces glucose from carbon dioxide and water under the catalytic action of chlorophyll in the presence of light.

physical change   A change that does not involve any alteration in chemical composition.

physical property   A property of a substance arrived at through observation of its smell, taste, color, density, and so on, which does not relate to chemical activity.

pi bond   A bond between p orbitals.

pile   A general term for a nuclear reactor; specifically, a graphite-moderated reactor in which uranium fuel is distributed throughout a "pile" of graphite blocks.

pitchblende   A massive variety of uraninite that contains a small amount of radium.

plasma   Very hot ionized gases.

polar covalent bond   A bond in which electrons are closer to one atom than to another. *See also* **polar molecule**.

polar dot structure   Representation of the arrangements of electrons around the atoms of a molecule in which the polar characteristics are shown by placing the electrons closer to the more electronegative atom.

polar molecule   A molecule that has differently charged areas because of unequal sharing of electrons.

polyatomic ion   A group of chemically united atoms that react as a unit and have an electric charge.

polymerization   The process of combining several molecules to form one large molecule (polymer). (a) Additional polymerization: The addition of unsaturated molecules to each other. (b) Condensation polymerization: The reaction of two molecules by loss of a molecule of water.

positron   A positively charged particle of electricity with about the same weight as the electron.

potential energy   Energy due to the position of a body or to the configuration of its particles.

precipitate   An insoluble compound formed in the chemical reaction between two or more substances in solution.

proteins   Large, complex organic molecules, with nitrogen an essential part, found in plants and animals.

proton   A subatomic particle found in the nucleus that has a positive charge.

qualitative analysis   A term applied to the methods and procedures used to determine any or all of the constituent parts of a substance.

quantitative analysis   A term applied to the methods and procedures used to determine the definite quantity or percentage of any or all of the constituent parts of a substance.

quark   A subatomic particle with a fractional charge of $1/3$ or $2/3$ the charge of an electron. Six quarks have been described.

quenching   Cooling a hot piece of metal rapidly, as in water or oil.

radiation   The emission of particles and rays from a radioactive source; usually alpha and beta particles and gamma rays.

radioactive   Referring to substances that have the ability to emit radiations (alpha or beta particles or gamma rays).

radioisotopes   Isotopes that are radioactive, such as uranium-235.

reactant   A substance involved in a reaction.

reaction   A chemical transformation or change. The four basic types are combination (synthesis), decomposition (analysis), single replacement or single displacement, and double replacement or double displacement.

reaction potential   The sum of the oxidation potential and reduction potential for a particular reaction.

reagent   Any chemical taking part in a reaction.

recrystallization   A series of crystallizations, repeated for the purpose of greater purification.

redox   A shortened name for a reaction that involves reduction and oxidation.

reducing agent   From an electron standpoint, a substance that loses its valence electrons to another element; a substance that is readily oxidized.

reducing sugar   A sugar that reduces copper or silver salts in an alkaline solution.

reduction   A chemical reaction that removes oxygen from a substance; a gain of electrons.

reduction potential   An electrode potential associated with a reduction half-reaction.

refraction (of light)   The bending of light rays as they pass from one material into another.

relative humidity   The ratio, expressed in percent, between the amount of water vapor in a given volume of air and the amount the same volume can hold when saturated at the same temperature.

resonance   The phenomenon in a molecular structure that exhibits properties between those of a single bond and those of a double bond and thus possesses two or more alternate structures.

reversible reaction   Any reaction that reaches an equilibrium, or that can be made to proceed from right to left as well as from left to right.

roasting   Heating an ore (usually a sulfide) in an excess of air to convert the ore to an oxide, which can then be reduced.

salt   A compound, such as NaCl, made up of a positive metallic ion and a negative nonmetallic ion or radical.

saturated solution   A solution that contains the maximum amount of solute under the existing temperature and pressure.

secondary cell   A cell (battery) that can be regenerated or "charged."

Second Law of Thermodynamics   *See under* **laws.**

shell   A level outside the nucleus of an atom that electrons can inhabit without expending energy.

sigma bond   A bond between *s* orbitals or between an *s* orbital and another kind of orbital.

significant figures   All the certain digits recorded in a measurement plus one uncertain digit.

slag   The product formed when the flux reacts with the impurities of an ore in a metallurgical process.

solid   A phase of matter that has a definite size and shape.

solubility   A measure of the amount of solute that will dissolve in a given quantity of solvent at a given temperature.

solute   The material that is dissolved to make a solution.

solution   A uniform mixture of a solute in a solvent.

solvent   The dispersing substance that allows the solute to go into solution.

specific gravity (mass)   The ratio between the mass of a certain volume of a substance and the mass of an equal volume of water (or, in the case of gases, an equal volume of air); expressed as a single number.

specific heat   The ratio between the number of calories needed to raise the temperature of a certain mass of a substance 1 degree on the Celsius scale and the number of calories needed to raise the temperature of the same mass of water 1 degree on the Celsius scale.

spectroscope   An instrument used to analyze light by separating it into its component wavelengths.

spectrum   The image formed when radiant energy is dispersed by a prism or grating into its various wavelengths.

spinthariscope   A device for viewing through a microscope the flashes of light made by particles from radioactive materials against a sensitized screen.

spontaneous combustion (ignition)   The process in which slow oxidation produces enough heat to raise the temperature of a substance to its kindling temperature.

stable   Referring to a substance not easily decomposed or dissociated.

standard conditions   An atmospheric pressure of 760 mm or torr or 1 atm (mercury pressure) and a temperature of 0°C (273 K) (abbreviation: STP).

stratosphere   The upper portion of the atmosphere, in which the temperature changes but little with altitude, and clouds of water never form.

strong acid (or base)   An acid (or a base) capable of a high degree of ionization in water solution.

Examples: sulfuric acid, sodium hydroxide.

structural (graphic) formula   A pictorial representation of the atomic arrangement of a molecule.

sublime   To vaporize directly from the solid to the gaseous state, and then condense back to the solid.

substance   A single kind of matter, element, or compound.

substitution products   Products formed by the substitution of other elements or radicals for hydrogen atoms in hydrocarbons.

sulfation   An accumulation of lead sulfate on the plates and at the bottom of a (lead) storage cell.

supersaturated solution   A solution that contains a greater quantity of solute than is normally possible at a given temperature.

suspension   A mixture of finely divided solid material in a liquid, from which the solid settles on standing.

symbol   A letter or letters representing an element of the periodic table. Examples: O, Mn.

synthesis   The chemical process of forming a substance from its individual parts.

Systeme International d'Unites   The modernized metric system of measurements universally used by scientists. There are seven base units: kilogram, meter, second, ampere, kelvin, mole, and candela.

temperature   The intensity or the degree of heat of a body, measured by a thermometer.

tempering   The heating and then rapid cooling of a metal to increase its hardness.

ternary   Referring to a compound composed of three different elements, such as H_2SO_4.

theory   An explanation used to interpret the "mechanics" of nature's actions; a theory is more fully developed than a hypothesis.

thermochemical equation   An equation that includes values for the calories absorbed or evolved.

thermoplastic   Capable of being softened by heat; may be remolded.

thermosetting   Capable of being permanently hardened by heat and pressure; resistant to the further effects of heat.

tincture   An alcoholic solution of a substance, such as a tincture of iodine.

torr   A unit of pressure defined as 1 mm Hg; 1 torr equals 133.32 pascals.

tracer   A minute quantity of radioactive isotope used in medicine and biology to study chemical changes within living tissues.

transmutation   Conversion of one element into another, either by bombardment or by radioactive disintegration.

tribasic acid   An acid that contains three replaceable hydrogen atoms in its molecule, such as H_3PO_4.

tritium   A very rare, unstable, "triple-weight" hydrogen isotope (H^3) that can be made synthetically.

Tyndall effect   The scattering of a beam of light as it passes through a colloidal material.

ultraviolet light   The portion of the spectrum that lies just beyond the violet; therefore of short wavelength.

U.S.P. (United States Pharmacopeia) chemicals   Chemicals certified as having a standard of purity that demonstrates their fitness for use in medicine.

valence   The combining power of an element; the number of electrons gained, lost, or borrowed in a chemical reaction.

valence electrons   The electrons in the outermost level or levels of an atom that determine its chemical properties.

van der Waals forces   Weak attractive forces existing between molecules.

vapor   The gaseous phase of a substance that normally exists as a solid or liquid at ordinary temperatures.

vapor density   The density of a gas in relation to another gas taken as a standard under identical conditions of temperature and pressure.

vapor pressure   The pressure exerted by a vapor given off by a confined liquid or solid when the vapor is in equilibrium with its liquid or solid form.

volatile   Easily changed to a gas or a vapor at relatively low pressure.

volt   A unit of electrical potential or voltage, equal to the difference of potential between two points in a conducting wire carrying a constant current of 1 ampere when the power dissipated between these two points is equal to 1 watt (abbreviation: V).

volume   The amount of three-dimensional space occupied by a substance.

VSEPR   The valence shell electron pair repulsion model. It expresses the non-90° variations in bond angles for p orbitals in the outer energy levels of atoms in molecules because of electron repulsions.

water of hydration   Water that is held in chemical combination in a hydrate and can be removed without essentially altering the composition of the substance. *See also* **Hydrate**.

weak acid (or base)   An acid (or base) capable of being only slightly ionized in an aqueous solution. Examples: acetic acid, ammonium hydroxide.

weak electrolyte   A substance that, when dissolved in water, ionizes only slightly and hence is a poor conductor of electricity.

weight   The measure of the force with which a body is attracted toward the earth by gravity.

work   The product of the force exerted on a body and the distance through which the force acts; expressed mathematically by the equation $W = Fs$, where W = work, F = force, and s = distance.

X rays   Penetrating radiations, of extremely short wavelength, emitted when a stream of electrons strikes a solid target in a vacuum tube.

zeolite   A natural or synthesized silicate used to soften water.

Index

Accuracy, 35–36
Acids, 169–178
  binary, 82
  concentration, 172–173
  ternary, 82
Air pressure, 94
Alcohols, 215–216
Aldehydes, 216–217
Alkali metals, 201
Alkaline earth metals, 201
Alkane series, 209–212
Alkene series, 212–213
Alkyne series, 213
Alloys, 203
Alpha particles, 44, 222, 225
  detection, 223–224
Aluminum, 201
Amines, 218–219
Amino Acids, 218–219
Ammonia, 197
  preparation, 233–234
Amphoteric substances, 177
Analysis reactions, 141
Anhydrides, 126–127
Anions, 180
Anode, 180
Antimony, 199
Aromatics, 213–214
Arrhenius theory, 170–171
Arsenic, 199
Atmospheric pressure, 94
Atomic:
  mass unit, 45
  radii, 58–59
  spectra, 47–50
  structure, 42–63
    components, 45–46
Atoms, 28, 42
  electric nature, 42–47

Barium, 201
Barometer, 94–95
Bases, 82, 169–178
Benzene series, 214
Beryllium, 201
Beta particles, 222–223, 225
  detection, 223–224
Bismuth, 199
Bohr Model, 44–45, 47
Boiling point, 121
Bonding, 64–76
Bonds, 64–76
  dissociation energy, 149–150
Boyle's Law, 98–99
Bronsted theory, 171
Bubble chamber, 224
Buffer solutions, 176

Calcium, 201
Carbon, 205–220
  dioxide, 206–208

preparation, 232
forms, 205–206
Carboxylic acids, 217
Cathode, 180
Cations, 180
Celsius scale, 33–34
Cesium, 201
Chain reaction, 227
Charles's Law, 97–98
Chemical:
  bonds, 64
  changes, 28–29
  energy, origin, 146
  equilibrium, 157–168
  formulas, 77–88
  properties, 28
  reactions, 141–151
    predicting, 141–145
    rates of, 152–156
    types, 141
Chlorine, preparation, 232
Chromatography, 235
Cloud chamber, 223–224
Collision theory of reaction rates, 153–154
Colloids, 131
Combination reactions, 141–142
Combined gas law, 99
Common ion effect, 165–166
Composition of matter, 27–28
Compounds, 28
  naming, 81–82
Concentration:
  changing, effect of, 161
  expressions, 131–134
Conjugate acids and bases, 171–172
Coordinate covalent bonds, 66–67
Covalent bonds, 66–67
Critical:
  mass, 227
  temperature and pressure, 121
Crystallization, 137–138
Cycloalkanes, 212

Dalton's Law of Partial Pressures, 100
Decomposition reactions, 141, 143
Density, molar mass, 109
Diagnostic test, 1–25
Diffusion, 96
Dilution, 134–135
Dimensional analysis, 35
Dipole, 66, 68
Displacement reactions, 141
Dissociation, 180
Distillation, 234

Double:
  bonds, 69
  replacement, 141, 143–144, 177
Dow process, 202

Effusion, 96
Electrochemical:
  cells, measuring potential, 235–236
  reactions, 184, 186
Electrochemistry, 179–192
Electrolysis, quantitative aspects, 186–190
Electrolytes, 179–180
Electrolytic cells, 184–186
Electromotive series, 180–181
Electrons, 43, 45
  configuration, 54–55
Electronegativity, 59, 61
Electroplating, 184
Electroscope, 224
Electrostatic repulsion, 70–71
Electrovalence bond, 65
Elements, 28
Empirical formulas, 208
Endothermic, 130
  reactions, 29–30, 146
Energy, 29–31
  conservation, 31
  forms, 30
  origin, 146
Enthalpy, 30–31, 166–167
  bond energies, 150
  changes, 146–147
Entropy, 145, 166–167
Equations:
  ionic, 87
  showing phases, 86–87
  writing and balancing, 85–86
Equilibrium:
  constant, 159, 162, 166–167
  heterogeneous systems, 162–165
Esters, 219
Ethanol, 215–216
Ethers, 218
Exothermic, 130
  reactions, 29–30, 146

Factor-label method of conversion, 35
Fission, 226–228
Flame tests, 240
Formula weight, 82–84
Formulas, 77–88
  writing, 77–79
Francium, 201
Free energy, 167
Fusion, 228

Gamma rays, 222–223, 225
  detection, 223–224
Gases, 27, 89–105
  characteristics, 94–96
  drying, 237
  identification, 238–239
  laws, 97–103
  pressure, measuring, 94–95
  properties, 96
  volume:
    measuring, 238
    molar mass, 108
Geiger counter, 224
Graham's Law, 97
Gram-molecular mass, 132
Graphite, 205

Haber process, 198
Half-cell, standard voltages, 181
Half-reactions, 180, 182–183
Halogen family, 196
Heat:
  measurements, 34
  of combustion, 142
  of formation, 142
  of fusion, 125
  of reaction, additivity, 147–149
  of vaporization, 125
Hund's Rule of Maximum Multiplicity, 53–54
Hybridization, 71–74
Hydrocarbon derivatives, 215–219
Hydrochloric acid, 196
Hydrofluric acid, 196
Hydrogen, 91–93
  bonds, 68–69
  chloride, preparation, 232
  peroxide, 125
  preparation, 92–93, 232
  replacement by a metal, 237
  sulfide, 195
    preparation, 233
  tests, 240
Hydrolysis reactions, 145

Ideal gas law, 101–103
Indicators, 173
Intermolecular forces of attraction, 68–69
Ionic:
  bonds, 65
  equations, 87
  substances, properties, 75
Ionization, 179–180, 221
  constants, 162–164
    water, 163–164
  energy, 61–62

351

Ions:
hydrated, 130
negative, 239–240
positive, 240
Iron, 202
Isomers, 215
Isotopes, 45–46
IUPAC naming system,
211–212, 214, 217

Joules, 30

Kelvin scale, 33–34
Ketones, 218
Kinetic:
energy, 30
-molecular theory, 96

Laboratory setups,
231–240
Law of:
conservation of:
mass and energy, 31
matter, 29
definite composition,
28, 85
mass action, 154–155,
157
multiple proportions, 85
Le Châtelier's Principle,
160–161
Length, measurements,
32
Leptons, 229
Lewis Dot Structure, 55
Liquids, 27, 119–140
distillation, 234
intermolecular
interaction, 119
kinetics, 119
properties, 75
Lithium, 201
London forces, 68

Magnesium, 201–202
Manometer, 95
Mass:
conservation of, 29
-mass problems, 111–113
measurements, 32
number, 45
spectroscopy, 50
volume relationships,
109–111
Matter, 27–29
Measurements, 31–38
Metallic:
bonds, 68
structures, 47
Metalloids, 203
Metals, 199–203
extraction, 200
properties, 199–200
qualitative tests, 240
Methanol, 215
Metric system, 31–33
Millikan oil drop
experiment, 43–44

Miscible, 130
Mixtures, 27–28
Molality, 133
Molar mass, 106–108
density, 109
gas volumes, 108
Molarity, 133
Mole:
concept, 106–108
relationships, 108
Molecular crystals,
properties, 75
Molecule, 64

Naming compounds,
81–82
Neutralization, 169
reaction, 176
Neutron, 44–45
Nitrate ion, 198
Nitric
acid, 197–198
oxide, preparation, 232
Nitrogen, 199
family, 196–199
Nitrous oxide, 198
Noble gas structure, 47
Nonmetallic structure, 47
Nonpolar covalent bond,
66
Non-standard state cell
potentials, 183–184
Normality, 134
Nuclear energy, 226–229
Nucleonics, 221–230
Nucleons, 45
Nucleus, 44–45

Orbitals, 51–53
Organic:
acids, 217
chemistry, 205–220
Oxidation, 91
number, 46–47, 79–80
-reduction, 179–192
Oxygen, 89–91
preparation, 89–90, 231
properties, 91
Ozone, 91

Periodic Table, 56–62
pH, 172–173
Phase:
changes, 119–140
diagrams, 121–122
equilibrium, 120
Phosphorus, 199
Photographic plate, 223
Photons, 48
Physical:
changes, 28–29
properties, 28
Pi bonds, 74
Polar covalent bond, 66
Polyatomic ions, 77
Potassium, 201
Potential energy, 30
Precision, 35–36

Pressure:
corrections of, 100–101
effect on equilibrium,
161
gases, measuring, 94–95
vs. Temperature, 99–100
Properties, 28
Protons, 44–45

Qualitative tests, 238–240
Quantum:
mechanics, 50–51
numbers, 51–53
Quarks, 228–229

Radioactive dating, 226
Radioactivity, 221–226
detection, 223–224
source, 224–226
Radium, 201
Rate of reactions, 152–156
activation energy, 154
collision theory, 153–154
measurement, 152
reaction rate law,
154–155
Reactant, excessive,
problems, 115–116
Reactions, 141–151
heats, additivity, 147–149
predicting, 141–145
rates of, 152–156
reversible, 157–160
types, 30–31, 141
Reactivity, 47
Redox equations,
balancing, 187–190
Reduction, 180 See also:
Oxidation-Reduction
Replacement reactions,
141, 143
Resonance structures, 70
Rubidium, 201
Rutherford experiment,
44

Salts, 169–178
Scientific notation, 34–35
Scintillation counter, 223
Semiconductors, 203
Sigma bonds, 74
Significant figures, 36–37
Silver:
bromide, 196
iodide, 196
Single replacement
reactions, 141, 143,
176
Sodium, 201
Solids, 27, 119–140
Solubility, 128–131
of compounds, 144
products, 164–165
Solutions, 28, 131
colligative properties,
135–137
Specific gravity, 132
Spectroscopy, 48–50

Standard:
reduction potentials, 182
state, 146
voltage, 181
States of matter, 27
Stoichiometry, 106–118
Strontium, 201
Structural formulas, 208
Subatomic particles,
228–229
Sublevels, 54–55
Substances, 27–28
Sulfur:
allotropic forms, 194
dioxide, 195–196
preparation, 233
family, 193–196
Sulfuric acid, 194–195
Suspensions, 131
Synthesis reactions,
141–142

Temperature:
equilibrium, effect of,
161
measurements, 33–34
Titration, 173–176, 236
Transition elements,
55–56
Transmutation, 224–225
Triple bonds, 69

Uncertainty, 35–36, 51
Unified mass unit, 45

Valence, 46–47
van der Waals forces, 68
Voltaic cell reactions, 182
Volume:
mass, 109–111
measurements, 32
- volume problems,
113–115
Volumetric analysis,
173–176
VSEPR, 70–71

Water, 122–127
calorimetry problems,
125
composition, 123
electrolysis, 185–186
heavy, 124
polarity, hydrogen
bonding, 127–128
properties, 125
purification, 122–123
reactions with
anhydrides, 126–127
solutions, 129–130
synthesis, 124
Wave mechanics, 50–54